Chitin Chemistry

CHITIN CHEMISTRY

GEORGE A.F. ROBERTS

Senior Lecturer in Dyeing
Nottingham Polytechnic

MACMILLAN

First published 1992 by
THE MACMILLAN PRESS LTD
Houndmills, Basingstoke, Hampshire RG21 2XS
and London
Companies and representatives
throughout the world

ISBN 0–333–52417–9

A catalogue record for this book is available
from the British Library.

Printed in Hong Kong

In memory of my parents

Isobel M.L. Roberts (née Gunion)
(1904–89)
and
George A.E. Roberts
(1901–89)

Contents

Foreword

Chitin is the second most important natural polysaccharide (cellulose being the first). But fundamental knowledge about chitin and its applications is frequently difficult to obtain.

Even though its original and spectacular properties have been recognised for a long time (it was first isolated by Braconnot in 1811), it seems that no great uses were made of these until comparatively recently.

The first review of chitin was presented by R. Muzzarelli in 1977. Subsequently, International Symposiums on Chitin and Chitosan kept the information up to date. *Chitin Chemistry* answers the needs of people working in this field as well as those who want to develop new products for specific applications. The book gives an exhaustive overview of chitin and its derivatives. Both the solid state and the solution properties of chitin and chitosan are discussed.

The book is a very useful source of knowledge for everyone involved with chitin and its developments.

Professor M. Rinaudo
University Joseph Fourier of Grenoble

Preface

Although the isolation of chitin by Braconnot in 1811 preceded Payen's isolation of cellulose by some 30 years, the development of the chemistry of chitin and its industrial applications has lagged far behind that of cellulose. However the 1970s saw the beginning of an upsurge of interest in chitin chemistry, spearheaded by the publication of Professor Muzzarelli's two books, *Natural Chelating Polymers* (1973) and *Chitin* (1977), and the organisation of the 1st International Conference on Chitin/Chitosan in 1977. In the intervening period the chemical behaviour of these polymers has been studied intensively and the object in writing this book is to present an up-to-date, ordered, coherent and critical account of the chemistry of chitin and chitosan.

As indicated by the title, the book is concerned with the chemistry of these polymers and no attempt has been made to deal with their biology except for two topics, the chitin–protein complexes of native chitin and the biodegradation of chitin and chitosan, whose inclusion seemed particularly relevant. Furthermore, despite the considerable interest that exists in the commercial utilisation of the two polymers, their possible applications have not been discussed apart from a few brief references. Although there is a large and ever-increasing number of potential uses proposed for these polymers and their derivatives, particularly in the patent literature, most would appear not to be in use at the present time and I did not feel in a position to find the few needles in the numerous haystacks. However it is my hope that this book, by presenting a comprehensive account of their chemistry, may point the way to the development of useful applications for these most intriguing polymers.

Acknowledgements

My interest in chitin developed out of an earlier interest in cellulose and I am indebted to Dr Hal S. Blair of Queen's University, Belfast, who first introduced me to cellulose chemistry, and to Mr Eugene D. Klug who further stimulated my interest in it during the period that I was fortunate to work with him at Hercules Inc. I am also indebted to my former research students – Drs Graham K. Moore, Julian G. Domszy, Beryl D. Gummow, Ghobad G. Maghami, Kathryn E. Taylor and Ing. Irineu Batista – whose work has contributed much to my understanding of chitin.

My thanks are due to Drs Joan C. Halfpenny and Neil A.A. MacFarlane, both of Nottingham Polytechnic, for interesting discussions and helpful comments on aspects of X-ray crystallography and of biology respectively, and to Dr Keith D. Parker of Leeds University for reading the complete manuscript and for his constructive comments.

My most sincere thanks are due to Mrs Frances A. Wood, Nottingham Polytechnic, not only for typing the manuscript, preparing the Figures, organising the Indexes and helping with correction of the proofs, but also for the valuable technical support that she has given my research students and myself over many years.

Glossary of Symbols and Abbreviations

a	exponent in the Mark–Houwink equation
AA	alginic acid
Am	Kuhn statistical chain element
BSA	bovine serum albumin
CD	circular dichroism
C.I.	Colour Index (published by the Society of Dyers and Colourists)
CMC	carboxymethyl cellulose
CP/MAS	cross polarisation/magic angle spinning
CS	cellulose sulphate
CSA	chondroitin sulphate A
CSC	chondroitin sulphate C
DCA	dichloroacetic acid
DCC	dicyclohexylcarbodiimide
DD	degree of deacetylation
DEAE cellulose	diethylaminoethyl cellulose
DIPEA	di-(prop-2-yl)ethylamine (di-*iso*propylethylamine)
DMAc	dimethylacetamide
DMF	dimethylformamide
DMSO	dimethylsulphoxide
DNA	deoxyribonucleic acid
DNase	deoxyribonuclease
DNFB	2,4-dinitrofluorobenzene
DP	degree of polymerisation
$\overline{DP}_n$	number-average degree of polymerisation
$\overline{DP}_v$	viscosity-average degree of polymerisation
$\overline{DP}_w$	weight-average degree of polymerisation

DS	degree of substitution
DxS	dextran sulphate
EDTA	ethylenediaminetetraacetic acid (Na$^+$ salt)
EO	ethylene oxide
ESR	electron spin resonance
FTIR	Fourier transform infra-red
GC	gas chromatography
GCD	glycidol (2,3-epoxypropanol)
GC/MS	gas chromatography/mass spectroscopy
GLC	gas–liquid chromatography
GPC	gel permeation chromatography
GTMAC	glycidyltrimethylammonium chloride
GUA	poly(galacturonic) acid
HA	hyaluronic acid
HECh	hydroxyethyl chitosan
HEP	heparin
HMPA	hexamethylphosphoramide
IPA	2-propanol (*iso*-propanol)
IR	infra-red
K	constant in the Mark–Houwink equation
LC	liquid chromatography
LVN	limiting viscosity number
$\bar{M}_n$	number-average molecular weight
$\bar{M}_v$	viscosity-average molecular weight
$\bar{M}_w$	weight-average molecular weight
MCA	monochloroacetic acid
MS	moles of substitution
MWD	molecular weight distribution
NADH	nicotinamide adenine dinucleotide (reduced form)
NASA	1-naphthylamine-4-sulphonic acid
NCA	*N*-carboxyanhydride
NMP	*N*-methyl-2-pyrrolidone
NMR	nuclear magnetic resonance
NSA	1-naphthol-4-sulphonic acid
Py	pyridine
RNA	ribonucleic acid
RNase	ribonuclease
SEM	scanning electron microscopy
TBA	2-methyl-2-propanol (*tert*-butanol)
TCA	trichloroacetic acid
TEA	triethylamine
THF	tetrahydrofuran
TLC	thin-layer chromatography
UV	ultra-violet

$[\eta]$	limiting viscosity number (formerly intrinsic viscosity)
$\eta_{(inh)}$	logarithmic viscosity number (formerly inherent viscosity)
$\eta_{sp/c}$	viscosity number (formerly reduced viscosity)

1

Structure of Chitin and Chitosan

1.1 INTRODUCTION

Chitin is one of the most abundant organic materials, being second only to cellulose in the amount produced annually by biosynthesis.[1] It occurs in animals, particularly in crustacea, molluscs and insects where it is an important constituent of the exoskeleton, and in certain fungi where it is the principal fibrillar polymer in the cell wall. Its occurrence has been reviewed by Jeuniaux[2] and by Richards[3] and a table of the distribution of chitin in living organisms, adapted from that given by Jeuniaux,[4] is given in Table 1.1.

Chitin is poly[β-(1$\rightarrow$4)-2-acetamido-2-deoxy-D-glucopyranose] and its idealised structure (**1.1**) is shown in Figure 1.1, from which it can be seen that it is structurally similar to cellulose (**1.2**) except that the C(2)-hydroxyl group of cellulose is replaced by an acetamido group. This similarity in structure is reflected in the similar roles played by the two polymers in nature, both acting as structural and defensive materials. Chitin is also structurally related to murein (**1.3**) which is the main structural polymer in the cell wall of bacteria.

The principal derivative of chitin is chitosan, produced by alkaline deacetylation of chitin. Chitosan also occurs naturally in some fungi[5] but its occurrence is much less widespread than is that of chitin. Chitosan is poly[β-(1$\rightarrow$4)-2-amino-2-deoxy-D-glucopyranose] and its idealised structure is (**1.4**).

1

TABLE 1.1 *Occurrence of chitin in living organisms*

| Organism | Structure | Chitin | | | Other components | |
		% Organic fraction	Crystal type	Inorganic	Inorganic	Organic
FUNGI						
Ascomyceta	cell walls and structural	traces–45	—	—		polysaccharides such as glucans or mannans
Basidiomyceta	membranes of mycelia,					
Phycomyceta	stalks and spores					
Imperfecti						
ALGAE						
Chlorophyceae	cell wall	+	—	—		cellulose
PROTOZOA						
Rhizopoda:						
Pelomyxa	cyst wall	+	—	—		—
Plagiopyxidae	shell	+	—	silica		—
Allogromia	shell	+	—	iron		proteins and lipids
Ciliata	cyst wall	+	—	—		proteins
CNIDARIA						
Hydrozoa:		3.2–30.3	α	—		proteins, sometimes tanned
Hydroidea	perisarc	+	α	—		—
Milleporina	coenosteum	+	α	CaCO₃		—
Siphonophora	pneumatophore	+	—	—		proteins
Anthozoa	'skeleton'	+	—	CaCO₃		proteins
Scyphozoa	podocyst	+	—	—		proteins
ASCHELMINTHES						
Rotifera	egg envelope (inner membrane)	14.6	—	—		proteins

Taxon	Structure				
Nematoda	egg capsule (middle membrane)	16.6	−	−	proteins
Acanthocephala	egg capsule	+	−	−	−
Priapulida	cuticle	+	−	−	tanned proteins
ENDOPROCTA	cuticle	+	−	−	tanned proteins
BRYOZOA	ectocyst	1.6–6.4	−	$CaCO_3$	proteins
PHORONIDA	tubes	13.5	−	−	proteins
BRACHIOPODA					
Articulata	stalk cuticle	3.8	−	−	−
Inarticulata	stalk cuticle	+	γ	−	collagen
	shell	29.0	β	$CaCO_3$	−
ECHIURIDA	hooked chaetae	+	−	−	−
ANNELIDA					
Polychaeta	chaetae	20.0–38.0	β	−	quinone-tanned proteins
	jaws (*Eunicidae*)	0.28	−	unidentified	proteins
Oligochaeta	chaetae; gizzard cuticle	+	β	−	−
All	peritrophic membrane	+	−	−	proteins
MOLLUSCA					
Polyplacophora	shell plates; mantle bristles	12.0	−	$CaCO_3$	proteins
	radula	+	−	iron	proteins
Gastropoda	shell (mother of pearl)	3.0–70.	−	$CaCO_3$	conchiolin
	radula	19.7	α	iron & silica	tanned proteins
	jaws	+	−	−	tanned proteins
	'stomacal plates' (*Opisthobranchia*)	36.8	−	−	tanned proteins

continued on page 4

TABLE 1.1 *continued*

Organism	*Structure*	*Chitin*		*Other components*	
		% Organic fraction	*Crystal type*	*Inorganic*	*Organic*
Cephalopoda	calcified shell	3.5–26.0	β	$CaCO_3$	conchiolin
	'pen' (Loligo, Octopus)	17.9	β	—	'conchagen'
	jaws and radula	19.5	α	—	tanned proteins
	stomach cuticle	–	γ	—	—
Lamellibranchia	shells: periostracum	0.7–3	–	$CaCO_3$	proteins
	shells: prisms	traces–0.2	–	$CaCO_3$	conchiolin
	shells: mother of pearl	0.1–1.2	–	$CaCO_3$	conchiolin
	shells: calcitostracum	0.2–8.3	–	$CaCO_3$	conchiolin
	gastric shield	17.3	–	—	—
ONYCHOPHORA	cuticle	+	–	—	proteins
ARTHROPODA					
Crustacea, *Diplopoda*	calcified cuticle	58.0–85.0	α	$CaCO_3$	arthropodins + sclerotins (10–32%)
	intersegmental membranes	48.0–80.0	α	—	arthropodins (23–51%)
Insecta, *Arachnida*, *Chilopoda*	hardened cuticle	20.0–60.0	α	—	arthropodins + sclerotins (40–76%)
	unhardened cuticle	20.0–60.0	α	—	arthropodins + (in some parts) resilin
All	peritrophic membrane	3.8–22.0	–	—	proteins (21–47%) + mucins
CHAETOGNATHA	grasping spines	+	–	—	—
POGONOPHORA	tubes	33.0	β	—	proteins (47%)
TUNICATA	peritrophic membrane	+	–	—	—

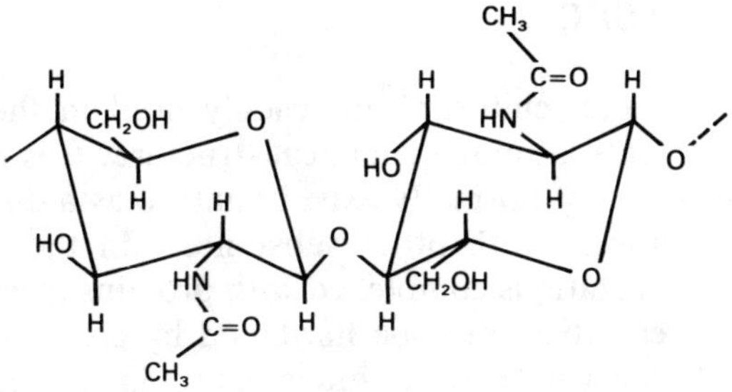

(1.1)

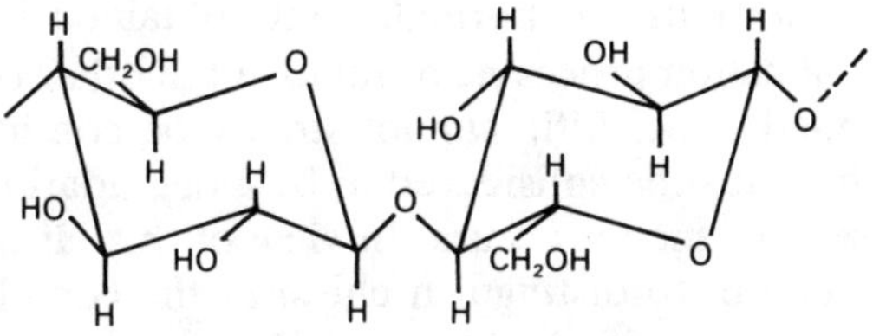

(1.2)

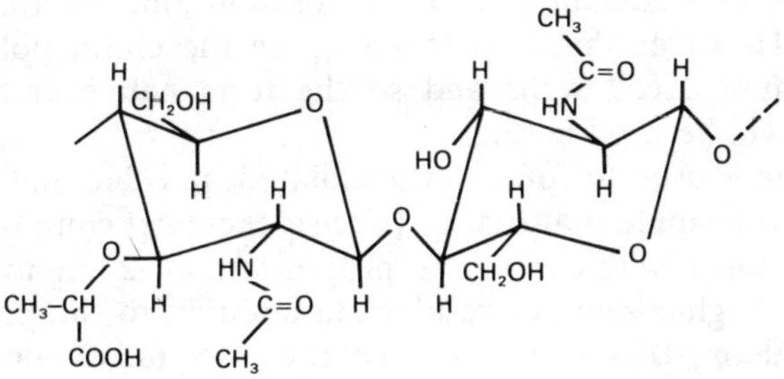

(1.3)

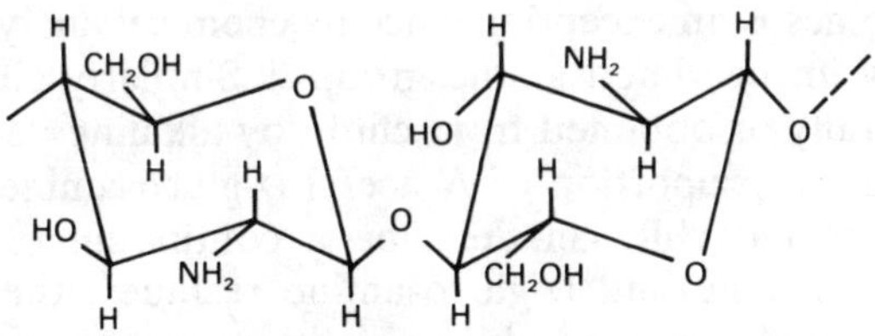

(1.4)

FIGURE 1.1   The structure of chitin (**1.1**) and chitosan (**1.4**), together with the structurally related polysaccharides cellulose (**1.2**) and murein (**1.3**)

1.2 NOMENCLATURE

The names 'chitin' and 'chitosan' are widely used in the literature but neither term represents a unique chemical structure. It is generally recognised that chitin does not normally exist in nature as a discrete substance but occurs as a complex with other substances. In the exoskeletons of insects and crustacea chitin is complexed with proteins (section 1.5), which in the case of insect cuticle may be hardened by crosslinking with polyhydroxyphenols (sclerotised), while fungal chitin is complexed with other polysaccharides including cellulose.[5] To date, pure chitin has only been identified[6] in the extracellular spines of the diatoms *Thalassiosira fluviatilis* and *Cyclotella cryptica*. Thus poly[β-(1→4)-2-acetamido-2-deoxy-D-glucopyranose] as an entity is normally only obtained in the laboratory after rigorous purification processes to remove the other components, and Richards has argued[3] that chitin cannot strictly be considered a 'natural' chemical entity but must be considered to be a degradation product of the various complexes in which it occurs. In view of this Hackman proposed[7] the term 'native chitin' to distinguish chitin in the complexed composite material from isolated, purified chitin, but Hunt has argued[8] that the term 'native chitin' suggests only differences in the conformation and macromolecular organisation of the polymer rather than its association with other molecules, and used the term 'chitin–protein complex' for chitin in its natural state. However this fails to recognise the chitin–polysaccharide complexes in fungal cell walls and so the term native chitin is to be preferred and will be used here.

Even if chitin is used to refer to the isolated, purified material, it will still not refer to a single material of precise chemical composition since the polymer chains will contain a proportion of 2-amino-2-deoxy-D-glucopyranose (D-glucosamine) residues in addition to the preponderant 2-acetamido-2-deoxy-D-glucopyranose (*N*-acetyl-D-glucosamine) residues. Whether these are present in the native chitin, or are introduced through hydrolysis of some of the acetamido groups during the isolation and purification processes, has not been established. Again the chitin obtained from diatom spines is an exception since its chemical analysis indicates the absence of any unacetylated amine groups.[6] Similarly chitosan, whether occurring naturally or obtained from chitin by alkaline deacetylation, will normally contain a proportion of *N*-acetyl-D-glucosamine residues. Thus the terms chitin and chitosan describe a continuum of copolymers of *N*-acetyl-D-glucosamine and D-glucosamine residues, the two being distinguished by insolubility or solubility in dilute aqueous acid solutions. As the properties are frequently dependent on the relative proportions of *N*-acetyl-D-glucosamine and D-glucosamine residues, as is the biodegradability and the biological role, a more precise system of nomenclature is desirable.

One attractive system would be to use the term chitin to refer to all isolated, purified material regardless of the extent of deacetylation, and to indicate the actual composition of the sample by giving the mole fraction of N-acetyl-D-glucosamine residues in brackets. Thus the pure homopolymer (**1.1**) would be chitin[1.0], homopolymer (**1.4**) would be chitin[0.0], and a typical commercial chitosan having 85% deacetylation would be chitin[0.15]. However there are a number of difficulties in applying such a system to the field of chitin chemistry as it exists today. Firstly the term chitosan is well known and accepted even if it is not clearly defined. Secondly there are a number of chitinolytic enzymes known whose degradative action decreases with decrease in the mole fraction of N-acetyl-D-glucosamine residues in the chain, and it would be confusing in the extreme to have a chitinase that would not degrade a sample of chitin, as would be the case if the substrate were chitin[0.0]. Finally, there is a large volume of published work on chitin/chitosan chemistry in which the composition of the material studied has not been determined, the polymer being merely described as chitin or chitosan depending on its provenance and on its insolubility or solubility in dilute aqueous organic acid solutions.

Because of these objections the convention adopted in this book is to use either chitin or chitosan, depending on the sample's solubility characteristics in dilute acid solution, followed by a figure giving the mole fraction of N-acetyl-D-glucosamine residues where this is known. Thus the three examples discussed above would be chitin[1.0], chitosan[0.0] and chitosan[0.15]. Where the sample's composition is not known the term chitin or chitosan, based on the same solubility criterion, will be used without any amending figure. Multiple numbers may be used for derivatives containing two or more chemically different N-acyl groups, the N-acetyl group concentration being given by the first figure with the others being given in the order of increasing molecular weight of the N-acyl group. Thus a derivative in which 15% of the amine groups are N-acetylated and 20% are N-hexanoylated would be N-hexanoylchitosan[0.15/0.20].

1.3 CHEMICAL STRUCTURE

1.3.1 Historical background

Chitin was first isolated in 1811 by Braconnot[9] as the alkali-resistant fraction from some of the higher fungi. The product, which he termed fungine, appears to have been impure, possibly containing other polysaccharides, the reported analysis indicating an approximately equal mixture of chitin and a non-nitrogeneous polyglucan. Braconnot also reported the formation of acetic acid from fungine and concluded that fungine was a new and quite distinct substance. In particular he was definite in his opinion that fungine differed from the woody material of plants.

In 1823 Odier isolated an insoluble residue, to which he gave the name chitin (*Greek*, χιτων, tunic or covering), from the elytrum of the cock-chafer beetle, or May bug, by repeated treatments with hot KOH solutions.[10] He failed to detect nitrogen in the elytra and concluded that chitin is related to vegetable rather than animal substances, commenting that it is very remarkable that the same substance should be found in the 'framework of insects' and in vegetables. It is a frequent misapprehension that this statement indicates that Odier recognised the similarity between chitin and the fungine of Bracconot, but a further statement that 'lignin (woody fibre) is the only proximate vegetable principle which can be compared to it' clearly shows that he did not. Furthermore there is no reference to Bracconot's paper in that of Odier and there is no evidence that the latter was aware of the earlier work. Odier also identified chitin as present in demineralised crab carapace and suggested that it is the basic material of the exoskeletons of all insects and, possibly, the arachnides.

The following year Children[11] published an English translation of Odier's paper, together with some additional results from his own studies on chitin.* Doubting the validity of Odier's negative results for nitrogen, which had been based on his failure to detect NH_3 on burning elytra, Children determined the elemental composition and found nitrogen to be present. He suggested that Odier's test could have failed if a volatile acid such as acetic acid was evolved simultaneously with the NH_3. This suggestion appears to have been made in ignorance of Bracconot's work in which the evolution of acetic acid was reported;[9] certainly no reference is made to it. However, since the samples burnt by Odier had only been extracted with boiling water and not with hot KOH solution, they would have contained a considerable amount of protein (section 1.5) which should have given a positive test for nitrogen on burning. Hence Children's explanation cannot be correct and the reason for Odier's negative result is still unresolved.

From the description of the purification process used to obtain chitin, multiple extractions with hot concentrated KOH solutions, it is very probable that both Odier and Children actually isolated chitosan rather than chitin. Indeed the elemental analysis[11] gives an empirical formula of approximately $C_{11}H_{17}O_7N_2$ which is considerably closer to that for the disaccharide repeat unit of chitosan ($C_{12}H_{22}O_8N_2$) than that of chitin ($C_{16}H_{26}O_{10}N_2$). However chitosan was first recognised and described in

---

* Not everyone agreed with the name 'chitin', as shown by one of the editorial comments. 'Every man has a right to name his own child, but we think M. Odier might have made a happier election. *Elytrine* would have been more significant and at least as euphonius.' This suggestion must have gained some currency, for the *Sydenham Society Lexicon of Medicine and Allied Sciences* (1879–1899) refers to elytrine as 'the form of chitin which composes the elytra of insects.'

1859 by Rouget[12] who reported that treatment of chitin with concentrated KOH solution under reflux gave a product, 'modified chitin', that was soluble in dilute solutions of organic acids and gave a different colour on treatment with an acidified iodine solution than did the original chitin. In 1894 Hoppe-Seyler described heating chitin from the shells of crabs, scorpions and spiders with KOH at 180°C to give a product readily soluble in dilute acetic acid and hydrochloric acid solution, and which could be precipitated from such solutions by addition of alkali.[13] Hoppe-Seyler, who did not refer to the 'modified chitin' of Rouget,[12] proposed the name chitosan for the product which was stated, surprisingly, to have the same nitrogen content as the original chitin.

Owing perhaps to Odier's conclusion[10] that the structural material in the exoskeletons of insects and in plants is similar, there was some confusion between cellulose, chitin and chitosan for much of the remainder of the 19th Century. This also arose partly from the tendency to use the term 'cellulose' as a group name and to apply it to all the structural materials in fungi. Thus Fremy stated[14] that there were three varieties of cellulose in the principal tissues of vegetables, one of which, called metacellulose (fungine) and found chiefly in agarics and lichens, could be identified by its insolubility in cuprammonium hydroxide. Although Schulze proposed[15] in 1891 that the name cellulose be restricted to that constituent of the cell wall that is not dissolved in dilute mineral acid or dilute alkali, is only slightly attacked by Schulze's reagent, is soluble in cuprammonium hydroxide, and yields only D-glucose (grape sugar) on hydrolysis, use of the name cellulose with reference to fungal chitin continued for several years. Thus in 1894 Winterstein published two papers[16, 17] dealing with 'fungus cellulose', the nitrogen-containing material obtained by him from various fungi by fusion with KOH at 180°C. Although on hydrolysis it gave the same mono-saccharide (D-glucosamine hydrochloride, see section 1.3.2) as did chitin,[18] Winterstein concluded that he had not established that 'fungus cellulose' and chitin were identical. In this he was technically correct since in view of the isolation procedure used the 'fungus cellulose' would have been chito-san and not chitin, as is shown by its reported solubility in dilute aqueous HCl. This latter fact caused Winterstein to conclude that the material could not be unaltered cellulose. In a third paper[19] Winterstein reported that the residue obtained after treating certain fungi with dilute H_2SO_4 and then dilute NaOH under reflux gave the same monosaccharide (D-glucosamine hydrochloride) on hydrolysis with HCl and that, just as in the case of chitin, acetic acid was produced during the hydrolysis. Moreover the residue behaved like chitin when fused with KOH for 1 hour at 180°C, giving acetic acid and a substance resembling chitosan. Despite this evi-dence, Winterstein still used the name 'cellulose' for the material.

Gilson[20, 21] also reported the presence of chitin in fungi, and its conver-sion to chitosan (mycosin). He noted its insolubility in cuprammonium hydroxide and that its elemental composition was in close agreement with previously reported analyses for animal chitin. Gilson suggested that the chitin in mushrooms has the same role as does the cellulose in the phanerogams and many of the cryptogams. He also pointed out that the chitin may be associated with other carbohydrate materials analogous or identical to those found in phanerogams.

Two years later Tarret[22] isolated 'fungin' from *Aspergillus niger* and confirmed its identity as chitin by hydrolysis with HCl and by alkali fusion with KOH. Although other fungi gave similar results, only in the case of *A. niger* was the chitin as pure as that obtained from crabs. Some 30 years later Karrer and co-workers[23, 24] examined the action of an extract from snails (*Helix pomatia*), containing a chitinase, on animal chitin[23] and fungal chitin[24] and found that the two substrates behaved similarly. Final confirmation that they have the same overall structure came from the similarity in the X-ray diffraction patterns of chitin from crustacea, fungi and insects[25, 26] and the identical IR spectra of chitin from insect and fungal sources.[27]

The accepted view today is that chitin from both animal and fungal sources is predominately poly[β-(1→4)-2-acetamido-2-deoxy-D-glucopyranose] (**1.1**), but that there may be differences in the mole fraction and distribution pattern of any D-glucosamine residues present, depending on the source. Furthermore animal-derived chitin may contain small amounts of bound aminoacid residues from the complexed proteins of the original native chitin, while chitin from fungal sources will not contain any such residues.

Four factors had to be determined in order to obtain the chemical structure of chitin and hence also of chitosan, These are:

(a) the component monosaccharide(s);
(b) the ring size – pyranose or furanose;
(c) the position of the glycoside bond between successive monosaccharide units;
(d) the stereochemistry of the glycoside bond.

Although investigations of the factors occurred in parallel in many instances, they will be discussed separately.

1.3.2 The component monosaccharide(s)

The configuration of 'glucosamine'/chitosamine

The first information regarding the chemical structure of chitin was obtained in 1876 when Ledderhose hydrolysed lobster shell chitin by heating it in concentrated HCl for 30 minutes.[18] Analysis of the crystals obtained on evaporation showed them to have a composition in agreement with that of a hexosamine hydrochloride salt. Ledderhose referred to it as a glycosamine hydrochloride and assigned structure (1.5) to it with the amine group at C(6). He later reported the production of acetic acid during the hydrolysis step.[28]

$$CHO.(CHOH)_4.CH_2NH_2 + HCl$$

(1.5)

The name 'glucosamine' was first applied to the sugar by Tiemann[29] although its configuration was not to be unequivocably established for over 55 years. Tiemann attributed the name to Ledderhose, stating that the latter had used the name in his original paper:[18]

> Mit dem Namen 'salzsaures Glucosamin' hat
> G. Ledderhose die schön kristallisierte, von
> ihm zuerst in chemisch reinem Zustande dargestellte
> substanz bezeichnet, welche bei dem Kochen von
> Chitin mit konzentrierten Salzsaure entseht.

This attribution is in error, for Ledderhose referred to the monosaccharide from chitin as a glycosamine and did not attempt to assign a specific configuration to it. Subsequently, although a number of research workers used the trivial name chitosamine in view of the absence of a definite configuration, most followed Tiemann's nomenclature.

The first evidence of its configuration was obtained by Tiemann[30] who prepared the phenylosazone derivative of 'glucosamine' and found it to be identical to that obtained previously by Fischer[31] from D-glucose and D-fructose (dextrose and laevulose respectively in the original paper). This proved that the configurations at carbon atoms C(3)–C(5) are identical to those of D-glucose and that the amine group is attached to C(2). These two facts were confirmed by Fischer and Leuchs by the synthesis of 'glucosamine' from D-arabinose.[32] However the question of the stereochemistry at the C(2) position was still unresolved, it being possible for 'glucosamine' to have the configuration of either D-glucose or D-mannose.

In 1912 Irvine and Hynd[33] attempted to resolve the problem by deamination of 'glucosamine' to a hexose by the reaction scheme given in Figure 1.2. Such a series of reaction indicates that 'glucosamine' has the D-glucose configuration, provided the authors' assumption that a Walden inversion does not occur during the deamination step is correct. However the authors themselves stated that no rigorous proof could be offered in support of this assumption and that 'the displacement of any group by any other group may be accompanied by inversion.' The wisdom of this caveat was demonstrated two years later when the same workers obtained[34] D-mannose by nitrous acid deamination of 'glucosamine' (Figure 1.3).

Having produced both D-glucose and D-mannose by deamination of 'glucosamine', the authors argued[34] that the first deamination process is the

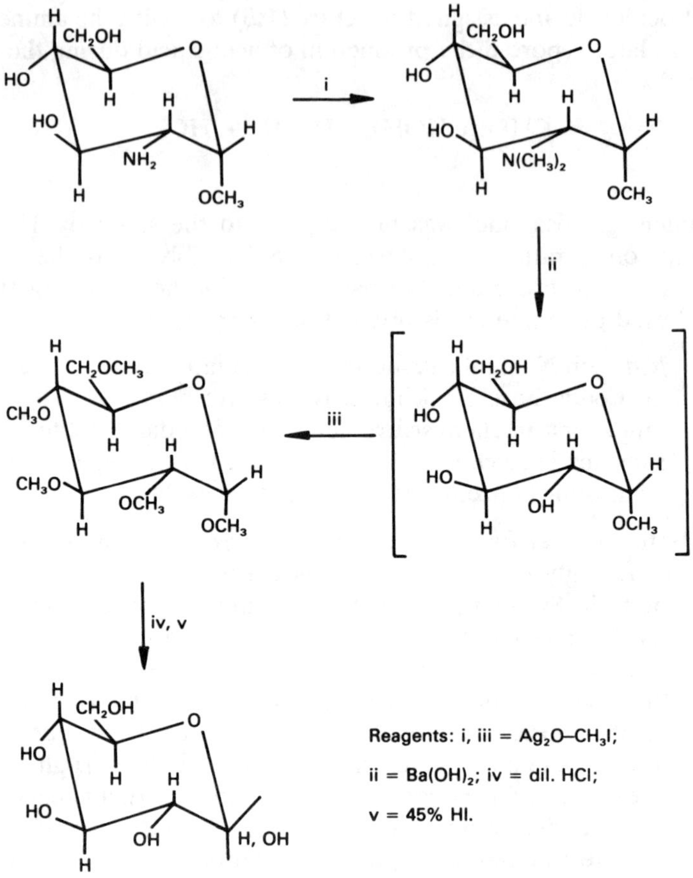

FIGURE 1.2   Deamination of methyl 2-amino-2-deoxy-α-D-glucopyranoside[33]

simpler one since 'the critical reaction is simple and confined to a group attached to one asymmetric C atom whilst in the second the fundamental reaction is complex and unusual and directly affects two asymmetric centres.' In their opinion the first reaction is the less likely to be accompanied by inversion and they therefore concluded that the bulk of the evidence was in support of 'glucosamine' having the D-glucose configuration.

The C(2) configuration was conclusively established in 1939 with the publication of three papers dealing with the problem. Neuberger and Pitt Rivers concluded,[35] on the basis of a comparison of the relative rates of hydrolysis of the methyl α- and β-D-glycosides of 'glucosamine' with those for methyl α-D-glucopyranoside and its β-anomer, that the methoxyl group in the methyl α-D-glycosaminide has a *cis* configuration with respect to the C(2)-amine group and hence 'glucosamine' must have the same C(2)

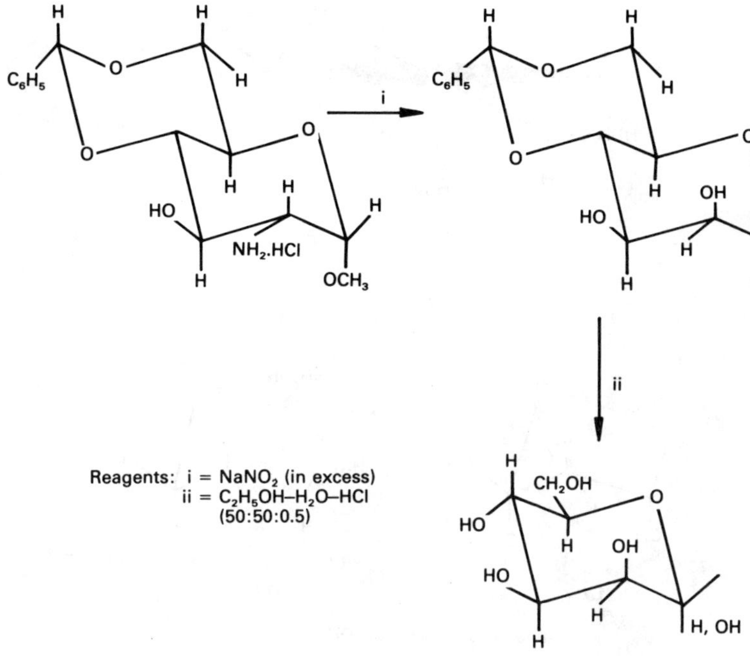

Reagents: i = NaNO₂ (in excess)
ii = C₂H₅OH–H₂O–HCl
(50:50:0.5)

FIGURE 1.3   Deamination of methyl 4, 6-O-benzylidine-2-amino-2-deoxy-α-D-glucopyranoside hydrochloride[34]

configuration as does D-glucose. This was supported by the optical behaviour of 'glucosamine hydrochloride', its glycosides and derivatives, which was found to resemble that of D-glucose and to differ from that of D-mannose.

Unequivocable proof by synthesis followed swiftly in a paper by Haworth et al.[36] Previous work by Haworth's group[37–39] on anhydro sugars had established the general rule that fission of the epoxide ring by NH_3 leads to the formation of two isomeric amino sugars, resulting from nucleophilic attack at either C atom of the epoxide ring. In each case, ring opening is accompanied by a Walden inversion at the C atom of the epoxide ring at which the attack occurs and to which the amine group therefore becomes attached. For example, 2,3-anhydro-D-allose (**1.6**) gives 2-amino-2-deoxy-D-altrose (**1.7**) and 3-amino-3-deoxy-D-glucose (**1.8**) (Figure 1.4).

Haworth et al.[36] found that methyl 4,6-di-O-methyl-2,3-anhydro-β-D-mannopyranoside (**1.9**), when heated under pressure with anhydrous CH_3OH–NH_3, gave two isomeric amines that were converted to the N-acetyl derivatives and then separated by fractional crystallisation. Haworth et al. established that the major component (**1.10**) (90%) was a derivative of 3-amino-3-deoxy-D-altrose and hence the minor component (~10%)

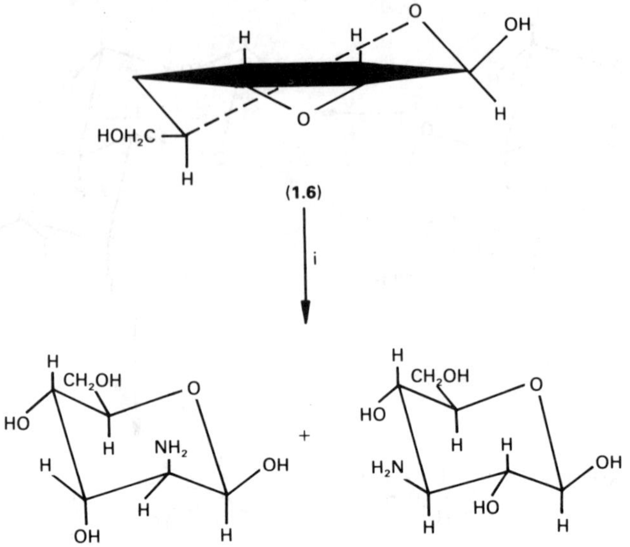

(1.6)

i

(1.7)   +   **(1.8)**

Reagents i = NH₃–CH₃OH

FIGURE 1.4   Stereochemistry of the epoxide ring opening of 2, 3-anhydro-D-allose on
reaction with anhydrous NH₃/CH₃OH

must be, on the basis of the previously established general rules,[37–39] a
derivative of 2-amino-2-deoxy-D-glucose, namely methyl 4,6-di-*O*-methyl-
2-acetamido-2-deoxy-β-D-glucopyranoside **(1.11)** (Figure 1.5). Methylation
of **(1.11)** gave methyl 3,4,6-tri-*O*-methyl-2-acetamido-2-deoxy-β-D-gluco-
pyranoside **(1.12)** which was identical in properties with the compound
prepared from 'glucosamine' by *N*-acetylation followed by *O*-methylation.
In the words of the authors[36] this synthesis 'established beyond reasonable
doubt that the biologically important hexosamine, glucosamine or chitosamine is configurationally related to D-glucose and not D-mannose.'

In the third paper Cox and Jeffrey[40] determined the crystal structures of
α-chitosamine hydrobromide and hydrochloride by X-ray crystallography
and showed that the configuration of chitosamine is the same as that of
D-glucose, the substituents on C(1) and C(2) of the α-form being in the *cis*
position with respect to each other.

Evidence for N-*acetylation of* D-*glucosamine*

As mentioned above, Ledderhose[28] reported the production of acetic acid
during acid hydrolysis of chitin. Other workers subsequently reported its

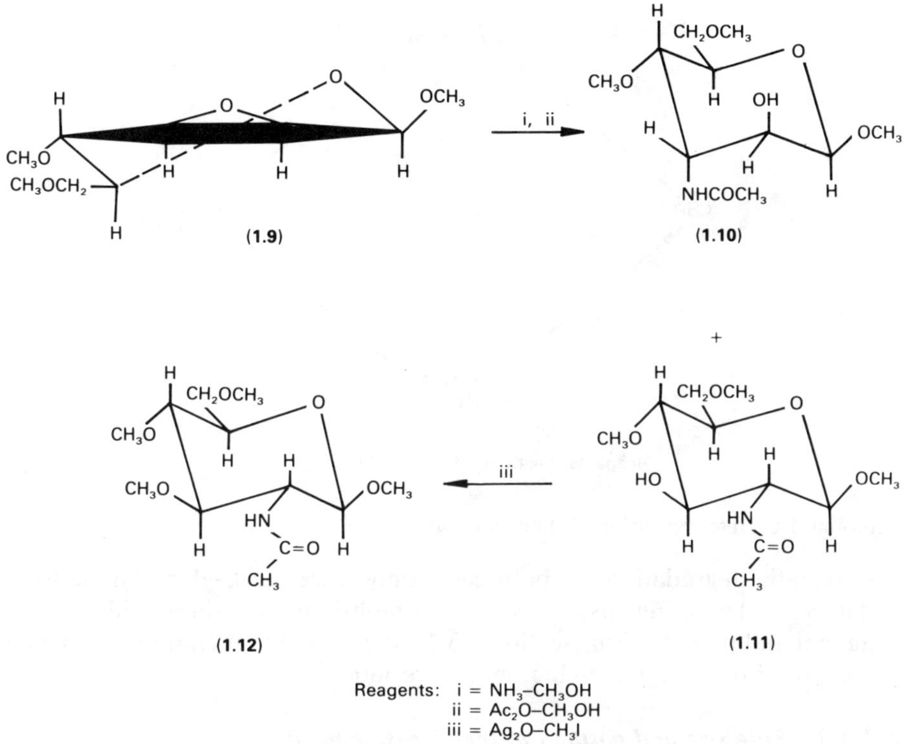

FIGURE 1.5 Synthesis of methyl 3, 4, 6-tri-O-methyl-2-acetamido-2-deoxy-β-D-glucopyranoside from methyl, 4, 6-di-O-methyl-2, 3-anhydro-β-D-mannopyranoside[36]

production both during acid hydrolysis[19, 41] and in the preparation of chitosan by KOH fusion of chitin.[13, 19, 41, 42] Indeed the cycle carried out by Hoppe-Seyler[13] (Figure 1.6) clearly demonstrates the relationship between chitin and chitosan but neither this cycle, nor the results of the other workers,[19, 41, 42] established whether chitin contains N-acetyl or O-acetyl groups.

This was resolved in 1902 when Fränkel and Kelly[43] carried out the acid hydrolysis of chitin under milder conditions than used previously (70% H_2SO_4, room temperature) and obtained N-acetyl-D-glucosamine, identical with that obtained[44] by N-acetylation of D-glucosamine. Ten years later Brach and von Fürth[45] determined the stoichiometry of D-glucosamine and acetic acid production, found it to be a 1:1 ratio and concluded that chitin is a 'polymerised monoacetylglucosamine'.

Karrer et al. obtained crystalline N-acetyl-D-glucosamine in 50% yield from lobster shell chitin[23] and in 80% yield from fungal chitin[24] by

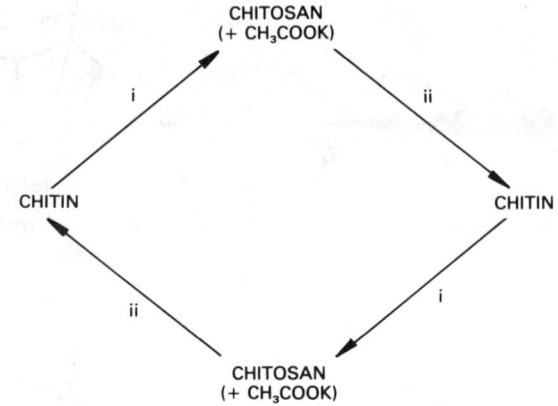

Reagents: i = KOH/180°C; ii = (CH₃CO)₂O

FIGURE 1.6   Interconversion of chitin and chitosan[13]

enzymatic degradation. In both cases only traces of D-glucosamine were detected. These results, as well as establishing the similar identity of animal and fungal chitin (section 1.3.1), confirmed that chitin is a polymer composed of *N*-acetyl-D-glucosamine residues.

1.3.3   Ring size and position of the glycoside bond

Contemporary work on the structure of cellulose

In dealing with this aspect of chitin structure it is of interest to consider the development of structural theories for cellulose, as these had a considerable influence on the structures proposed for chitin.

In 1914 Denham and Woodhouse reported[46] the isolation, in about 10% yield, of a new tri-*O*-methyl-D-glucose compound from the acid hydrolysis of a partially *O*-methylated cellulose (24–26% –OCH₃). They subsequently determined it to be 2,3,6-tri-*O*-methyl-D-glucose,[47] a conclusion supported by Haworth and Leitch[48] and confirmed by Irvine and Hirst.[49] These latter workers improved the methylation process for cellulose and obtained[50] a yield of greater than 90% of a mixture of methyl 2,3,6-tri-*O*-methyl-α-D-glucoside and methyl 2,3,6-tri-*O*-methyl-β-D-glucoside on methanolysis of an almost completely *O*-methylated cellulose (43.8% –OCH₃). This mixture, on normal hydrolysis with aqueous acid, yielded 2,3,6-tri-*O*-methyl-D-glucose as the sole product, except for a small quantity of di-*O*-methyl-D-glucose corresponding exactly to the deficiency in the methoxyl content of the methyl cellulose used. Since monosaccharides were considered at that time to exist in the 5-membered (furanose) ring form, Irvine and Hirst[50] argued that the quantitative formation of

2,3,6-tri-*O*-methyl-D-glucose from *O*-methyl cellulose shows that the C(1) –OH and C(5) –OH groups of the anhydro-D-glucose residues in cellulose must be protected from *O*-methylation through being substituted by other anhydro-D-glucose residues.

However in 1926 Haworth's group[51] demonstrated that D-glucose tends to form a 6-membered (pyranose) ring and pointed out[52] that the constitutional formulae allocated to di- and trisaccharides, and also to starch and cellulose (in all of which, furanose ring structures had previously been assumed) would have to be altered. In the particular case of cellobiose, this change from a furanose to a pyranose ring form would also require the adoption of a (1→4) glucoside link rather than the then accepted (1→5) link, and this was subsequently confirmed.[53]

Also in 1926 Sponsler and Dore[54] compared the X-ray diffraction pattern obtained from ramie fibres with a number of structures previously proposed for cellulose on the basis of chemical evidence. They concluded that the most satisfactory agreement with the experimental data was obtained by assuming a pyranose ring form for the anhydro-D-glucose residues and based on this they proposed a most peculiar structure in which the D-glucose residues were linked by alternating (1→1) and (4→4) bonds. This was criticised by Haworth[55] on the grounds that the (4→4) link, being an ether link, would be expected to be resistant to acid hydrolysis but no such disaccharide had been obtained on hydrolysing cellulose. Furthermore such a structure does not contain the cellobiose unit, although cellobiose octaacetate had been obtained in up to 50% yield on acetolysis of cellulose, so it would have to be formed from D-glucose molecules during the hydrolytic process. Not surprisingly, Haworth considered this to be most unlikely.

In 1928 Meyer and Mark published a classic paper[56] on the structure of cellulose, based on X-ray diffraction studies, in which the chemical and diffraction data were reconciled. In this it was argued that Sponsler and Dore had made errors in interpreting their diffraction data and that the anhydro-D-glucose residues are in the form of pyranose rings linked β-(1→4), in agreement with the chemical evidence, and also that the cellulose molecule has a twofold screw axis with each pyranose ring being rotated 180° relative to the preceding one.

This was rapidly followed by another paper[57] in which the same authors assigned an analogous structure to chitin, with the anhydro-*N*-acetyl-D-glucosamine residues in the pyranose form linked β-(1→4) and with each ring rotated 180° relative to the preceding one. In assigning this structure to chitin, Meyer and Mark had to make several assumptions regarding the chemistry of chitin. The configuration at C(2) had not been established (section 1.3.2) and chitobiose, the chitin-derived disaccharide analogous to cellobiose which had been so important in elucidating the structure of cellulose, had not yet even been isolated (see later in this section). Thus neither ring size, nor the position and stereochemistry of the glycosidic link, nor the stereochemistry of the monosaccharide itself had been established. These factors were only determined by chemical and biochemical means after this structure, which was in general correct, had been proposed by Meyer and Mark from the X-ray diffraction data and from consideration of the similar biological roles of cellulose and chitin which suggested similar structures.

Ring size and glycoside bond position in chitin

Bergmann et al. were the first to isolate a crystalline disaccharide, as the octaacetate, by acetolysis of chitin.[58] They assigned it the name chitobiose and expressed the opinion that it might be as useful in the study of chitin as cellobiose had been in the study of cellulose. Its preparation was subsequently reported by Zechmeister and Tóth[59, 60] who also isolated chitotriose undecaacetate from the acetolysis mixture. In a second paper Bergmann et al. investigated the chemical behaviour of chitobiose[61] and their results, coupled with the assumptions that the primary hydroxyl groups are unsubstituted and that the sugar residues have the pyranose ring form, enabled them to propose structure (1.13) for chitobiose, the stereochemistry of the glycoside bond being unresolved.

$$HOCH_2\text{--}CH\text{--}CHOH\text{--}CHOH\text{--}CH(NH_2)\text{--}CHO \qquad O \qquad HOCH_2\text{--}CH\text{--}CH\text{--}CHOH\text{--}CH(NH_2)\text{--}CH(OH)O$$

(1.13)

Zilliken et al.[62] later obtained di-N-acetylchitobiose as a pure crystalline material, m.p. 245–247°C, whereas Bergmann et al.[58, 61] had described their product as an amorphous material with m.p. > 185°C. These later workers confirmed the stoichiometry of its oxidation by NaOI and also reported the Morgan–Elson reaction to be negative,[62] behaviour that had been shown to be typical of N-acetyl-D-glucosamines in which the C(4) –OH is substituted.[63]

Among further attempts at hydrolysis and fractionation of chitin[64–70] only one[64] has had, as the major objective, the preparation of a polymer–homologous series of oligosaccharides in order to investigate changes in properties with increase in chain length. Barker et al.,[64] following the work of Horowitz et al.,[69] hydrolysed chitosan rather than chitin so that the hydrolysis reaction could be carried out homogeneously. However they introduced a major development, selective N-acetylation of the oligosaccharide mixture prior to fractionation. This enabled them to isolate the oligosaccharide components up to the hepta-N-acetylchitoheptose (**1.14**, $n = 5$).

(1.14)

That these N-acetylchitosaccharides formed a polymer-homologous series was shown by:

(a) the rectilinear relationship for the plot of $[M]_D/n$ *versus* $(n - 1)/n$, where $[M]_D$ is the molecular rotation and n is the degree of polymerisation (*DP*), as previously predicted by Freudenberg *et al.*[71] for an homologous series of oligosaccharides when $n > 1$;

(b) the rectilinear relationship between $\log(1/R_F - 1)$ *versus* n, where R_F is the mobility in paper chromatography, as previously found for an homologous series of oligosaccharides.[72]

The presence of a (1→4) link between the penultimate sugar residue and the reducing end residue was shown by the negative results obtained on subjecting the oligosaccharides to the Morgan–Elson test.[73] Substitution of the C(4) –OH group eliminates any response to the test by preventing the amino sugar adopting the furanose form,[63] hence the failure of the oligosaccharides to respond to the test indicates that the C(4) position of the reducing sugar residue of each member of the series is involved in the glycoside link. This had previously only been demonstrated for di-N-acetylchitobiose.[61, 62]

Periodate oxidation of di-N-acetylchitobiose showed the rapid consumption of 1 mole of IO_4^- per mole of substrate, followed by the very much slower consumption of a further 1 mole of IO_4^-, and during this second stage reaction 1 mole of HCHO is released. This pattern of oxidation is only compatible with a structure in which both N-acetyl-D-glucosamine residues have a pyranose configuration and in which they are linked (1→4'). The initial rapid reaction involves cleavage of the C(3) –C(4) bond of the non-reducing sugar residue, while the slower reaction involves cleavage of the C(5') –C(6') bond of the open chain form produced during mutarotation of the reducing residue (**1.15**).

(1.15)

Direct chemical evidence from chitin or chitosan is limited. Jeanloz and Forchielli[74] determined the periodate consumption, formic acid production and, in the case of chitosan, the amount of ammonia liberated, during periodate oxidation of chitin and chitosan[0.2]. Comparison with the theoretical values for the various possible linking arrangements for pyranose residues showed that the experimental results agreed most closely with those for a (1→4) linked structure. Wolfrom et al.[75] prepared an O-methyl chitin which on hydrolysis yielded crystalline 3, 6-di-O-methyl-2-acetamido-2-deoxy-α-D-glucopyranose. This was taken as proof of the (1→4) linkage in chitin, although the same product would have been obtained if the anhydro-N-acetyl-D-glucosamine residues had the furanose ring form and were linked (1→5) (discussed earlier in this section).

More recently Saito et al.[76] examined chitin in the solid state by CP/MAS NMR spectroscopy and found that the ^{13}C NMR spectrum is in agreement with a (1→4) linked structure. Similarly the ^{13}C NMR spectrum of chitin during hydrolysis in anhydrous hydrogen fluoride[68] shows the N-acetyl-D-glucosamine residues to be linked (1→4), a result confirmed by Gagnaire et al.[77] in their NMR studies of chitin dissolved in DMAc–LiSCN and DMAc–LiCl. Di-N-acetylchitobiose, whose structure had previously been rigorously confirmed, was used as a reference for all the peak assignments in this later work.[77]

1.3.4 Stereochemistry of the glycoside bond

In deriving their proposed structure for chitin (section 1.3.3) Meyer and Mark assumed a β-configuration for the C(1) anomeric centres in chitin,[57] in part because of the similarity in the repeat distances for cellulose and chitin as determined by X-ray crystallography. The first direct evidence in support of this was obtained some 10 years later when Zechmeister and Tóth[78, 79] found that a chitinase extracted from snails (Helix pomatia) and freed from β-glucosidase and α-galactosidase by chromatography, cleaved the glycoside bond of phenyl 2-acetamido-2-deoxy-β-D-glucopyranoside but not that of the α-isomer. Similar behaviour was observed for the chitinase obtained from emulsin.[80] It was also found that the crude snail chitinase consists of two components, a chitinase and a chitobiase. The latter component did not degrade chitin but did hydrolyse oligomers of chitin.

The stereochemistry of the C(1) anomeric centre has also been investigated by IR spectroscopy of the polymer–homologous series up to hepta-N-acetylchitoheptose.[64] All exhibited an absorption band at 884–900 cm^{-1}, previously shown[81] to be indicative of β-D-glucopyranosides. The spectra of N-acetyl-D-glucosamine, di-N-acetylchitobiose and tri-N-acetylchitotriose also showed an absorption band in the region of 850 cm^{-1}, indicative of α-D-glucopyranosides[81] but the presence of this band in these spectra was attributed to the fact that in the crystalline state the reducing sugar residue of these compounds has the α-configuration, a conclusion that is supported by the downward mutarotation of their solutions.[64]

The presence of the β-glycoside link in chitin has also been shown by ^{13}C NMR spectroscopic studies of solutions of chitin in DMAc–LiSCN and DMAc–LiCl[77] and anhydrous HF,[68] as well as in the solid state using CP/MAS techniques.[76] It has also been demonstrated in chitosan, and hence in the chitin precursor, by ^{13}C NMR spectroscopy of the oligosaccharide series up to the pentadecamer.[70] In these studies the ^{13}C signals are assigned on the basis of the ^{13}C shifts in the spectra of di-N-acetylchitobiose and other chitooligosaccharides.

1.4 PHYSICAL STRUCTURE

1.4.1 Introduction

Chitin has a highly ordered, crystalline structure as evidenced by X-ray diffraction studies. It has been found in three polymorphic forms, α-, β- and γ-chitin,[82] which differ in the arrangement of the chains within the crystalline regions. In α-chitin the chains are anti-parallel, in β-chitin they are parallel, and in γ-chitin two chains are 'up' to each chain 'down'[82] (Figure 1.7).

α-chitin   β-chitin   γ-chitin

FIGURE 1.7   Arrangement of the polymer chains in the three forms of chitin

The most abundant form is α-chitin, which also appears to be the most stable since both the β- and γ-chitin may be converted into the α-form by suitable treatments. Thus β-chitin is converted to the α-form on precipitation from solution in formic acid,[82] or on treatment with cold 6M HCl.[82, 83] The ease of this solid state transformation depends on the extent of crystallinity of the sample, occurring very readily in the case of β-chitin from *Loligo* pen and *Aphrodite* chaetae but less readily with the more crystalline β-chitin from pogonophore tubes and the spines of the diatoms *Thalassiosira fluviatilis* and *Cyclotella cryptica*.[83] Treatment of γ-chitin with a saturated aqueous solution of LiSCN converts it to α-chitin.[83] However both β- and γ-chitin are stable to boiling 1.25M NaOH solution[84] and to 1.57M HCl, both of which are used extensively in the preparation of chitin from native chitin (section 2.2). The β → α and γ → α transformations appear to be irreversible.

Rudall has suggested[85] that the three polymorphic forms are related to the diversity of function. α-Chitin is found where extreme hardness is required, as in arthropod cuticle, and is frequently associated with sclerotised protein or inorganic materials, or both. β-Chitin and γ-chitin are found where flexibility and toughness are required. All three forms have been shown to be present in the squid *Loligo*; α-chitin in the relatively thin oesophageal cuticle lining the stomach, β-chitin in the skeletal pen, and γ-chitin in the thick cuticle lining the stomach.[82] This indicates that the different forms relate to function rather than to taxonomic grouping.

In the numerous X-ray crystallographic studies, there is some variation in the assignment of axes in the unit cells of α- and β-chitin. Meyer and Pankow[86] and Carlstrom[87] designated b as the fibre axis in their work on α-chitin, as had Meyer and Mark,[57] while Dweltz[88] and Minke and Blackwell[89] designated c. In the case of β-chitin Dweltz initially designated c as the fibre axis[90] but changed this to b in a subsequent paper,[91] while Blackwell changed from b initially[92] to c subsequently.[93]

Interchanging the b and c axes is permissible in the case of α-chitin because of the symmetry of the orthorhombic cell ($P2_12_12_1$ – section 1.4.2) which possesses three sets of non-intersecting, mutually perpendicular, twofold screw axes. However it is not permissible in respect of β-chitin in which the unit cell is monoclinic (section 1.4.3) which is conventionally defined[94] as having $a \neq b \neq c$, $\alpha = \gamma = 90°$, $\beta \neq 90°$. Thus to reverse the b and c axes imposes a change in angles, making $\beta = 90°$ and $\gamma \neq 90°$, which is not in accord with the conventional crystallographic definition. Furthermore the space group for β-chitin is $P2_1$ which requires b to be the unique axis and therefore, since the twofold screw axis is in the direction of the fibre axis, b must be the fibre axis. Although the *International Tables for X-ray Crystallography* gives a second setting for the space group $P2_1$ in which c is the unique axis, the former setting is used unless a comparison is being made, in the same paper, with hexagonal or rhombohedral unit cells where c is the unique axis.[95]

Walton and Blackwell have commented[96] that 'traditionally, b is taken as the fibre axis for cellulose and chitin; in most other polymer work, c is so used.' However it is not simply a tradition; cellulose I (native cellulose) has a monoclinic cell with the space group $P2_1$ and the fibre axis parallel to the 2_1 screw axis, hence b must be the unique axis. Furthermore a number of synthetic polymers, including poly(vinylidene chloride), poly(vinylidene fluoride), nylons 3, 4, 6, 8 and 10 and several poly(amino acid)s, all of which have monoclinic unit cells and space group $P2_1$, take b as the fibre axis.[97] Thus since crystallographic practice requires b to be taken as the fibre axis for β-chitin, the same assignment has been made for α-chitin in the current work and the literature data has been transposed, where necessary, to conform to this.

1.4.2 Structure of α-chitin

The structure of α-chitin has been investigated more extensively than that of either the β- or γ-form, because it is the more common polymorphic form. The earliest X-ray investigation was that of Gonell[98] whose results (Table 1.2) formed the basis for the paper of Meyer and Mark in which a structure analogous to that of cellulose was proposed,[57] based on the similarity in the fibre repeat distances. It should be remembered that since very many fewer X-ray reflections are obtained from a polysaccharide sample than from a mono- or disaccharide single crystal, it is not possible to determine directly the complete spatial arrangement of all the atoms in even a simple polysaccharide unit cell. Instead, comparisons are made between theoretical and observed intensities, the former being calculated from possible structures selected on the basis of the available chemical and biochemical evidence. The extent of compatibility of the calculated and observed intensities gives an indication of the correctness of the selected model structure.[99]

TABLE 1.2   *Unit cell dimensions (Å) for α-chitin*

a	b	c	
11.58	10.44	19.42	Gonell[98]
9.40	10.46	19.25	Meyer and Pankow[86]
9.25	10.46	19.25	Clark and Smith[100]
9.40	10.26	19.25	Lotmar and Picken[108]
4.76	10.28	18.85	Carlstrom[87]
4.69	10.43	19.13	Dweltz[88]
4.74±0.01	10.32±0.02	18.86±0.01	Minke and Blackwell[89]

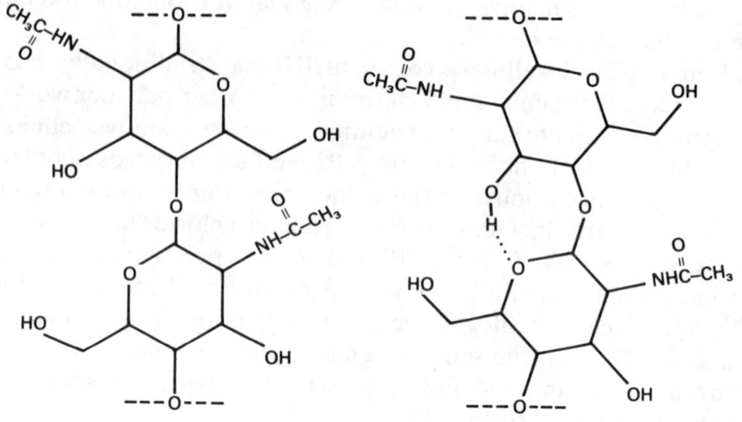

FIGURE 1.8   (a) Straight and (b) bent conformations for the disaccharide repeat unit in the chitin chain

Other early workers[86, 100] established the unit cell as orthorhombic with cell dimensions[86] $b = 10.46$ Å, $c = 19.25$ Å and a either 9.4 Å[86] or 9.25 Å,[100] but the first detailed structure analysis was that of Carlstrom[87] who concurred with an orthorhombic unit cell but obtained different dimensions, namely $a = 4.76$ Å, $b = 10.28$ Å, and $c = 18.85$ Å (Table 1.2). Carlstrom's data also indicated the space group to be either $P22_12_1$ or $P2_12_12_1$, both of which require an antiparallel arrangement of chains, and the plane of the pyranose rings to be almost parallel to the (100) plane. Of the two possible space groups the latter, $P2_12_12_1$, was eventually assigned to α-chitin by Carlstrom.

A major difference in the structure proposed by Carlstrom, compared with previous models,[57, 86] is that it is based on a 'bent' chain, similar to that proposed by Hermans et al. for cellulose[101, 102] (see Figure 1.8) rather than on a 'straight' chain. Carlstrom pointed out[87] that the 'straight' chain structure for cellulose, and by analogy for chitin, is unlikely owing to steric hindrance factors. Furthermore the 'bent' chain gives a C(3') –OH . . . O(5) distance of 2.68 Å, suitable for the formation of an intramolecular hydrogen bond, and a repeat distance along the fibre axis of 10.28 Å which agrees with the X-ray diffraction results.

A second difference between earlier models and that of Carlstrom is that the side groups are taken into consideration in the latter. The –NHCO.CH$_3$ groups are assumed to be planar and, from polarised IR spectroscopic results,[87, 103] to be predominantly perpendicular to the fibre axis. Carlstrom further proposed that because of the limited extent of rotation of the C(2)–N bond the N–H group has to point approximately towards a C(7) =O group of an adjacent chain along the *a* axis. This fact, together with the $C(2_1)N$–H . . . O=C(7_3) distance of 2.69 Å and IR evidence[87] of strong N–H . . . O=C hydrogen bonding, led Carlstrom to propose the presence of interchain hydrogen bonding in the direction of the *a* axis (see Figure 1.9 for the system used to identify atoms in the different chains in the unit cell). Although Carlstrom found that optically derived Fourier transforms based on his proposed structure (Figure 1.10) had intensity distributions similar to the observed X-ray intensities, he was careful to point out that because of the limited resolution of the diffraction patterns his model could only be regarded as a good approximation to the structure.

Several attempts were made over the following 20 years to improve on Carlstrom's model, most notably by Pearson *et al.*,[104] Ramakrishnan and Prasad,[105] and Haleem and Parker.[106] Pearson *et al.* carried out a detailed IR spectroscopic investigation and in general the results agreed with the Carlstrom model. However they pointed out that no free O–H stretching absorption band can be observed, so that it is necessary to modify the Carlstrom model to take account of this. After considering various arrangements for the CH_2OH group, the authors proposed a bifurcated hydrogen bonding system in which the C(6′)OH forms an intrachain donor

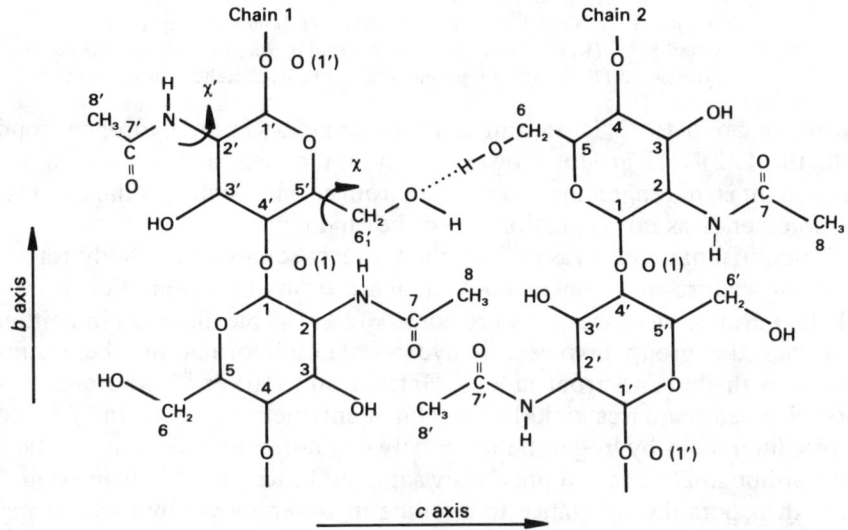

FIGURE 1.9 System used to identify atoms in the different chains in α-chitin. The chain directly below chain 1, that is the adjacent chain along the *a* axis, is designated chain 3. Interchain bonding is indicated by the appropriate subscripts on the atom numbers, thus the hydrogen bond depicted is described as $O(6_2)H$ . . . $O(6_1')$

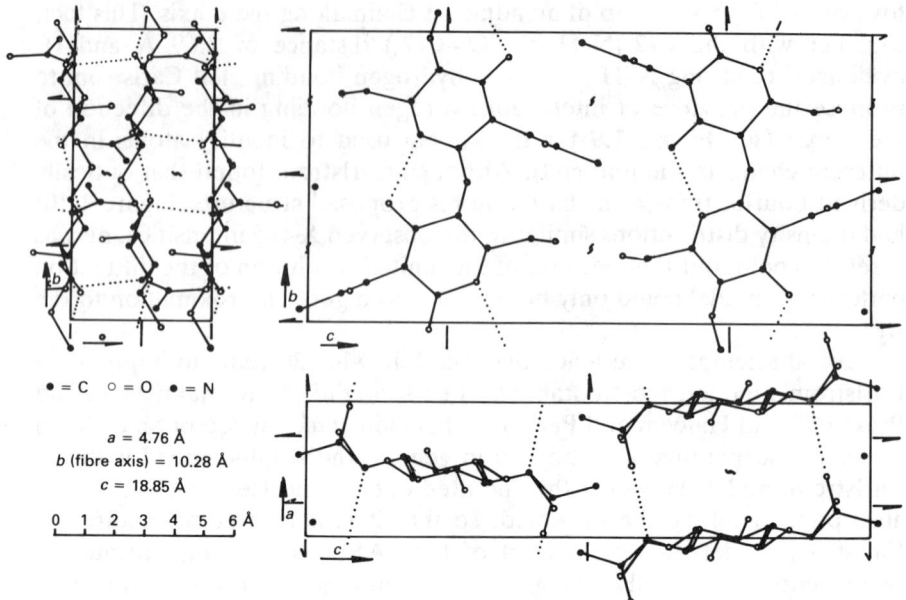

● = C   ○ = O   ● = N

a = 4.76 Å
b (fibre axis) = 10.28 Å
c = 18.85 Å

0 1 2 3 4 5 6 Å

FIGURE 1.10   The structure of α-chitin proposed by Carlstrom[87] showing the three main
projections of the unit cell. The C(3′)O–H . . . O(5) intrachain and
C(2₁)N–H . . . O=C(7₃) interchain hydrogen bonds are represented by
dotted lines. (Reproduced from the *Journal of Biophysical and Biochemical
Cytology*, **3** (1957) 679, by permission of the Rockefeller University Press)

hydrogen bond to O(1) and an intramolecular acceptor hydrogen bond
with the C(2)N–H group. However they noted that such a bonding arrangement is in disagreement with Carlstrom's conclusion that the CH₂OH
group extends as far as possible from the chain.[87]

Ramakrishnan and Prasad[105] applied a least-squares rigid-body refinement procedure and concluded that apart from the orientation of the
CH₂OH group, of which two were considered possible although in neither
case was the group involved in hydrogen bond formation, the results
agreed with the Carlstrom model. Haleem and Parker,[106] who also employed a least-squares rigid-body refinement method, were the first to
depict interchain hydrogen bonding between adjacent chains along the *c*
axis, although Dweltz had previously suggested such an interchain bond.[90]
In both papers the resistance to swelling in water shown by α-chitin was
attributed to this interchain bond. As with Ramakrishnan and Prasad[105]
the model of Haleem and Parker involves two possible positions for the
CH₂OH group but in this later model the group is involved in a hydrogen
bond in either position, both arrangements being energetically equivalent.
Although the CH₂OH groups were considered to be arranged somewhat at
random between the two possible positions, the particular arrangement in

any given hydrogen bond must affect the arrangement of neighbouring hydrogen bonds.

In 1978 Minke and Blackwell[89] and Blackwell et al.[93] pointed out a number of deficiencies in the Carlstrom model, their major criticisms being:

(a) the CH_2OH side chains are not hydrogen-bonded although IR spectroscopic studies show that all the hydroxyl groups form donor hydrogen bonds;
(b) the presence of two amide I peaks suggests that the arrangement of the amide groups, all of which are in the same environment in Carlstrom's model, cannot be correct;
(c) a medium intensity 001 meridional reflection is observed which is forbidden by the $P2_12_12_1$ space group;
(d) there is no interchain hydrogen bonding in the direction of the c axis which is surprising, in view of the lack of swelling of α-chitin in water, since β-chitin, where such interchain hydrogen bonding has been shown to be absent,[92] swells readily in water.

Minke and Blackwell[89] therefore concluded that although the conformation and approximate mode of packing of the chains in Carlstrom's model are correct, a conclusion supported by Marchessault and Sarko who found reasonable agreement between observed X-ray intensities and those calculated for the model,[99] the arrangement of the side chains and hence of the hydrogen-bonding network was not established.

These workers[89, 93] applied least-squares rigid-body refinement methods to their X-ray diffraction data using models with $P2_12_12_1$ symmetry (a2) and $P2_1$ symmetry, in the latter case both parallel (P_1) and antiparallel (a1) chain arrangements being considered. The $P2_1$ models were refined in terms of nine parameters and the $P2_12_12_1$ model in terms of six.

No acceptable hydrogen-bonded model with $P2_12_12_1$ symmetry could be found initially and the P_1 model was also eliminated. The refined a1 model ($R'' = 0.161$), which was not significantly different from the a2 model except that χ and χ' may have different values for the two chains (see Figure 1.9), allows the formation of an intramolecular $O(6')H \ldots O(7)$ bond for one chain and an intermolecular $O(6_2)H \ldots O(6_1')$ bond for the other. Although this conformation gives an increase in the R'' value to 0.190, it does allow the formation of a completely hydrogen-bonded structure.

This a1 model has intra- and intermolecular donor hydrogen bonds for the CH_2OH groups of chain 1 and chain 2 respectively, but an equally acceptable conformation is that which has inter- and intramolecular donor hydrogen bonds for the CH_2OH groups of chain 1 and chain 2 respectively; the model is degenerate. Since either arrangement can be adopted without influencing the choice of arrangement in neighbouring unit cells, a one-to-one statistical mixture of the two hydrogen-bonding arrangements was proposed as the most probable structure in large crystallites. Such a structure may be modelled by replacing the $O(6)$ atoms of each chain by two half oxygens and introduction of this one-to-one statistical mixture of inter- and intramolecular hydrogen bonds for the CH_2OH groups produces $P2_12_12_1$ symmetry (Figure 1.11).

Two projections of the structure are given in Figure 1.11. The unit cell contains disaccharide sections of two chains with full intramolecular $C(3')-OH \ldots O(5)$ and intermolecular $C(2_1)N-H \ldots O=C(7_3)$ hydrogen bonding and a 50/50 statistical mixture of $O(6_1')H \ldots O(7_1)$ / $O(6_2)H \ldots O(6_1')$ and $O(6_2)H \ldots O(7_2')$ / $O(6_1')H \ldots O(6_2)$ intra-/intermolecular pairs of hydrogen bonds. The intramolecular $CH_2OH \ldots$

$$O=C \text{ and intermolecular } CH_2OH \ldots \overset{H}{OCH_2} \text{ hydrogen bond lengths are}$$

2.85 Å and 2.79 Å respectively, while the intermolecular $N-H \ldots O=C$ hydrogen bonds linking the chains in sheets along the a axis are 2.73 Å.

This model answers the criticisms of the Carlstrom model listed above. It is fully hydrogen-bonded with all the hydroxyl groups being involved in donor hydrogen bonds, and has intermolecular hydrogen bonds along both the a and c axes to prevent swelling in water. An exact 50/50 mixture of the two patterns of hydrogen bonding corresponds to $P2_12_12_1$ symmetry and furthermore, since all the amide groups are involved in intermolecular $C(2_1)NH \ldots O(7_3)$ hydrogen bonds and half are also involved in the intramolecular $O(6')H \ldots O(7)$ hydrogen bonds, splitting of the amide I band would be expected, in agreement with the IR spectrum of α-chitin.

Finally the one-to-one statistical mixture of hydrogen-bonding patterns required for $P2_12_12_1$ symmetry will be disrupted by crystal defects and edge effects, thereby reducing the structure's symmetry and giving rise to the observed 001 reflection whose intensity will be inversely related to the crystalline perfection of the sample.

Support for this structure has come from polarised IR and Raman spectroscopy.[107] The $O(6')H \ldots O(7)$ hydrogen bond would be expected to alter the direction of these $C=O$ bonds, compared with those carbonyl groups involved only in intermolecular hydrogen bonds, thereby giving their absorption band a somewhat larger parallel component. In agreement with this, the dichroic ratio for the singly hydrogen-bonded carbonyl group absorbing at 1660 cm^{-1} was found to be 0.18 while that for the doubly hydrogen-bonded group at 1624 cm^{-1} was 0.26. Furthermore there is splitting of the N-H (stretching) band at 3270 cm^{-1} and of the amide II band at 1560 cm^{-1}, which in the case of the amide II band is considerably enhanced on partial deuteration by immersion in D_2O. Both observations were considered as supporting the structure proposed for α-chitin by Blackwell and co-workers.[89, 93]

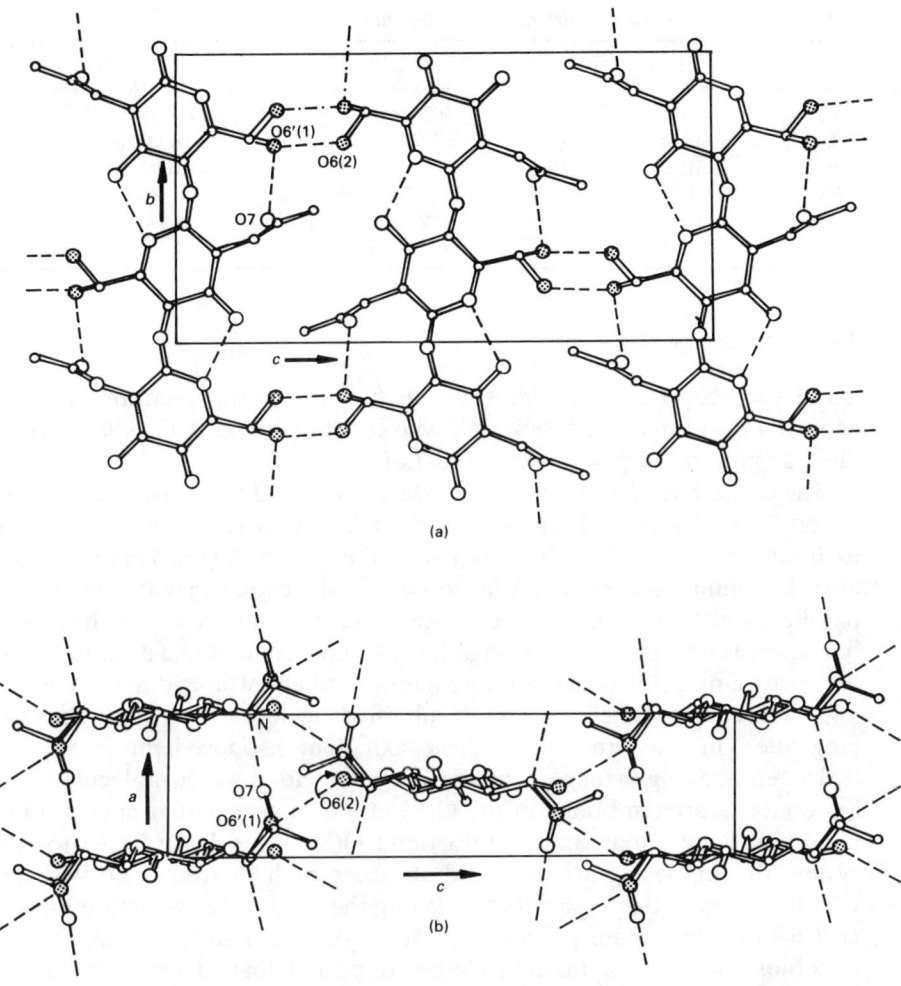

(a)

(b)

FIGURE 1.11   The structure of α-chitin proposed by Blackwell *et al.*[89, 93]
(a) The *bc* projection showing the intrachain C(3')O–H . . . O(5) bonds
together with the C(6₂)O–H . . . O(6₁') and the intrachain
C(6₁')O–H . . . O(7₁) bonds (– – – –). The alternative interchain
C(6₁')O–H . . . O(6₂) and intrachain C(6₂)O–H . . . O(7₂') bonds are
also shown (– · – ·).
(b) The *ac* projection showing the C(2₁)N–H . . . O=C(7₃) and
C(6₂)O–H . . . O(6₁') interchain bonds.
The axes designated *b* and *c* have been interchanged from the original in
order to conform with the text (section 1.4.1). (Reproduced from the
Journal of Molecular Biology, **120** (1978) 179, by permission of the authors
and of Academic Press Inc.)

TABLE 1.3   *Unit cell dimensions (Å) for β-chitin*

a	b	c	β		
4.7	10.3	10.5	90°		Dweltz[90]
4.85	10.38	9.26	97.5°		Blackwell *et al.*[109]
4.80	10.32	9.83	112°		Dweltz *et al.*[91]
4.85	10.38	9.26	97.5°	(anhydrous)	⎫
4.8	10.4	10.5	97°	(monohydrate)	⎬ Blackwell[92]
4.8	10.4	11.1	97°	(dihydrate)	⎭

1.4.3   Structure of β-chitin

The existence of a second crystal structure for chitin, β-chitin, was first reported by Lotmar and Picken[108] who identified it in *Aphrodite aculeate* chaetae and in the pen of the squid *Loligo*.

The earliest attempt to determine the structure of the β-form was that of Dweltz[90] who reported the unit cell to be approximately half that of α-chitin with a = 4.70 Å, b = 10.3 Å, and c = 10.5 Å (see Table 1.3). He also determined the space group to be $P2_1$, indicating that the chains are parallel, and that similarly to α-chitin there are intermolecular hydrogen bonds between amide groups on adjacent chains along the a axis, the bond distance being 2.82 Å. In addition he proposed an intraresidue O(3)H . . . O(7) hydrogen bond and intermolecular hydrogen bonding, through water molecules, in the direction of the c axis. The proposed intermolecular hydrogen-bonding arrangement was complex with a water molecule forming donor hydrogen bonds to the O(3) atoms of two similar chains translated along the a axis, that is $O(3_1)$ and $O(3_3)$ – see Figure 1.12 for the system of numbering atoms – and an acceptor hydrogen bond with the $O(6_2)$H group of the adjacent chain along the c axis. This structure may be criticised on two main points, the use of the 'straight' structure for the chitobiose unit rather than the 'bent' one, and that at least one of the hydrogen bond lengths is outside the normal range of values for this type of bond.

Blackwell *et al.*[92, 109, 110] studied β-chitin from pogonophore tubes and also from the spines of the marine diatoms *Thalassiosira fluviatilis* and *Cyclotella cryptica*. Although agreeing with Dweltz[90] that the unit cell is monoclinic and has the space group $P2_1$ slightly different cell dimensions were obtained[109] (Table 1.3). The increase in the unit cell dimensions along the a axis, compared with that for α-chitin,[87] indicates a greater spacing between the chains in this direction. This agrees with the increased frequency of the N–H (stretching) band at 3293 cm^{-1} and of the amide I band at 1631 cm^{-1}, compared with 3265 cm^{-1} and 1621 cm^{-1} respectively for α-chitin, these increased values being indicative of a longer $C(2_1)$NH . . . $O=C(7_3)$ hydrogen bond in β-chitin. Another difference is that the amide I

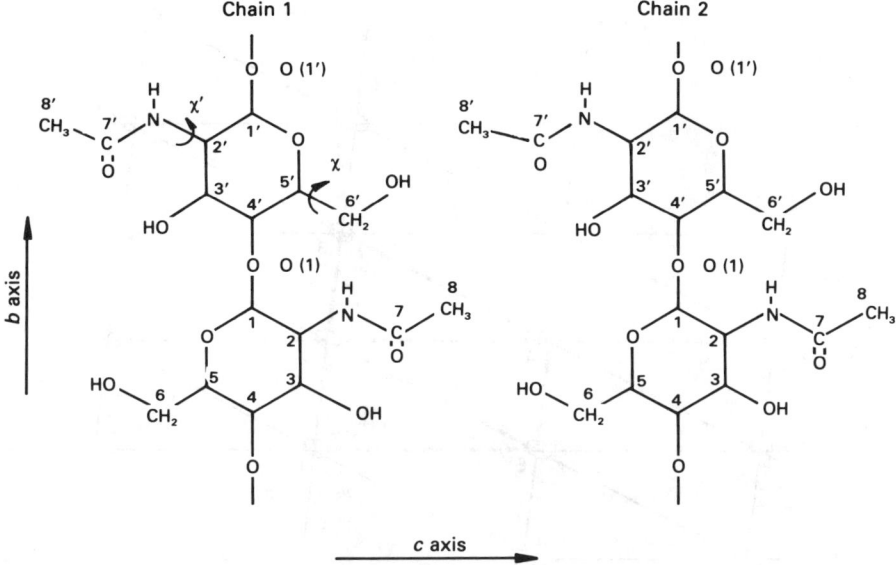

FIGURE 1.12   System used to identify atoms in the different chains in β-chitin. The chain
directly below chain 1, that is the adjacent chain along the *a* axis, is
designated chain 3. Interchain bonding is indicated by the appropriate
subscripts on the atom numbers, for example $C(2_1)N-H \ldots O=C(7_3)$

band is a single peak indicating that all the amide groups are in the same
environment.

In a paper published between the first two of Blackwell *et al.*,[92, 109] Dweltz
et al.[91] reported on an X-ray diffraction study of the β-chitin from *Thalassiosira
fluviatilis*, from which they concluded the unit cell to be monoclinic with the
space group P2₁ and cell dimensions $a = 4.80$ Å, $b = 10.32$ Å, $c = 9.83$ Å, and
$\beta = 112°$. This was shown to be similar to that proposed by Blackwell *et al.*,[109]
the essential difference between the two being that a basal *ac* plane diagonal in
the unit cell of Dweltz *et al.*[91] becomes a basal cell edge in the unit cell of
Blackwell *et al.*[109] (Figure 1.13).

In a second paper, Blackwell reported[92] that diaphanol-bleached
β-chitin (diaphanol is a solution of ClO_2 in 50% glacial acetic acid) from
pogonophore tubes formed two distinct hydrates as well as the anhydrous
form, and that the interchain spacing along the *c* axis increases from 9.26 Å
to 10.5 Å to 11.1 Å with increasing extent of hydration. Models showed
that with a unit cell dimension of $c = 9.1$ Å approximately, it is not possible
to fit water molecules between the chains without an unacceptable increase
in *c*. With the inclusion of one molecule of water per *N*-acetyl-D-
glucosamine residue, hydrogen bonded between the $O(6_1')H$ and $O(3_2')H$
of two adjacent chains along the *c* axis, there was an increase in *c* to 10.4 Å

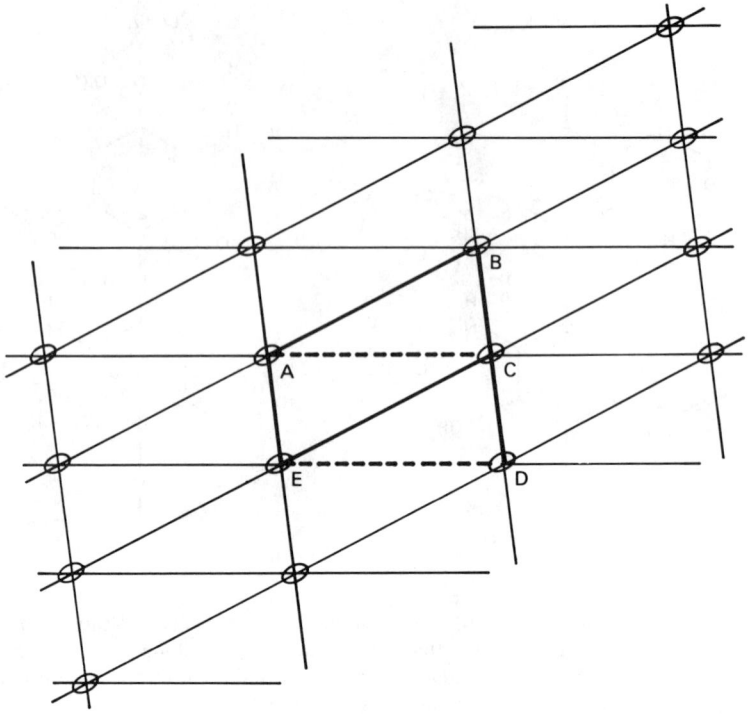

FIGURE 1.13   Relationship between the unit cell – ABCE – proposed by Dweltz et al.[91]
and the unit cell – ACDE – proposed by Blackwell et al.[109]

approximately, while increasing the interchain spacing along the c axis to 11.1 Å allowed the incorporation of two water molecules per residue. Blackwell concluded that the form having $c = 9.26$ Å is the anhydrous form while those having $c = 10.5$ Å and $c = 11.7$ Å are the mono- and dihydrate respectively.

In the same paper Blackwell reported on the occurrence of two structural phases, A and B, in a number of β-chitin samples. However diaphanol-bleached β-chitin, such as was used in studies on the hydrate structures, consists of pure phase A material.

Blackwell et al.[93, 110] refined the structure of β-chitin (anhydrous phase A) using the rigid-body least-squares method and this enabled the hydrogen-bonding pattern to be established. Figure 1.14 shows that the $C(7)=O$ groups are involved in two interchain hydrogen bonds involving two adjacent chains along the a axis. One is formed between $C(7_3)=O$ . . . H–N–C(2_1) and the second between $C(7_3)=O$ . . . HO$(6_1')$C. Thus the hydrogen bonds are either intrachain [$C(3_1')$OH . . . O(5_1)] or intrasheet, and there are no intersheet hydrogen bonds between adjacent chains along the c axis.

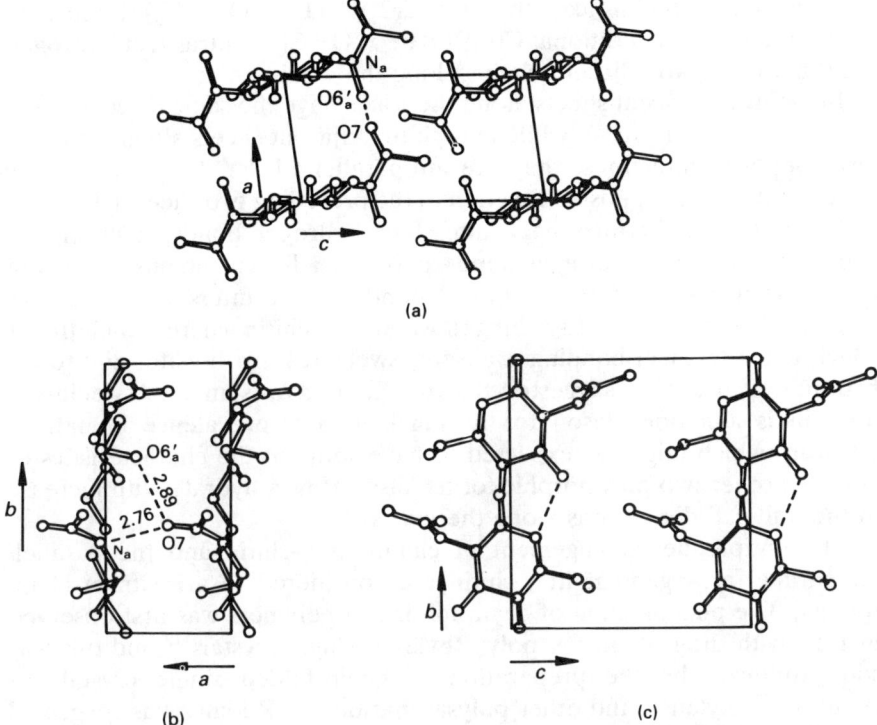

(a)

(b)   (c)

FIGURE 1.14   The structure of anhydrous β-chitin proposed by Blackwell et al.[93, 110]
              (a) The ac projection showing the $C(2_1)N–H \ldots O=C(7_3)$ interchain bond.
              (b) The ab projection showing the $C(2_1)N–H \ldots O=C(7_3)$ and the
              $C(6_1')O–H \ldots O=C(7_3)$ interchain bonds.
              (c) The bc projection showing the $C(3_1')O–H \ldots O(5_1)$ intrachain bond.
              (Reproduced from Biopolymers, 14 (1975) 1592, by permission of the
              authors and of John Wiley and Sons, Inc.)

1.4.4   Structure of γ-chitin

Very few studies have been carried out on γ-chitin since Rudall[82] first
reported a third structure for chitin. His X-ray diffraction studies showed
that the chains repeat in groups of three ('two up, one down') along the
c axis (Figure 1.7) with[96] $a = 4.7$ Å, $b = 10.3$ Å, $c = 28.4$ Å, $\beta = 90°$,
and space group $P2_1$. Recently it has been suggested that γ-chitin may be a
distorted version of either α- or β-chitin rather than a true third polymorphic form.[111]

1.4.5   Inter-relationships between α–, β– and γ-chitin

In chitin the chains are arranged in sheets or stacks, the chains in any one
sheet having the same direction or 'sense' and being hydrogen-bonded

together along the *a* axis through $C(2_1)N–H \ldots O=C(7_3)$ bonds. In β-chitin there is an additional $C(6_1')OH \ldots O=C(7_3)$ intrasheet hydrogen bond between two adjacent chains along the *a* axis.

In β-chitin adjacent sheets along the *c* axis have the same direction; the sheets are parallel (↑ ↑), while in α-chitin adjacent sheets along the *c* axis have opposite directions, they are antiparallel (↑ ↓). In γ-chitin every third sheet has the opposite direction to the preceding two sheets (↑ ↑ ↓).

In addition to the intrasheet, interchain hydrogen bonds, α-chitin also contains intersheet hydrogen bonds between adjacent chains along the *c* axis through $C(6_2)–OH \ldots O(6_1')$ bonds. These intersheet bonds are responsible for the lack of swelling in water of α-chitin whereas β-chitin, in which this intersheet bonding is absent, swells readily in water and forms hydrates. It has been suggested that the greater resistance to swelling of α-chitin is a major reason for its much greater prevalence in nature. γ-Chitin, which might be expected to have some of the characteristics of both the other two polymorphic forms, also forms a hydrate with increase in the unit cell dimensions along the *c* axis.[83]

The antiparallel arrangement of chains in α-chitin and the parallel/ antiparallel arrangement in γ-chitin are considered to arise from chain folding. The phenomenon of chain folding in polymers was first observed in 1957 with the isolation of poly(ethylene) single crystals[112] and this was soon followed by the preparation of chain-folded single crystals of cellulose,[113] xylan[114] and other polysaccharides.[115] Rudall[82] has suggested that the γ-chitin system could be produced by chain folding as in Figure 1.15a, and a related arrangement (Figure 1.15b) has been proposed for α-chitin.[116] If the initial direction of the new chain has the same sense as the final direction of the preceding chain the 'two up, one down' arrange-

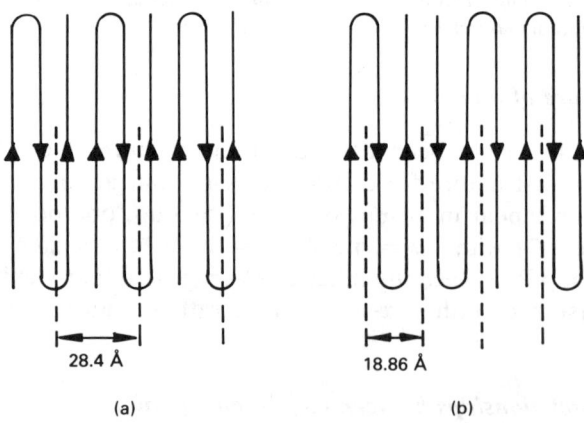

28.4 Å

(a)

18.86 Å

(b)

FIGURE 1.15   Proposed chain folding arrangements in (a) γ-chitin[82] and (b) α-chitin[116]

ment of γ-chitin is obtained, while if the initial direction of the new chain is opposite in sense to that of the final section of the preceding chain the 'one up, one down' arrangement of α-chitin is obtained. These have been referred to as 'parallel threes' and 'antiparallel threes' respectively.[96] In support of this concept of chain folding, sheets that are three chains thick have been observed in α-chitin–protein complexes[116] (see section 1.5.4).

As stated previously (section 1.4.1) α-chitin is the most stable of the three forms, as is demonstrated by the solid state conversion of β-chitin to α-chitin on treatment with 6M HCl.[82, 83] During this morphological change the swollen material contracts by about 50% in length and Rudall has suggested[117] that the β→α transformation occurs through the parallel chains of β-chitin folding back on themselves as illustrated in Figure 1.16.

This transformation has also been followed by solid state CP/MAS ^{13}C spectroscopy,[118] the two forms being distinguishable by the fact that the two signals assigned to the C(3) and C(5) carbon atoms appear as a doublet in the spectrum of α-chitin but are superimposed and appear as a singlet in the spectrum of the β-form.[118, 119] The spectrum for the α-chitin produced by this transformation is characterised by peaks with broader bases, which suggests that a considerable amorphous content is created during the solid state transformation.[118]

1.5 CHITIN–PROTEIN COMPLEXES

1.5.1 Introduction

As stated in section 1.2, chitin from animal sources usually occurs in association with protein which functions as a lower modulus matrix surrounding the chitin. The protein may be sclerotised, crosslinked by o-dihydric phenols, but in many instances it is not and studies on chitin–protein complexes are usually carried out on unsclerotised material. Its biological role is thought to be that of a defence mechanism against the action of chitinases, and Jeuniaux has reported a membrane containing approximately 75% chitin and 15% protein that is resistant to the action of chitinases unless the protein is removed, after which the chitin is rapidly attacked.[120] The protein, once sclerotised, may also prevent excess hydration.[121]

The extent of interaction between chitin and protein molecules differs within a given sample. Hackman and co-workers[122–124] have shown that there is only weak bonding between chitin and some of the protein in the cuticle of several insects and of the crab Cancer pagurus L. Hackman and Goldberg[124] introduced the use of a sequence of extractions with aqueous solutions of increasing extractive power. The series is:

(a) cold water at pH 7.0 for 48 h, removes soluble, unbound protein;

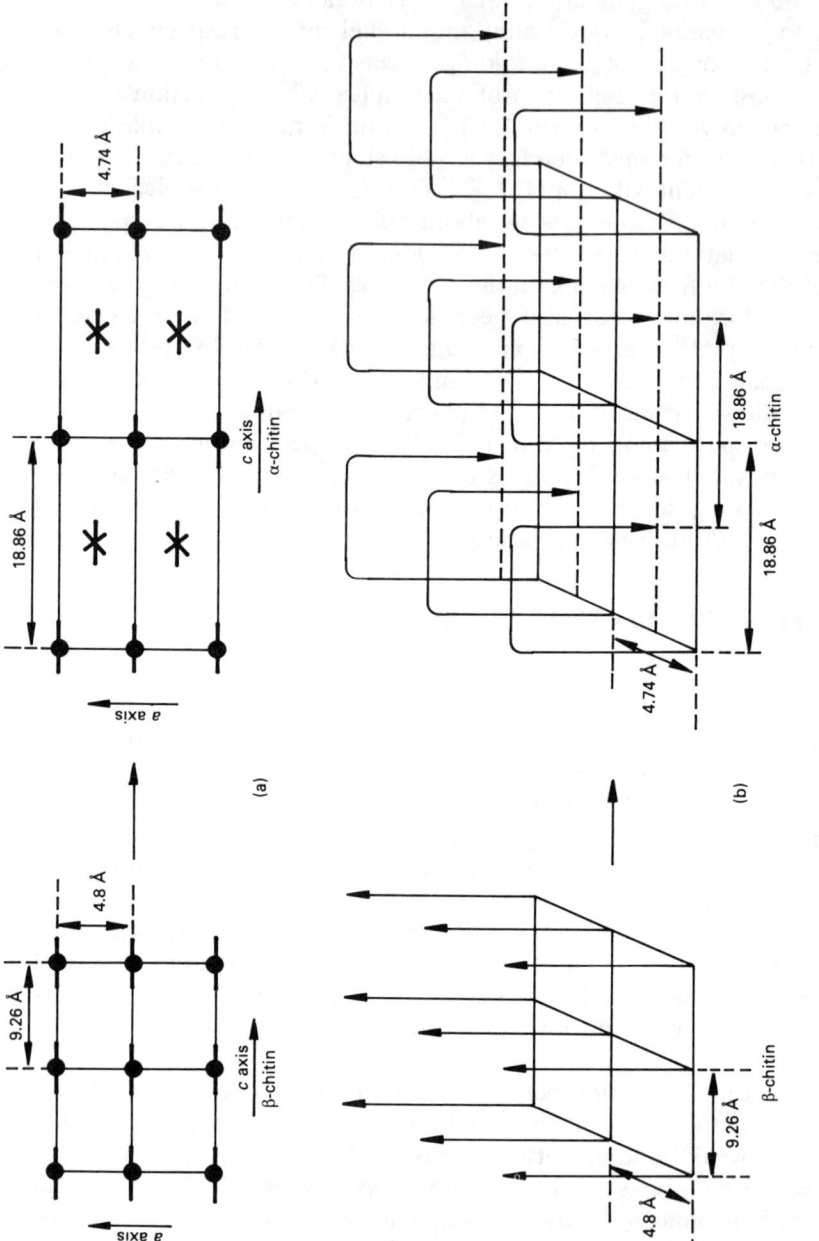

FIGURE 1.16 (a) Schematic representation of change in unit cells on β-chitin → α-chitin solid state transformation – chains of opposite sense are denoted by circles and crosses; (b) diagram of possible mechanism for the β-chitin → α-chitin solid state transformation involving chain folding

(b) cold 0.17M Na_2SO_4 at pH 7.0 for 48 h, removes protein bound by weak forces such as van der Waals' forces;
(c) cold 7M urea at pH 7.0 for 48 h, removes hydrogen-bonded protein;
(d) cold 0.01M NaOH for 5 h, removes electrostatically bound protein;
(e) 1.0M NaOH at 50–60°C for 5 h, removes strongly bound protein, presumably covalently bonded to the chitin.

Applying this series to the larval cuticle of the beetle *Agrianome spinicollis* showed that it contained a total of 63% protein that could be divided into 14% water soluble, 2% bound by van der Waals' forces, 25% bound by hydrogen bonds, 3% by ionic bonds, and the remainder (56%) strongly bound. The question is whether the residual chitin–protein complex remaining after stage (d) is a true proteoglycan with the protein molecules linked through covalent bonds to the chitin chains.

1.5.2 Evidence for covalent bonds between chitin and protein

Much of the evidence for the presence of a true proteoglycan structure in native chitins comes from precipitation studies. Puparia of the blowfly *Sarcophaga falculata*,[125] cuticle of the crab *Cancer pagurus*,[123] and larval cuticle of *Agrianome spinicollis*[124] have been dispersed in aqueous LiSCN solutions at 170°C and reprecipitated without separating the chitin and protein components. In a more detailed study[7] LiSCN solutions of four chitin–protein complexes were fractionated by progressive dilution with acetone to give a series of fractions which all contained both chitin and protein. No protein-free fraction was obtained. The resistance of the protein component of each fraction to extraction by water, electrolyte solution, aqueous urea and aqueous phenol, together with the fact that the protein content of various native chitins could be substantially reduced but never completely eliminated by repeated extractions with hot NaOH solutions, was cited as additional evidence of covalent bonding between chitin and protein chains.[7] Furthermore the fact that fractions were obtained, rather than a single precipitate, was taken as being indicative of the polydisperse nature of the complexes.

Doubts concerning the usefulness of these precipitation studies have been raised by the results of Attwood and Zola who examined the solution obtained on dissolving the chitin–protein complex from *Loligo* pen in hot aqueous LiSCN.[126] Its behaviour on ultracentrifugation suggested it to be a dispersion of particulate aggregates rather than a true molecular solution and therefore, as Hunt has pointed out,[8] reprecipitation of a chitin–protein complex, rather than separation of the two components, is not conclusive proof of covalent bonding between them.

However later work by Lipke and Geoghegan showed that treatment of native chitin with *N*-bromosuccinimide, followed by a proteinase then chitinase, yielded peptidochitodextrins, together with unsubstituted chitin oligomers, with the latter predominating.[127] Since both the unsubstituted and the peptidylated chitin oligomers were soluble, the objection raised to the precipitation results involving the much higher molecular weight chitin–protein complexes does not apply. Finally Hunt and Huckerby[128] obtained chitin–protein complexes by precipitation from solutions of native chitin, from both *Loligo* pen and *Lutaria* siphon sheath, in anhydrous formic acid. On the assumption that the process of dissolving in anhydrous formic acid will eventually sever all non-covalent bonding, Hunt and Huckerby concluded that the precipitates were true chitin proteoglycans with covalent bonds between the chitin and protein chains.

1.5.3   Nature of the chitin–protein link

Hackman subjected chitin samples from puparia of *Lucilia cuprina* (Wied.), larval cuticles of *Agrianome spinicollis*, the carapace of *Scylla serrate*, cuttlefish shell and *Loligo* pen, all previously deproteinised by treatment with 1M NaOH at 100°C, to hydrolysis in 5.7M HCl at 105°C in sealed tubes. Examination of the hydrolysates by ionophoresis and paper chromatography revealed the presence of aspartic acid and histidine, in addition to D-glucosamine, and Hackman concluded that these two amino acids were the ones involved in the chitin–protein covalent link.[7]

Later workers have detected other amino acids in the traces of protein left after exhaustive extraction of native chitins with hot NaOH solution. Attwood and Zola identified valine, methionine, tyrosine, phenylalanine, ornithine and lysine in chitins prepared from *Loligo* pen and *Calliphora* larval cuticle. In addition, the former chitin yielded histidine but no aspartic acid while the latter yielded histidine and aspartic acid together with threonine, serine, glutamic acid, glycine, alanine, isoleucine and leucine. In neither case was either histidine or aspartic acid predominant.[126] Brine and Austin examined chitins from five different sources and found the main residual amino acids to be aspartic acid, glycine and serine.[129] Thus despite the results of Hackman[7] there does not appear to be a universal chitin proteoglycan.

A number of possible bonding arrangements between the chitin and protein chains have been proposed and these include (Figure 1.17): formation of an amide group between an unacetylated amine group of chitin and a carboxylic acid group in the protein chain,[7] the latter being either the C-terminal carboxylic acid group or the side group of an acidic amino acid (**1.16**); bonding through a Schiff's base structure[122] (**1.17**) or through an N-glycosidic structure involving the amide group of asparagine[8] (**1.18**) or an O-glycosidic structure involving serine[8] (**1.19**); or through the carboxylic acid groups of a small number of N-acetylmuramic acid units in the chain and the amine group of a terminal alanyl residue[82, 130] (**1.20**).

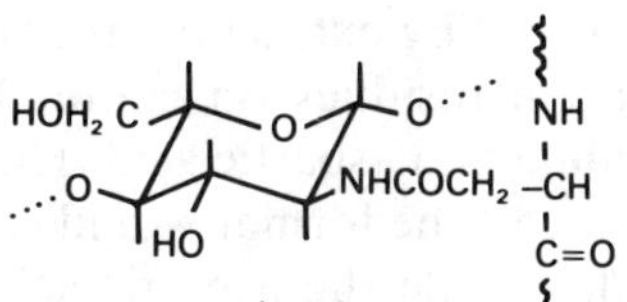

(1.16)

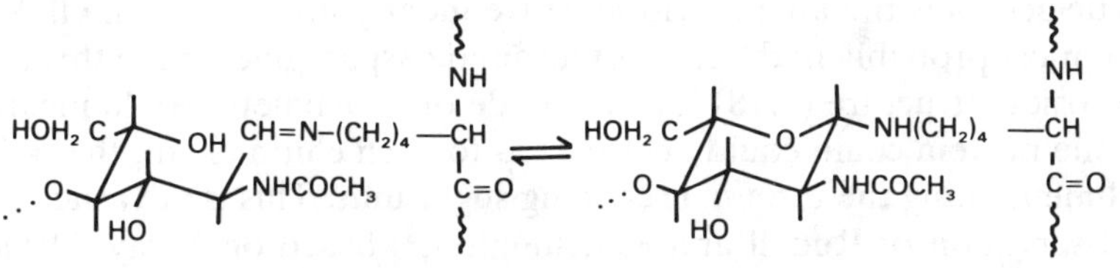

(1.17)

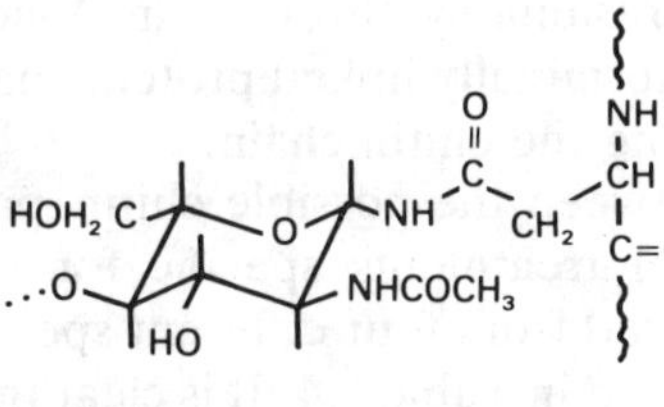

(1.18) 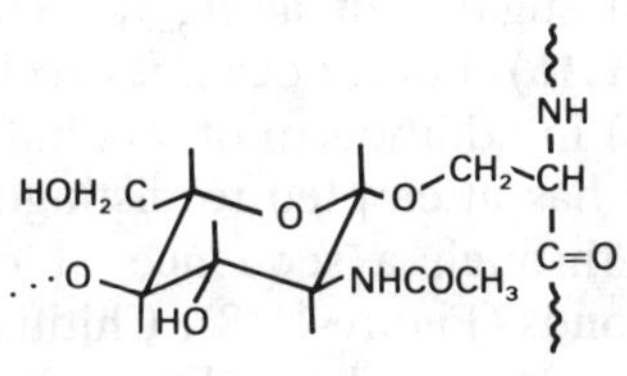   (1.19)

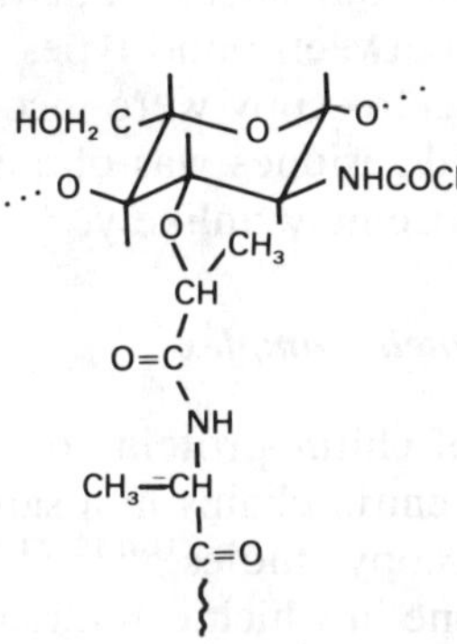

(1.20)

FIGURE 1.17   Possible bonds involved in the chitin–protein link

Hackman initially proposed[122] a Schiff's base type of bond (**1.17**) on the basis of studies of the adsorption of water-soluble protein onto chitin from aqueous solution but, since this is the only one of the structures (**1.16**)–(**1.20**) likely to be formed under the conditions of the experiment, the validity of this conclusion in respect of a biologically produced chitin proteoglycan is questionable. Further doubt arises from the fact that this

adsorbed protein could be removed by extraction at pH = 9, which is not
the case with the normal protein residues on chitin. Based on his subsequent detection of only histidine and aspartic acid[7] Hackman proposed a
type (1.16) link, which in the case of the former would require involvement
of a C-terminal histidine residue while the aspartic acid residue could be
located in any position in the protein chain.

On the basis of the then available information Hunt concluded[8] that
aspartic acid was the amino acid most frequently involved in the link with
chitin, most probably in the form of its amide asparagine linked through an
N-glycosidic structure (1.18). Such a mode of attachment would mean that
only one protein chain could be attached to each chitin chain, the point of
attachment being the terminal reducing sugar unit. This is at variance with
the observation of Rudall and Kenchington,[83] based on X-ray diffraction
studies (section 1.5.4), that there are regular repeats of bound protein at
31 Å intervals along the chain. This represents a protein chain linked to
every sixth sugar unit along the chain, presumably through an N-acyl
structure (1.16), but it does not preclude a terminally linked protein chain
as in (1.18) in addition to others linked along the chitin chain.

Brine[131] has attempted to distinguish between the possible chitin–protein links through a sequence of chemical treatments specific for the
different bonds (Figure 1.18). Chitins prepared from four different species
of crab were examined and the results are given in Table 1.4. It is clear that
amide links are predominant in all four species but the hydroxylamine
reagent used does not distinguish between bond types (1.16) and (1.18),
and attempts to do so using IR spectroscopy were inconclusive. No indication of any N-acetylmuramic acid residues was obtained, showing that
the suggested[82, 130] link (1.20) is extremely unlikely.

1.5.4 Morphology of the chitin–protein complex

Despite the convincing evidence of chitin–protein covalent bonds, other
evidence indicates that not all the chitin chains in a sample are linked to
protein molecules. Electron microscopy studies[116, 121, 132] have shown two
morphological types of complex; one in which cylindrical chitin fibrils are
imbedded in a protein matrix in an approximately hexagonal packing
arrangement, and the other in which the chitin is arranged in layers
interspersed with layers of protein. The first morphology has been reported
for *Megarhyssa nortoni* ovipositor,[116] *Megarhyssa lunator* ovipositor,[121, 132]
Sirex ovipositor[121] and *Aphrodite* chaetae,[121] and the second for blowfly
larval cuticle.[116] In both morphological types the absence of intimate
mixing of chitin and protein chains is shown by the absence of any staining
in the chitin-containing regions, indicating that protein chains are only
bound to chitin chains at the chitin/protein interface.

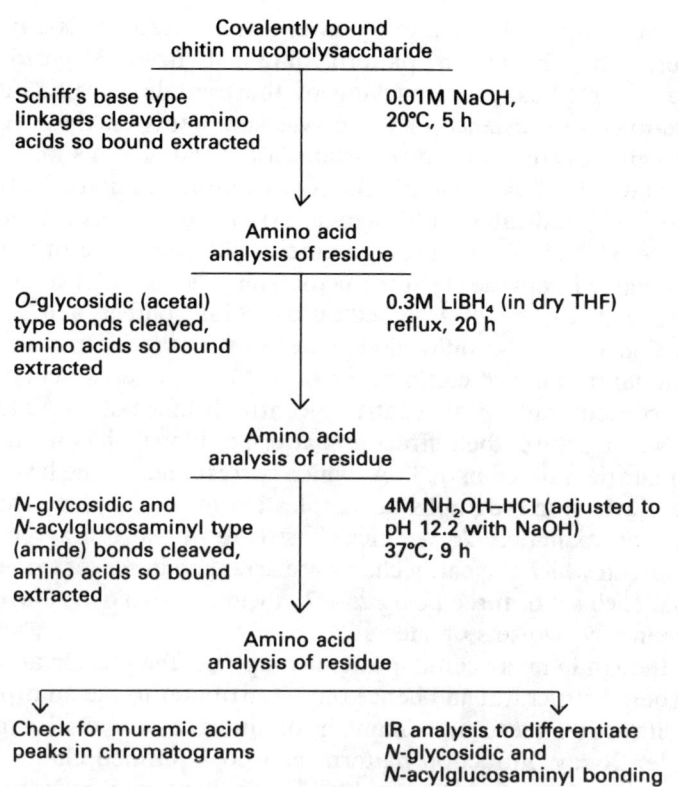

FIGURE 1.18   Sequential chemical treatments to differentiate types of covalent bonding in
the chitin mucopolysaccharide. (Reproduced from reference 131)

TABLE 1.4   *Covalent bonding type fractions of protein in chitin*[131]

	Molar percent of bound protein			
Bond type	*Horseshoe crab*	*Blue crab*	*Stone crab*	*Red crab*
Double covalent (Schiff's base)	5.8	22.6	3.5	18.7
Acetal (C-1, *O*-Glycosidic)	3.6	16.0	23.2	29.0
Amide (C-1, *N*-Glycosidic or C-2, *N*-Acylglucosaminyl)	58.3	54.9	51.1	43.5
Other (Residual, strong)	32.3	6.5	22.2	8.8

These proposed morphologies are supported by X-ray diffraction studies. The diffraction pattern obtained from *Megarhyssa nortoni* indicates[96, 116] hexagonal packing of the crystalline chitin fibrils with a centre-to-centre distance of 69 Å. Measurements on the electron micrographs give a centre-to-centre distance of 75–80 Å and a fibril diameter of approximately 50 Å. Similarly the X-ray diffraction pattern of *Megarhyssa lunator*[121, 132] indicates an hexagonal arrangement with a centre-to-centre distance of 72.5 Å, in close agreement with the value of approximately 75 Å obtained from the electron micrograph. The diameter of the chitin fibrils was calculated to be 28 Å, based on the chitin content, and 38 Å based on diffraction data. The diffraction pattern of *Aphrodite* chaetae[121] indicates hexagonal packing of chitin fibrils of 38 Å diameter, calculated from the chitin content, and with a centre-to-centre distance of 70 Å. In the case of the layer structure, the diffraction pattern of blowfly larval cuticle[116] shows an equatorial reflection of 33 Å, which corresponds to the layer separation, while electron micrographs of samples cut in the *bc* plane show layers of chitin approximately 25 Å thick. These results are in agreement with a structure in which the chitin chains are arranged in sets of three antiparallel sheets, each set of three being 25–27 Å wide, with a 6–8 Å layer of protein between any two sets of sheets.[96]

Although in many chitin–protein complexes the protein appears to have no ordered structure, and hence only contributes to the amorphous halo in the diffraction pattern, a number of intact native chitins give a more complex X-ray diffraction pattern than does purified chitin. Lotmar and Picken were the first to suggest[108] that the extra reflections in these diffraction patterns are due to ordered protein, rather than to modification of the three-dimensional arrangement of the chitin chains as had been previously suggested.[133] Subsequently Rudall examined the diffraction patterns of a number of intact chitin–protein complexes[82, 83] and noted the presence of layer lines showing axial repeats at approximately 31 Å in most samples, although some repeated at 41 Å or 62 Å intervals. These represent a repeat in the protein structure every 3, 4 or 6 chitin disaccharide repeat units respectively. However even complexes that contain structured protein will also contain a proportion that has no tertiary structure – the so-called 'satellite' protein[83] and it has been demonstrated for at least one sample, *Megarhyssa lunator* ovipositor,[121] that removal of up to 15% of loosely bound and hydrogen-bonded protein by the progressive extraction process[124] does not cause any change in the X-ray diffraction pattern.

In nearly all the native chitin samples examined the patterns corresponding to α- or β-chitin are recognisable in the patterns of the intact chitin–protein complexes, indicating that there is no change in the arrangement of the chitin chains on removal of protein. This supports the concept of two discrete phases linked only at the interface, as indicated in the electron micrographs.[116, 121, 132] Only with the oesophagel cutical of the squid *Loligo* was there any indication that the packing of the chitin chains in the purified chitin was different from that in the intact chitin–protein complex.[82] It has been suggested[8] that in this case the chitin and protein polymer chains are intimately associated, thereby affecting the packing of the chitin chains which spontaneously adopt the α-chitin arrangement on removal of the protein. Unfortunately no electron micrographs of stained sections of this cuticle appear to have been reported.

Blackwell and Weith[121, 132] have attempted to use the additional reflections present in the diffraction patterns of intact chitin–protein complexes to gain an insight into the structure of the ordered protein. They have proposed an elegant model for the complexes found in *Megarhyssa lunator* and *Sirex* ovipositors and in *Aphrodite* chaetae. Despite the differences in protein content and in protein subunit structure of these three samples, each has a core–sheath structure in which the protein subunits form an approximately 6_1 helical array around the chitin fibril (Figure 1.19). Support for the sixfold symmetry of the protein subunits has come from Fourier reconstruction of the electron micrographs of stained cross-sections.[134]

Closer inspection of the X-ray diffraction patterns disclosed minor differences between the structures of the three samples.[121] The layer lines in the pattern for *Megarhyssa lunator* show distinct fanning and splitting into components, indicating a much longer repeat than 30.6 Å, and suggesting that the protein helix has approximately six subunits per turn but is a non-integral helix repeating after five turns to give a 153 Å repeat. The layer lines in the pattern for *Sirex* are not appreciably split, indicating that the helix symmetry is close to being exactly 6_1, while those in the pattern for *Aphrodite* chaetae are split much more than for *Megarhyssa*, indicating greater deviation from an exact 6_1 symmetry for the helix of protein subunits. Despite these differences, the general principles of the model were considered to hold for all three samples.

1.6 PHYSICAL STRUCTURE OF CHITOSAN

1.6.1 Introduction

Considerably less work has been carried out on the physical structure of chitosan than on that of chitin (section 1.4). This disparity reflects the much greater importance of chitin in nature but, with the growing interest in chitosan as a specialty polymer, there is an increasing interest in this aspect of its structure. However studies in this area are complicated by the various routes available for preparing chitosan samples and these include:

(a) direct information by heterogeneous deacctylation of chitin without subsequent solubilisation and regeneration;

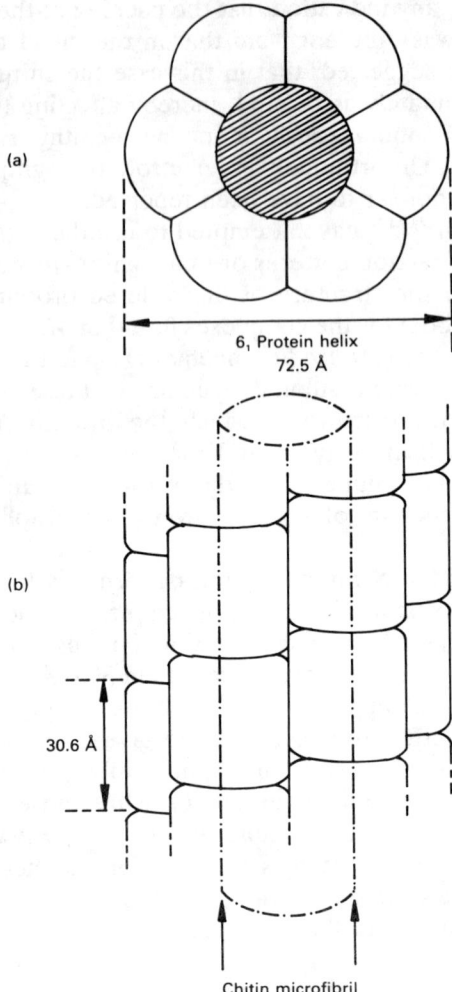

FIGURE 1.19   (a) Cross-sectional and (b) longitudinal views of the chitin–protein helix
proposed by Blackwell and Weith.[121] The protein subunits form a 6_1 helix
surrounding the chitin microfibril core

(b) regeneration from solution by solvent evaporation, followed by steep-
    ing the as-cast film in alkali – the post-treatment process;
(c) regeneration from solution by coagulation on immersion in alkali – the
    precipitation process;

and it is possible that the crystal structure will depend, to some extent, on
the particular method used. Furthermore, although only α-chitin has been
used to date as the starting material for route (a), both β- and γ-chitin

could also be used and these would be expected to give chitosans of different structures because of the different arrangements of chains in the starting chitins (section 1.4.5). Even with regenerated material it currently appears doubtful that there is a single unique crystal structure for chitosan. Instead there are a number of structures of differing stabilities and extents of hydration, the most stable of which may be considered to be the correct crystal structure of chitosan.

1.6.2   Chitosan crystal structure

Clark and Smith[100] followed the deacetylation of α-chitin from lobster tendon by X-ray crystallography and determined the unit cell to be ortho-rhombic, with $a = 8.9$ Å, $b = 10.25$ Å and $c = 17.0$ Å. They pointed out that a monoclinic structure with $\beta = 88°$ was an alternative possibility but favoured the orthorhombic structure (Table 1.5). They also found that the products obtained by alkali fusion of chitin with KOH at 180°C and by treatment with concentrated aqueous NaOH solutions are identical, giving the same diffraction pattern and the same analysis.

Clark and Smith stated that details of the crystal structure would be given in a subsequent paper but this did not materialise, hence the next reported work was that of Samuels some 45 years later.[135] Samuels pre-pared a film of chitosan[0.17] from a solution in aqueous formic acid, using the precipitation process. Although he was unable to obtain an unambigu-ous identification of the crystal structure he suggested an orthorhombic unit cell with $a = 4.4$ Å, $b = 10.30$ Å and $c = 10.0$ Å (Table 1.5). Samuels referred to this as the form II structure to distinguish it from the form I structure that he obtained for the as-cast film. However he appears not to

TABLE 1.5   *Unit cell dimensions (Å) for chitosan*

a	b	c	β	Crystal type	Method of preparation[a]	Reference
8.9	10.25	17.0	90°	Orthorhombic	1	Clark and Smith[100]
4.4	10.30	10.0	90°		2	Samuels[135]
4.46	10.30	8.63	96.3°		2	Sakurai et al.[136]
5.82	10.30	8.37	99.2°		3	
8.63	10.30	8.92	96.3°	Monoclinic	2	Sakurai et al.[138]
8.37	10.30	11.64	99.2°		3	
8.67	10.24	8.96	92.6°		3'	
8.24	10.39	16.48	90°	Orthorhombic	3"	Ogawa et al.[140]
8.07	10.34	8.44	90°		4	Cartier et al.[142]

[a] 1 – deacetylation of α-chitin; 2 – precipitation process; 3 – post-treatment process; 3' – post-treatment process, film cast from 380 g dm^{-3} solution; 3" – post-treatment, film stretched 300% then annealed in water ≥ 190°C; 4 – precipitation from solution at 125°C as single crystals.

have realised that the as-cast film was chitosan formate and that his form I and form II structures represented different chemical entities and not two different crystal structures for chitosan. The chemical difference between the two films was subsequently elegantly demonstrated by Sakurai *et al.*[136]

These latter workers studied the crystal structures of chitosan films prepared by the post-treatment and precipitation processes. The material used was stated to be chitosan[0.0] on the basis of back-titration of excess HCl in a chitosan–hydrochloric acid solution[137] but the IR spectrum reproduced in the paper indicates a considerably higher degree of *N*-acetylation. Sakurai *et al.*[136] concluded that the crystal structures formed by the two preparation processes were both monoclinic but with different unit cell dimensions (Table 1.5) and that the sample prepared by coagulation was closer in structure to β-chitin[91] than to the form II structure of Samuels.[135] The unit cell volume for the post-treatment sample is larger than that for the precipitated sample, primarily because of the larger value for *a*. One possible explanation for this is that the arrangement of chains in the unit cell of the former sample occurs while they are in the salt form, and little change in their positions is likely to occur under the mild conditions of the neutralisation treatment, while in the precipitation process the chains are deprotonated before packing together in the unit cell. It is reasonable to expect that uncharged chains would pack more closely together than would charged chains, which must accommodate the counter ions in the structure.

In the second paper in the series Sakurai *et al.*[138] calculated the number of chains in the unit cell for four structures: as-cast chitosan formate, chitosan by the post-treatment process, chitosan by the precipitation process, and as-cast chitosan butyrate, and determined these to contain 2, 1, 1, and 2 chains respectively. This suggests that treatment of the as-cast chitosan formate film with dilute aqueous NaOH solution results in a 50% decrease approximately, in the unit cell volume and a decrease in the number of chains per unit cell of from 2 to 1. Such a drastic rearrangement of chains was considered unlikely in view of the mild conditions of the alkali treatment and the previously determined unit cell parameters were amended so as to double the unit cell volume and to include 2 chains. This was achieved by doubling the length of the repeat distance along the *a* axis, followed by reversing the *a* and *c* axis, which is permissible.

Another chitosan film was prepared by the post-treatment process from a 380 g dm^{-3} solution in pure formic acid, such a solution being considered by the authors to have liquid crystal characteristics. This film gave unit cell dimensions very close to those of the film prepared from a much more dilute solution by the precipitation process, rather than that prepared by the post-treatment process. The unit cell volume for this latter sample was some 25% greater than that for either of the other two and the authors suggested that these represent the stable crystal form for chitosan.

In a third paper models for chain packing in the *ac* projection were examined by calculating the structure factors followed by least squares analysis.[139] The three models examined were:

(a) chains at the corners and the centre of the unit cell;
(b) chains at the corners and at the centres of the *a* axis;
(c) chains at the corners and at the centres of the *c* axis.

The calculated *R* values indicated that model (c) is the most appropriate for both the samples prepared by the precipitation process and that prepared from the concentrated, liquid crystalline solution, while model (a) is the most suited for the sample prepared by the post-treatment of the film cast from the more dilute solution. The proposed crystal structures are shown in schematic form in Figure 1.20.

Ogawa *et al.*[140] studied chitosan samples prepared by post-treatment of films cast from solution in 0.2M acetic acid, followed by stretching 300% in water at 95°C and annealing at ≥ 190°C in water while kept at constant length. The films produced gave very sharp diffraction patterns indicative of a high degree of crystallinity. The unannealed film, when examined at 75% RH, showed a diffraction pattern similar to that reported by Clark and Smith[100] for unregenerated, deacetylated α-chitin, but gave a very diffuse pattern when recorded under vacuum. This indicates that the unit cell contains water molecules, a possibility originally suggested by Averbach[141] who found that the position of the (002) peak [(020) in the original in which *c* was taken as the fibre axis – see section 1.4.1] depended on the water content and hence postulated that water molecules are loosely bound between the chitosan chains along the *c* axis. In contrast to this, two annealed films of highly deacetylated chitosan (reported to be chitosan[0.005] and chitosan[0.0] based on elemental analysis) showed no change in diffraction pattern with change in relative humidity over the range 0–100% RH. Furthermore the (002) reflection was absent from the patterns. These results were considered to indicate that there are no water molecules present in the unit cell and this was confirmed by thermal analysis.[140]

There were some slight differences between the diffraction patterns obtained from the two samples and only reflections present in both patterns were used to determine the crystal structure. The analysis indicated an orthorhombic unit cell with $a = 8.24$ Å, $b = 10.39$ Å and $c = 16.48$ Å, very close to the cell dimensions reported by Clark and Smith.[100] Ogawa *et al.*[140] calculated that the unit cell contains 4 chains, double the number found by Sakurai *et al.*[138] This is in agreement with the greater volume (~80% larger) of the unit cell proposed by the former workers and the absence of any water molecules in the structure.

Cartier *et al.*[142] have recently reported the preparation of single crystals of chitosan. The lamellar single crystals, which were in the form of square

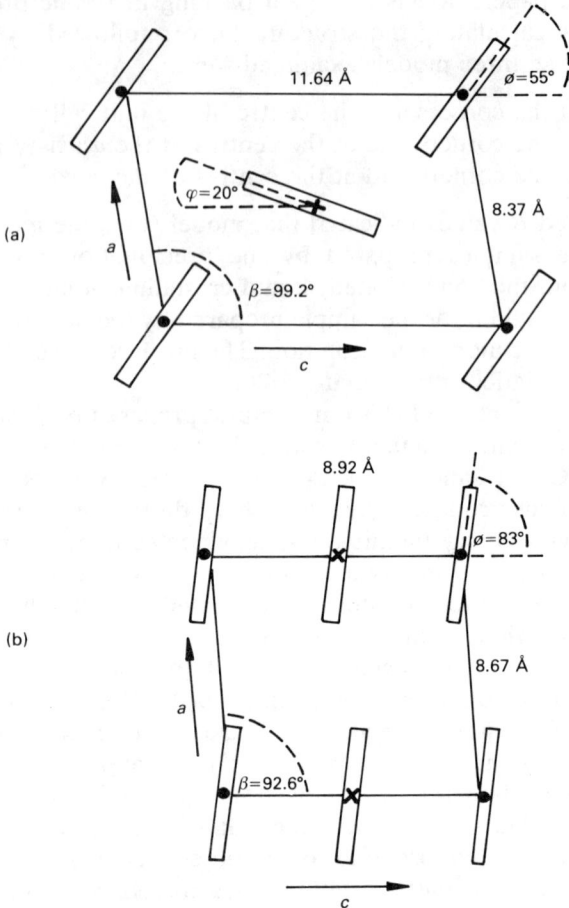

FIGURE 1.20 Schematic representation of the *ac* projection of the two unit cells for
chitosan proposed by Sakurai *et al.*[139]
(a) Chitosan film prepared by post-treatment process.
(b) Chitosan films prepared by precipitation process and from concentrated
solution (380 g dm^{-3}) by post-treatment process. Chains of opposite sense
are denoted by circles and crosses

platelets with diagonals of 0.2–0.5 μm, were produced at 125°C by addition
of ammonia to a solution of low molecular weight chitosan ($\overline{DP} = 35$; $\overline{DP}_w/$
$\overline{DP}_n = 1.14$). The thickness of the platelets is approximately 120 Å, which
is close to the chain length of the chitosan molecules and indicates that
chain folding does not occur.

Electron diffraction studies showed the unit cell to be orthorhombic,
space group P2$_1$2$_1$2$_1$, with $a = 8.07$ Å, $b = 10.34$ Å, $c = 8.44$ Å. These
dimensions, which together with the measured density of 1.47 g cm^{-3} mean

that the unit cell contains 2 chains, are similar to the values obtained by Sakurai et al.[138] for chitosan samples prepared by the precipitation process and by the post-treatment of film cast from a very concentrated chitosan solution (Table 1.5), both unit cells containing 2 chains. However these unit cells are monoclinic rather than orthorhombic, having $\beta = 96.3°$ and 92.6° respectively.[138] Cartier et al.[142] concluded that the crystal structure of the single crystals corresponds quite closely to that obtained by Ogawa et al.[140] for stretched, annealed chitosan films, all the reflections having medium or strong intensities in the X-ray diffraction diagrams obtained by the latter workers being observed in the electron diffraction patterns obtained by Cartier et al.[142] Furthermore neither unit cell contains any water molecules. The main difference between the two is the length of the c axis and Cartier et al. suggested that the approximate doubling of the c axis by Ogawa et al. is incorrect, being based on the presence of a single, very weak reflection at 3.40 Å on the third layer line of the X-ray diffraction pattern. Although such a reflection was not accessible in the work of Cartier et al., all 60 visible diffraction spots in the electron diffraction diagrams could be accounted for if c has a value of 8.44 Å. If Cartier et al. are correct in respect of the c axis dimension, then the 'unit cell' reported by Ogawa et al. represents a double cell and this would account for it containing four chains.

The cell proposed by Cartier et al. is very similar to that of cellulose IV$_{II}$[143] and mannan I,[144] particularly the latter, and must be regarded as the most stable structure for chitosan obtained to date. However it is reasonable to assume that the unit cell having a value of $c = 17.0$ Å, obtained by Clark and Smith[100] for chitosan prepared by heterogeneous deacetylation of α-chitin, is correct and does not represent a double cell. In this case the chain spacings in the chitosan will be largely governed by their spacings in the α-chitin precursor, a view first expressed by Darmon and Rudall[103] and subsequently endorsed by Averbach,[141] so that considerable similarity in the crystal structures would be anticipated. The substantial increase in the a axis on the conversion from chitin to chitosan (Table 1.5) is due to the removal of the N-acetyl groups that hold together adjacent chains along the a axis through $C(2_1)N–H \ldots O=C(7_3)$ interchain hydrogen bonds. Destruction of these interchain bonds allows the chains to move apart in the direction of the a axis and a distance of 8.9 Å between adjacent chains is in keeping with the analogous spacings of 7.85 Å in cellulose I and 9.80 Å in cellulose II[99] where neither structure contains interchain hydrogen bonds between anhydro-D-glucose residues in this direction.

REFERENCES

1. M.V. Tracey, Rev. Pure Appl. Chem., 7 (1957) 1.
2. C. Jeuniaux, Chitine et Chitinolyse, Masson, Paris, 1963.

3. A.G. Richards, *The Integument of Arthropods*, University of Minnesota Press, Minneapolis, 1951.
4. C. Jeuniaux, in *Comprehensive Biochemistry*, M. Florkin and E. Stotz (eds), Elsevier, Amsterdam, 1971, p. 595.
5. J. Ruiz-Herrera, in *Proceedings of 1st International Conference on Chitin/ Chitosan (1977)*, R.A.A. Muzzarelli and E.R. Pariser (eds), MIT Sea Grant Program Report MITSG 78–7, 1978, p. 11.
6. M. Falk, D.G. Smith, J. McLachlan and A.G. McInnes, *Can. J. Chem.*, **44** (1966) 2269.
7. R.H. Hackman, *Aust. J. Biol. Sci.*, **13** (1960) 568.
8. S. Hunt, *Polysaccharide–Protein Complexes in Invertebrates*, Academic Press, London, 1970, p. 129.
9. H. Braconnot, *Ann. Chim. (Paris)*, **79** (1811) 265.
10. A. Odier, *Mém. Soc. Histoire Nat. Paris*, **1** (1823) 29.
11. J.G. Children, *Zool. J.*, **1** (1824) 101.
12. C. Rouget, *Compt. Rend.*, **48** (1859) 792.
13. F. Hoppe-Seyler, *Ber.*, **27** (1894) 3329.
14. E. Fremy, *Compt. Rend.*, **83** (1877) 1136.
15. E. Schulze, *Ber.*, **24** (1891) 2277.
16. E. Winterstein, *Ber.*, **27** (1894) 3113.
17. E. Winterstein, *Ber.*, **27** (1894) 3508.
18. G. Ledderhose, *Ber.*, **9** (1876) 1200.
19. E. Winterstein, *Ber.*, **28** (1895) 167.
20. E. Gilson, *La Cellule*, **11** (1894) 7.
21. E. Gilson, *Compt. Rend.*, **120** (1895) 1000.
22. C. Tarret, *Bull. Soc. Chim. France*, **17** (1897) 921.
23. P. Karrer and A. Hofmann, *Helv. Chim. Acta*, **12** (1929) 616.
24. P. Karrer and G. von Francois, *Helv. Chim. Acta*, **12** (1929) 986.
25. G. van Iterson, K.H. Meyer and W. Lotmar, *Rec. Trav. Chim.*, **55** (1936) 61.
26. A.N.J. Heyn, *Protoplasma*, **25** (1936) 372.
27. K. Heller, L. Claus and J. Huber, *Zeit. Naturforsch.*, **14** (1959) 476.
28. G. Ledderhose, *Zeit. Physiol. Chemie*, **2** (1878) 213; **4** (1879) 139.
29. F. Tiemann, *Ber.*, **17** (1884) 241.
30. F. Tiemann, *Ber.*, **19** (1886) 49.
31. E. Fischer, *Ber.*, **17** (1884) 579.
32. E. Fischer and H. Leuchs, *Ber.*, **36** (1903) 24.
33. J.C. Irvine and A. Hynd, *J. Chem. Soc.*, (1912) 1128.
34. J.C. Irvine and A. Hynd, *J. Chem. Soc.*, (1914) 698.
35. A. Neuberger and R. Pitt Rivers, *J. Chem. Soc.*, (1939) 122.
36. W.N. Haworth, W.H.G. Lake and S. Peat, *J. Chem. Soc.*, (1939) 271.
37. S. Peat and L.F. Wiggins, *J. Chem. Soc.*, (1938) 1088.
38. W.H.G. Lake and S. Peat, *J. Chem. Soc.*, (1938) 1417.
39. S. Peat and L.F. Wiggins, *J. Chem. Soc.*, (1938) 1810.
40. E.G. Cox and G.A. Jeffrey, *Nature*, **143** (1939) 894.
41. F. Hoppe-Seyler, *Ber.*, **28** (1895) 82.
42. T. Araki, *Zeit. Physiol. Chemie*, **20** (1895) 498.
43. S. Fränkel and A. Kelly, *Monatsh.*, **23** (1902) 123.
44. R. Breuer, *Ber.*, **31** (1898) 2793.
45. H. Brach and O. von Fürth, *Biochem. Zeit.*, **38** (1912) 468.
46. W.S. Denham and H. Woodhouse, *J. Chem. Soc.*, **105** (1914) 2357.
47. W.S. Denham and H. Woodhouse, *J. Chem. Soc.*, **111** (1917) 244.
48. W.N. Haworth and G.C. Leitch, *J. Chem. Soc.*, **113** (1918) 118.

49. J.C. Irvine and E.L. Hirst, *J. Chem. Soc.*, **121** (1922) 1213.
50. J.C. Irvine and E.L. Hirst, *J. Chem. Soc.*, **123** (1923) 518.
51. W. Charlton, W.N. Haworth and S. Peat, *J. Chem. Soc.*, **128** (1926) 89.
52. W.N. Haworth, E.L. Hirst and E.J. Miller, *J. Chem. Soc.*, (1927) 2436.
53. W.N. Haworth, C.W. Long and J.H.G. Plant, *J. Chem. Soc.*, (1927) 2809.
54. H. Sponsler and W.H. Dore, *Coll. Symp. Mon.*, **4** (1926) 174.
55. W.N. Haworth, *Helv. Chim. Acta*, **11** (1928) 547.
56. K.H. Meyer and H. Mark, *Ber.*, **61** (1928) 593.
57. K.H. Meyer and H. Mark, *Ber.*, **61** (1928) 1936.
58. M. Bergmann, L. Zervas and E. Silberkweit, *Naturwiss.*, **19** (1931) 20.
59. L. Zechmeister and G. Tóth, *Ber.*, **64** (1931) 2028.
60. L. Zechmeister and G. Tóth, *Ber.*, **65** (1932) 161.
61. M. Bergmann, L. Zervas and E. Silberkweit, *Ber.*, **64** (1931) 2436.
62. F. Zilliken, G.A. Braun, C.S. Rose and P. Gyorgy, *J. Amer. Chem. Soc.*, **77** (1955) 1296.
63. R. Kuhn, A. Ganke and H.H. Baer, *Chem. Ber.*, **87** (1954) 1138.
64. S.A. Barker, A.B. Foster, M. Stacey and J.M. Webber, *J. Chem. Soc.*, (1958) 2218.
65. J.A. Rupley, *Biochem. Biophys. Acta*, **83** (1964) 245.
66. B. Capon and R.L. Foster, *J. Chem. Soc. C*, (1970) 1654.
67. A. Domard and A. Gadelle, in *Chitin in Nature and Technology*, R.A.A. Muzzarelli, C. Jeuniaux and G.W. Gooday (eds), Plenum Press, New York, 1986, p. 295.
68. C. Bosso, J. Defaye, A. Domard, A. Gadelle and C. Pedersen, *Carbohyd. Res.*, **156** (1986) 57.
69. S.T. Horowitz, S. Roseman and H.J. Blumenthal, *J. Amer. Chem. Soc.*, **79** (1957) 5046.
70. A. Domard and N. Cartier, *Int. J. Biol. Macromol.*, **11** (1989) 297.
71. K. Freudenberg, K. Friedrich and J. Bumann, *Ann.*, **494** (1932) 41.
72. D. French and G.M. Wild, *J. Amer. Chem. Soc.*, **75** (1953) 2612.
73. W.T.J. Morgan and L.A. Elson, *Biochem. J.*, **28** (1934) 988.
74. R. Jeanloz and E. Forchielli, *Helv. Chim. Acta*, **33** (1950) 1690.
75. M.L. Wolfrom, J.R. Vercellotti and D. Horton, *J. Org. Chem.*, **29** (1964) 547.
76. H. Saito, R. Tabeta and S. Hirano, in *Chitin and Chitosan*, S. Hirano and S. Tokura (eds), The Japanese Society of Chitin and Chitosan, Tottori, 1982, p. 71.
77. D. Gagnaire, J. Saint-Germain and M. Vincendon, *Makromol. Chem.*, **183** (1982) 593.
78. L. Zechmeister and G. Tóth, *Naturwiss.*, **27** (1939) 367.
79. L. Zechmeister and G. Tóth, *Enzymologia*, **7** (1939) 170.
80. L. Zechmeister and G. Tóth, *Enzymologia*, **7** (1939) 165.
81. S.A. Barker, E.J. Bourne, M. Stacey and D.H. Whiffen, *J. Chem. Soc.*, (1954) 171.
82. K.M. Rudall, *Adv. Insect Physiol.*, **1** (1963) 257.
83. K.M. Rudall and W. Kenchington, *Biol. Rev.*, **49** (1973) 597.
84. K.M. Rudall, *J. Pol. Sci. C*, **28** (1969) 83.
85. K.M. Rudall, *Symp. Soc. Exper. Biol.*, **9** (1955) 49.
86. K.H. Meyer and G.W. Pankow, *Helv. Chim. Acta*, **18** (1935) 589.
87. D. Carlstrom, *J. Biophys. Biochem. Cytol.*, **3** (1957) 669.
88. N.E. Dweltz, *Biochem. Biophys. Acta*, **44** (1960) 416.
89. R. Minke and J. Blackwell, *J. Mol. Biol.*, **120** (1978) 167.

90. N.E. Dweltz, *Biochem. Biophys. Acta*, **51** (1961) 283.
91. N.E. Dweltz, J.R. Colvin and A.G. McInnes, *Can. J. Chem.*, **46** (1968) 1513.
92. J. Blackwell, *Biopolymers*, **7** (1969) 281.
93. J. Blackwell, R. Minke and K.H. Gardner, in ref. 5, p. 108.
94. W.H. Bragg, *An Introduction to Crystal Analysis*, G. Bell and Sons, London, 1928.
95. *International Tables for X-ray Crystallography*, N.F.M. Henry and K. Lovedale (eds), Kynoch Press, 1976, vol. I.
96. A.G. Walton and J. Blackwell, *Biopolymers*, Academic Press, New York, 1973, p. 467.
97. R.L. Miller, in *Polymer Handbook*, J. Brandrup and E.H. Immerg (eds), Wiley, New York, 2nd edn, 1975, p. III–1.
98. H.A. Gonell, *Zeit. Physiol. Chem.*, **152** (1926) 18.
99. R.H. Marchessault and A. Sarko, *Adv. Carb. Chem.*, **22** (1967) 421.
100. G.L. Clark and A.F. Smith, *J. Phys. Chem.*, **40** (1936) 863.
101. P.H. Hermans, J. de Booys and C.J. Maan, *Kolloid-Zeit.*, **102** (1943) 169.
102. P.H. Hermans, *Physics and Chemistry of Cellulose Fibres*, Elsevier, Amsterdam, 1949, p. 13.
103. S.E. Darmon and K.M. Rudall, *Disc. Faraday Soc.*, **9** (1950) 251.
104. F.G. Pearson, R.H. Marchessault and C.Y. Liang, *J. Pol. Sci.*, **43** (1960) 101.
105. C. Ramakrishnan and N. Prasad, *Biochem. Biophys. Acta*, **261** (1972) 123.
106. M.A. Haleem and K.D. Parker, *Zeit. Naturforsch.*, **31c** (1976) 383.
107. R. Iwamoto, M. Miya and S. Mima, in ref. 76, p. 82.
108. W. Lotmar and L.E.R. Picken, *Experientia*, **6** (1950) 58.
109. J. Blackwell, K.D. Parker and K.M. Rudall, *J. Mol. Biol.*, **28** (1967) 282.
110. K.H. Gardner and J. Blackwell, *Biopolymers*, **14** (1975) 1581.
111. J. Blackwell, in *Methods in Enzymology*, W.A. Wood and S.T. Kellogg (eds), Academic Press, New York, 1988, vol. 161, p. 435.
112. A. Keller, *Phil. Mag.*, **2** (1957) 1171.
113. R. St. J. Manley, *Nature*, **189** (1961) 390.
114. R.H. Marchessault, F.F. Morehead, N.M. Walter, C.P.J. Glaudemans and T.E. Timell, *J. Pol. Sci.*, **51** (1961) S66.
115. B.G. Rånby and R.W.J. Noe, *J. Pol. Sci.*, **51** (1961) 337.
116. K.M. Rudall, in *Conformation of Biopolymers*, G.N. Ramachandran (ed.), Academic Press, New York, 1967, vol. 2, p. 751.
117. K.M. Rudall, *Sci. Basis Medicine Ann. Rev.*, (1962) 203.
118. M. Vincendon, J.C. Roux, H. Chanzy, S. Tanner and P. Belton, in *Chitin and Chitosan*, G. Skjåk-Braek, T. Anthonsen and P. Sandford (eds), Elsevier, London, 1989, p. 437.
119. M. Takai, Y. Shimizu, J. Hayashi, Y. Uraki and S. Tokura, in ref. 118, p. 431.
120. C. Jeuniaux, *Arch. Int. Physiol. Biochem.*, **67** (1959) 516.
121. J. Blackwell and M.A. Weith, in *Chitin, Chitosan and Related Enzymes*, J.P. Zikakis (ed.), Academic Press, New York, 1984, p. 257.
122. R.H. Hackman, *Aust. J. Biol. Sci.*, **8** (1955) 530.
123. A.B. Foster and R.H. Hackman, *Nature*, **180** (1957) 40.
124. R.H. Hackman and M. Goldberg, *J. Insect Physiol.*, **2** (1958) 221.
125. A.R. Trim, *Biochem. J.*, **35** (1941) 1088.
126. M.M. Attwood and H. Zola, *Comp. Biochem. Physiol.*, **20** (1967) 993.
127. H. Lipke and T. Geoghegan, *Biochem. J.*, **125** (1971) 703.
128. S. Hunt and T.N. Huckerby, *Comp. Biochem. Physiol. Biochem.*, **88B** (1987) 1107.

129. C.J. Brine and P.R. Austin, *Comp. Biochem. Physiol.*, **70B** (1981) 173.
130. E.J. Houk, G.W. Griffiths, N.E. Hadjokas and S.D. Beck, *Science*, **198** (1977) 401.
131. C.J. Brine, in ref. 76, p. 105.
132. J. Blackwell and M.A. Weith, *J. Mol. Biol.*, **137** (1980) 49.
133. G. Frankel and K.M. Rudall, *Proc. Roy. Sci.*, **B134** (1940) 111.
134. J. Blackwell, L.T. Germinario and M.A. Weith, in *The Chemistry and Biology of Mineralised Connective Tissues*, A. Veis (ed.), Elsevier North-Holland Inc., 1981, p. 465.
135. R.J. Samuels, *J. Pol. Sci., Pol. Phys. Ed.*, **19** (1981) 1081.
136. K. Sakurai, M. Takagi and T. Takahashi, *Sen-i Gakkaishi*, **40** (1984) T-246.
137. T. Nakajima, K. Sugai and Y. Ito, *Kobunshi Ronbunshu*, **37** (1980) 705.
138. K. Sakurai, T. Shibano, K. Kimura and T. Takahashi, *Sen-i Gakkaishi*, **41** (1985) T-361.
139. K. Sakurai, T. Shibano and T. Takahashi, *Fukui Daigaku Kogakubu Kenkyui Hokoku*, **33** (1985) 71.
140. K. Ogawa, S. Hirano, T. Miyanishi, T. Yui and T. Watanabe, *Macromolecules*, **17** (1984) 973.
141. B.L. Averbach, *Report MITSG 75–17*, National Technical Information Service, U.S. Department. (Quoted by Ogawa *et al.*, ref. 140.)
142. N. Cartier, A. Domard and H. Chanzy, *Int. J. Biol. Macromol.*, **12** (1990) 289.
143. A. Buléon and H. Chanzy, *J. Pol. Sci., Polym. Phys. Ed.*, **18** (1980) 1209.
144. I. Nieduszynski and R.H. Marchessault, *Can J. Chem.*, **50** (1972) 2130.

2

Preparation of Chitin and Chitosan

2.1 SOURCES OF CHITIN

Despite the great natural abundance of chitin, much of the annual production is not readily accessible for utilisation as a raw material. In this it differs markedly from cellulose. A critical evaluation of a number of potential sources of chitin has been given by Allan *et al.*[1] Their global estimate of the total annually accessible chitin was 150×10^3 t, of which 56×10^3 t was from krill, 39×10^3 t from shellfish (crab, shrimp, prawn, lobster and crayfish), 32×10^3 t from fungi, 22×10^3 t from clams and oysters and 1×10^3 t from squid. This total is much less than the estimated total of chitin produced annually by biosynthesis – one species alone, the marine copepods, is estimated to produce 10^9 t of chitin annually[2] – and a number of factors limiting the extent to which different sources may be utilised were discussed.[1]

The amount available from crustacea such as crab, shrimp and lobster is obviously restricted by the demand for what may be regarded as a luxury food item, while with krill the primary product is protein and the extent to which this can be marketed in its own right limits the chitin available from this source.

The use of clam and oyster shell is inhibited by the large quantities of inorganic material that must be removed (up to 90% dry weight) and by alternative uses for the ground shell itself.[3, 4] Fungal sources are considered to have a number of advantages over crustacean sources. These include a raw material that is consistent in composition, is available throughout the year and does not require a demineralisation step. However although chitin is present in the vast majority of fungi – the only main classes which do not contain chitin being the Schizomycetes, Myxomycetes and Tricho-mycetes[5] – it is usually present in association with other polysaccharides which must normally be removed.

Two marine diatoms, *Cyclotella cryptica* and *Thalassiosira fluviatilis*, have been shown to be a source of pure chitin that is not associated with protein.[6] They therefore represent an attractive source of chitin but for the fact that in both batch and continuous culture they were found to be slow growing and to produce low-density cultures, 5×10^4 litres of culture being required to yield 1 kg chitin.[1]

Currently all the chitin produced commercially would appear to be derived from crab, shrimp and prawn exoskeletons obtained as waste from the seafood processing industry. The amounts available from these sources, to which may be added crayfish exoskeletons which have been reported to represent a very extensive source of chitin,[7] are sufficient to meet the present demand for chitin and chitosan, and commercial exploitation of other potential sources is unlikely to take place for some years.

2.2 PURIFICATION

2.2.1 Introduction

The main sources of material for the laboratory preparation of chitin are also the exoskeletons of various crustacea, principally crab, and shrimp. In these the chitin is closely associated with proteins, inorganic material which is mainly $CaCO_3$, and pigments and lipids. Various procedures have been adopted to remove these impurities and no standard process has been developed. Demineralisation is most frequently carried out by treatment with HCl and deproteinisation by treatment with NaOH, but other methods may be used and the order in which these two steps are carried out has varied with different workers, although in most instances depro-teinisation has been carried out prior to demineralisation. The choice of processing conditions may be governed to some extent by the purpose for which the chitin is required, since partial deacetylation during deproteini-sation is not a disadvantage if the chitin is subsequently to be converted to chitosan, while some hydrolysis of the polymer chain during the deminer-alisation process can be tolerated if the chitin is to be used in the form of particles or converted to microcrystalline chitin (section 5.4.1).

2.2.2 Deproteinisation

Early patents[8, 9] claim the use of a wide range of agents for this step including NaOH, Na_2CO_3, $NaHCO_3$, KOH, K_2CO_3, $Ca(OH)_2$, Na_2SO_3, $NaHSO_3$, $CaHSO_3$, Na_3PO_4 and Na_2S, but NaOH is the preferred agent in the literature.

TABLE 2.1   *Conditions employed for deproteinisation of chitin-containing waste materials*

Material source	NaOH concentration	Temperature	Number of treatments	Total time of treatment (h)	Reference
Shrimp	0.25M	65°C	1	1	11
Crab	0.5M	65°C	1	2	12
Prawn	i 0.125M	100°C	1	0.5	13
	ii 0.75M	100°C	1	NS	
Krill	0.875M	90–95°C	1	2	14
Crab	1.0M	80°C	1	3	15
Crab	1.0M	100°C	1	36	16
Lobster	1.0M	100°C	5	60	17
Crab	1.0M	100°C	3	72	18
Lobster	1.25M	80–85°C	2	1	19
Crab	1.25M	85–90°C	3	1.5–2.25	20
Prawn	1.25M	100°C	1	0.5	21
Crab	1.25M	100°C	1	24	22
Crab	2.5M	Room temp.	3	72	23
Lobster	2.5M	100°C	1	2.5	24

NS = Not stated in the reference.

The mildest alkaline treatments reported are those utilising Na_2CO_3 at concentrations of up to 0.1M, together with a soap, at ~100°C for 4 hours,[8, 9] while the most severe involves treating the material, lobster shell, with 5M NaOH at ~100°C for 4 hours.[10] However most treatments have involved the use of approximately 1M NaOH but with wide variation in temperature and duration of treatment (see Table 2.1).

The use of enzymes for removal of protein has been examined by a number of workers. Although the use of 'certain putrificative bacteria' was claimed in the early patents,[8, 9] the first report in the literature of the use of enzymes for this purpose is by Giles *et al.*[25] more than 20 years later. In this work samples of lobster shell were incubated at 37°C for up to 8 days in a 0.5 wt-% pepsin solution, prior to demineralisation. It was concluded that enzymatic treatment of shells did not offer any substantial advantage over the shorter chemical treatments. Broussignac[20] suggested the use of enzymes such as pepsin or trypsin if the chitin is required to be as fully N-acetylated as possible but no experimental details were given.

Takeda and Abe[26] investigated the use of proteolytic enzymes such as tuna proteinase, papain or a bacterial proteinase, for removal of protein from shell previously demineralised with EDTA. The treatments left up to 5% protein which could be removed by boiling in a solution of sodium dodecylbenzenesulphonate. In a study of the preparation of chitin from krill waste it was found that treatment with an excess of a bacterial proteinase for 120 hours at pH 6.5 and 55°C enabled a much larger proportion of the pigments to be extracted with $CHCl_3$.[14]

Shimahara et al.[16] have studied the use of proteolytic bacteria specifically cultured to provide good proteolytic activity but not chitinolytic activity. The extent of removal of protein using the strain *Pseudomonas maltophilia* LC 102 varied with the species of crustacea with demineralised *Renaeus japonicus* carapace, the thinnest carapace examined, giving the best results. Although samples were incubated for up to 240 hours at 30°C, no further decrease in protein content occurred after 72 hours even if the chitin was transferred to a freshly innoculated bath. This indicates that the residual protein, ranging from 1% to 7% approximately, is inaccessible to the proteinase involved, in agreement with other workers using different enzymes.[25, 26]

2.2.3 Demineralisation

Demineralisation by a variety of acids including HCl, HNO_3, H_2SO_3, CH_3COOH and HCOOH is claimed by two early patents[8, 9] but apart from one use[10] of HNO_3 and one use[24] of HCOOH, all other workers have used HCl for acid demineralisation. The conditions used by a number of workers are given in Table 2.2. Despite patent claims[8, 9] that demineralisation with acid may be carried out at temperatures from 0 to 100°C, temperatures no higher than room temperature have been used in all

TABLE 2.2 *Conditions employed for demineralisation of chitin-containing waste materials*

Material source	HCl concentration	Temperature	Time (h)	Reference
Shrimp	0.275M	RT	16	8,9
Shrimp	0.5M	NS	NS	11
Krill	0.6M	RT	2	14
Crab	0.65M	RT	24	20
Crab	1.0M	RT	12	15
Crab	1.0M	RT	NS	18
Prawn	1.25M	RT	1	13
Crab	1.57M	RT	5	12
Lobster	1.57M	RT	11–14	25
Prawn	1.57M	20–22°C	1–3	21
Crab	2.0M	RT	48	16
Lobster	2.0M	RT	i 5 ⎫ ii 48 ⎭	17
Crab	11.0M	−20°C	4	23

RT = Room temperature.
NS = Not stated in the reference.

cases, even in the examples specified in the patents, in order to minimise hydrolysis of the polymer chain. However, as can be seen from Table 2.2, there is considerable variation in the HCl concentrations and the lengths of time of treatment used by different research workers.

Extraction with EDTA at alkaline pH values has been used as a non-degradative demineralisation process. Foster and Hackman[27] used EDTA at pH 9, followed by a further treatment at pH 3, to give a product having approximately 1.15% inorganic material, principally silica. The effect of particle size on the rate of demineralisation was particularly noticeable in this treatment. Other workers[26, 28] have employed EDTA at pH 10.

2.2.4 *Decolouration*

The exoskeletons of crustacea contain colouring matter, principally carotenoids, the main components being astacene, astaxanthin, canthaxanthin, lutein and β-carotene.[29] They do not appear to be complexed with either the inorganic material or the protein since treatments which remove these components do not remove the carotenoids.[30] However they may be removed by extracting the shell with ethanol or acetone after demineralisation by either acid or EDTA treatment. Warm 50% aqueous acetic acid simultaneously demineralises the shell and extracts the carotenoids.[31]

Alternatively, the colouring matters may be destroyed by bleaching and the use of $KMnO_4$, $NaOCl$, SO_2, $NaHSO_3$, $Na_2S_2O_4$ or H_2O_2 have been claimed in the literature.[8, 9]

2.3 COMPARISON OF PROCESSING CONDITIONS

Very little in the way of evaluation of different processing conditions has been carried out, despite the numerous variations that have been employed. One problem is that, at least up to the present time, chain degradation can only be assessed by deacetylation of the chitin to an extent at which it becomes soluble in dilute acid solutions (that is, chitosan) and determination of the viscosity under standard conditions. However the deacetylation process is severe (see section 2.5) and hence liable to cause chain degradation, so that any results obtained in this way must be interpreted with caution.

In general, it may be assumed that the milder the treatment the higher the viscosity of the solution prepared from the deacetylated chitin. Thus Lusena and Rose[32] found that chitosan prepared from chitin demineralised with HCl at 5°C gave more viscous solutions than that prepared from chitin demineralised at room temperature (LVN = 215 cm^3 g^{-1} and 124 cm^3 g^{-1} respectively) and that allowing the latter chitin sample to stand overnight at room temperature in the demineralisation medium, prior to deacetylation, reduced the LVN of the chitosan produced to 40 cm^3 g^{-1}.

However some apparently contradictory results have been reported for the effects of acid treatment variation in the demineralisation of prawn waste.[13] The concentration of HCl used varied from 0.75M to 2.0M and the treatment times from 0.5 hours to more than 3 hours, but there was no simple correlation between the severity of the demineralisation treatment and the extent of chain hydrolysis, as demonstrated by the viscosity of solutions of the samples after deacetylation. The results, in a rearranged form, are given in Table 2.3. For any given demineralisation time, the solution viscosity of the chitosan subsequently produced from the chitin increases with increase in the concentration of the HCl used, up to 1.25M. At higher concentrations, the viscosity decreases with increasing HCl concentrations. Furthermore, at HCl concentrations less than 1.25M the viscosity increases with increase in the time of demineralisation, or rises and then levels off in the case of 0.75M HCl. However the concentration of acid-soluble ash remaining in the chitin after demineralisation is extremely high in the case of samples treated with 0.75M and 1.0M HCl, or with more concentrated HCl solutions for the shorter treatment times, and decreases with increase in either HCl concentration or time of treatment. These high ash concentrations will presumably also be present in the deacetylated products and would be expected to affect the viscosity in two ways: by reducing the amount of chitosan actually present in the nominal 10 g dm^{-3} solution and by the introduction of appreciable concentrations of low-molecular-weight electrolyte into the solution through reaction with the acetic acid. Both factors would tend to reduce the viscosity of the solution and could account for the viscosity results in Table 2.3.

Similar results, also from prawn waste, were reported by other workers[21] who found that as the time of demineralisation in 1.57M HCl increased, the solution viscosity of the chitosan obtained from the resultant chitin decreased. However for a given time of demineralisation, 60 minutes, reducing the HCl concentration from 1.57M to 1.25M subsequently produced a chitosan having a lower solution viscosity value. Again the residual ash content of the latter sample was higher.

In the same paper[21] the effect of the position in the process sequence of a peroxide bleaching step was examined, bleaching being carried out by steeping the material overnight at room temperature in 5 g dm^{-3} H$_2$O$_2$.* The relative viscosities of the chitosans produced from these chitins decreased in the order: unbleached > bleached after demineralisation > bleached after deproteinisation > bleached after deacetylation, the relative viscosities being 226.1, 151.5, 98.2 and 1.35 respectively. These results

---

* Hydrogen peroxide is normally sold on a volume strength basis; 100 volume hydrogen peroxide contains 300 g dm^{-3} H$_2$O$_2$.

TABLE 2.3 *Effects of acid concentration and treatment time on the ash content of chitin from prawn waste and the viscosity of the resultant chitosan[13]*

HCl concentration	Treatment time (minutes)								α^a	
	30		60		120		180			
	Viscosity (cps)[b]	Ash (%)[c]	Viscosity (cps)[b]	Ash (%)[c]	Viscosity (cps)[b]	Ash (%)[c]	Viscosity (cps)[b]	Ash (%)[c]	Viscosity (cps)[b]	Ash (%)[c]
0.75M	14.63	48.44	16.84	46.30	18.86	41.52	18.45	39.44	18.02	37.50
1.0M	32.03	43.69	36.56	38.28	38.19	33.86	39.42	23.95	40.03	20.32
1.25M	106.85	24.34	97.07	18.82	58.05	6.33	46.44	2.97	40.89	1.31
1.50M	49.28	15.34	43.95	7.90	40.06	3.14	38.54	1.46	34.58	1.31
2.0M	37.66	2.71	31.52	1.76	26.94	1.03	17.79	0.65	17.20	0.54

[a] The length of time represented by α was not defined.[13]

[b] Viscosity, in centepoises, for a 10 g dm^{-3} solution in 1 vol.-% aqueous acetic acid after deacetylation in 50 wt-% NaOH at 100°C for 2 hours.

[c] Acid-soluble ash in chitin prior to deacetylation.

strongly suggest that the unacetylated amine groups are the sites for attack by the H_2O_2.

Work has also been carried out in the author's laboratory on the effects on the properties of the final chitosan of variations in the conditions used for deproteinisation of the native chitin.[33] Both Na_2CO_3 and NaOH solutions were used for the deproteinisation step and in general the solutions of the eventual chitosan end products showed an increase in viscosity and clarity, and a decrease in colour, with increase in severity of the preliminary treatment, indicating that like the demineralisation treatment the deproteinisation step may affect the properties of the final product, despite the fact that the subsequent deacetylation step involves a much more severe alkaline treatment (section 2.5).

Finally Shimahara et al.[16] have found differences between the chitins produced from Renaeus japonicus by enzymatic and by alkaline deproteinisation. The former had approximately 10% deacetylation while the latter had 21% deacetylation. The X-ray diffraction patterns showed that the NaOH treated material, containing only 0.2% residual protein, was more crystalline than the enzyme deproteinised sample (0.6% residual protein) both being more crystalline than the demineralised starting material (20.8% protein). This difference in crystallinity may explain the fact that the NaOH treated material dissolved more slowly in DMAc–LiCl (50 g dm^{-3}) than did the chitin treated with proteolytic bacteria.

2.4 CHITIN ISOLATES

2.4.1 Introduction

It has been recognised for some time that chitin, even after demineralisation and deproteinisation treatments, is not a simple chemical entity and that samples vary in chemical and physical properties depending on the source material and the purification process. In view of this Austin et al.[34] have proposed the term 'chitin isolate' for such material and consider that these differences arise from variation in the amount and nature of the residual protein, in the extent to which anomerisation of the β-glycosidic bonds has occurred, and in the extent of deacetylation.

2.4.2 Residual protein

The amount of protein remaining after deproteinisation varies considerably and values of from 0.2% to more than 5% may be found in the literature.[16, 34] Several studies have been carried out on the nature of the chitin–protein complex to ascertain the nature of the bond between the two components and the particular amino acids involved in the chitin–protein

link, and no amino acid residue can be identified as the sole link between
the two components (section 1.5).

2.4.3 Anomerisation

Much of the evidence in support of this comes from studies on model
compounds, particularly the alkylglycosides of N-acetyl-D-glucosamine.
While the opposing steric and electrostatic effects in D-glucose result in an
α:β ratio of 37:63 at equilibrium,[35, 36] the effect of the N-acetamido group
in N-acetyl-D-glucosamine is to shift this equilibrium ratio to 68:32.[37]
Furthermore di-N-acetylchitobiose monohydrate crystallises with an α:β
ratio of 90:10[38] while IR studies on oligomers of N-acetyl-D-glucosamine
have shown[39] that as the DP increases, the tendency of the terminal
reducing unit to form the β-isomer decreases. This evidence has led Austin
et al.[34] to suggest that, given favourable conditions, there would be a
considerable tendency for the β-glycosidic links in chitin to anomerise to
the α-configuration. They proposed that under acid conditions anomerisation and hydrolysis are in competition, the particular reaction that takes
place depending on whether it is the C(5)– or the C(1)–oxygen that is
protonated (Figure 2.1).

Experimental evidence cited by them in support of this is the observation that the optical rotation of ethyl 2-acetamido-2-deoxy-α-D-glucoside in
1.0M HCl at 53°C did not change for at least 2.5 hours, after which there
was a decrease with further increase in time.[34] It was argued that since
hydrolysis to N-acetyl-D-glucosamine would decrease the optical rotation
and anomerisation would increase it, the rates of the two processes are
approximately equal in this system and it is only after anomerisation
is substantially complete that the continuing hydrolysis reaction causes
the optical rotation to begin a steady decrease. At room temperature, the
same system showed no change in optical rotation over a period of approximately 720 hours, indicating that anomerisation did not reach completion
within this time at room temperature.

There is no direct evidence of anomerisation in chitin itself but supporting indirect evidence claimed[34] includes the differences in optical behaviour of various chitin samples dissolved in the DMAc–LiCl solvent
system[40, 41] and the slower than expected rates of hydrolysis of some chitosans,[42, 43] the hydrolysis rates for α-D-glucosides being slower in general
than those of the β-anomer.[44]

Austin et al. have also claimed[34] that the results of Falk et al.,[6] who
followed the deuterium exchange reaction of chitin by IR spectroscopy and
noticed the simultaneous disappearance and appearance, respectively, of
the anomeric C(1)–H deformation bands at 891 ± 7 cm^{-1} and 844 ± 8 cm^{-1},
support the concept of anomerisation in chitin. This is in fact not the case,
the error arising from a mistake regarding the reaction conditions. The

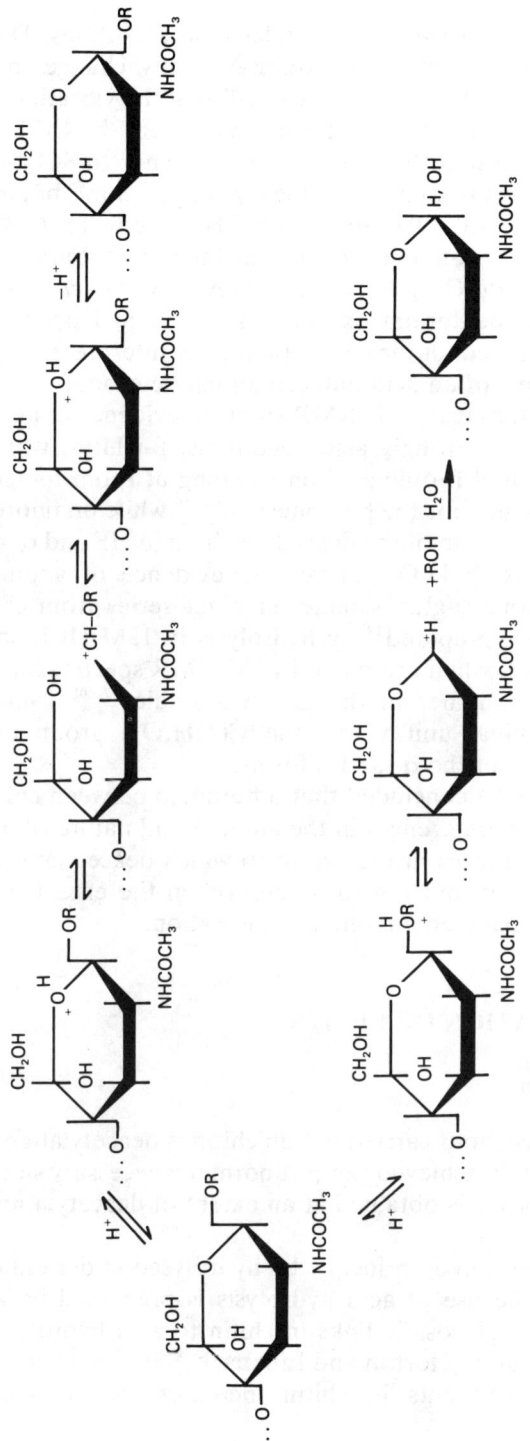

FIGURE 2.1   Pathways for acid catalysed anomerisation and hydrolysis of chitin[34]

deuterium exchange was carried out under mild conditions, D_2O at 21°C, so it would not be expected to give rise to any change in the chitin macrostructure. Hence Falk *et al.* concluded[6] that the vibrational modes of the anomeric C–H must contain a contribution from the C(3)–OH and/or C(6)–OH groups and that the shift from 891 ± 7 cm^{-1} to 844 ± 8 cm^{-1} does not represent a change from the β- to the α-configuration[45] but is due to the conversion of the C(3)–OH and C(6)–OH groups to C(3)–OD and C(6)–OD and the concomitant changes in their contributions to the vibrational modes of the C(1)–H bond. However Austin *et al.* erroneously assumed[34] that the deuterium exchange reaction had been carried out under strongly acid conditions and therefore interpreted the spectral changes as indicative of an acid-induced anomeric change.

More recent studies using ^{13}C NMR show no evidence of anomerisation occurring even under strongly acid conditions. β-Chitin, which was dissolved in concentrated formic acid on standing at room temperature for 14 days, gave the signal for the β-anomer only,[46] while on fluorohydrolysis the ^{13}C NMR spectra both of the degrading chitin in HF and of the isolated oligomeric products in D_2O, showed no evidence of anomerisation.[47] Finally the analogous D-glucosamine oligomer series from chitobiose to chitopentadecanose, prepared[48] by hydrolysis in 12M HCl, showed only β-(1→4)-linked units when examined by ^{13}C NMR spectroscopy. The only evidence of any α-anomer in the last two studies[47, 48] came from the reducing end terminal unit where the C(1)H,OH group exists as an equilibrium mixture of the α- and β-forms.

It must therefore be concluded that differences between chitin samples may be attributed to differences in the amount and nature of any residual protein and to differences in the extent to which deacetylation and chain cleavage has occurred, but not to differences in the extent to which the β-glycosidic bonds have undergone anomerisation.

2.5 DEACETYLATION OF CHITIN

2.5.1 Introduction

One of the main reactions carried out on chitin is deacetylation. Complete deacetylation is rarely achieved nor is it normally necessary since solubility in dilute aqueous acids is obtained at an extent of deacetylation of ~60% or above.

Although amides may in principle be hydrolysed under either acidic or basic conditions, the use of acid hydrolysis is precluded because of the susceptibility of the glycosidic links in chitin to acid hydrolysis. Furthermore, as pointed out by Horton and Lineback,[49] the *trans* arrangement of the C(2)–C(3) substituents in chitin increases the resistance of the

C(2)–acetamido group to alkaline hydrolysis, so that severe treatments are required to bring about deacetylation. At the same time care must be taken to avoid, or at least minimise, any accompanying degradation of the polymer chain.

2.5.2 Deacetylation by alkali fusion

This represents the most severe method used for deacetylation. Fusion with KOH at 180°C, preferably in a nitrogen atmosphere, has been used by a number of workers,[24, 39, 49–54] and although up to 95% deacetylation may be achieved[49] the product had a *DP* of only approximately 20. However the procedure included a purification step, carried out three times, that involved dissolving in 0.1M HCl at 50°C and precipitating out the chitosan hydrochloride salt by addition, at this temperature, of concentrated HCl. Thus part of the large reduction in chain length may be due to acid hydrolysis during purification rather than degradation during deacetylation.

2.5.3 Deacetylation with aqueous alkali

This is the most commonly used method for the deacetylation of chitin but no standard conditions have been established. The most frequently used alkali is NaOH, but KOH has been used in some instances and LiOH, Ca(OH)$_2$ and Na$_3$PO$_4$ have also been claimed to be suitable.[8, 9]

Two early patents established the main principles of the process.[8, 9] These stated that the extent of deacetylation is governed by the alkali concentration, the temperature, the time of reaction, and the particle size and density.[8, 9] The higher the concentration of alkali used the lower the temperature and/or the shorter the time of treatment required. Thus an acid soluble product may be produced by treatment with:

(a) 5 wt-% NaOH at 150°C for 24 hours; or
(b) 40 wt-% NaOH at 100°C for 18 hours; or
(c) 50 wt-% NaOH at 100°C for 1 hour.

It was further claimed that while treatment with 50 wt-% NaOH at 100°C for 1 hour gave a product having 82% deacetylation, extending the reaction time to 48 hours enabled almost 100% deacetylation to be achieved. However this was at the expense of a considerable decrease in the solution viscosity, indicating chain degradation (Table 2.4).

TABLE 2.4   *The effect of time on the quality of deacetylated chitin prepared with 50.9 wt-% NaOH at 100°C and a NaOH solution:chitin ratio of 10:1*[8]

Time (hours)	1	2.5	6.0	16.0	48.0
Viscosity of 50 g dm^{-3} soln (poises)	630	63	46	36	11

The patents[8, 9] stressed the need to exclude air from the reaction if a high-molecular-weight product was to be obtained, but the use of H_2O_2 to bring about controlled chain cleavage to produce lower molecular weight chitosan was also described. Results for its use under both homogeneous and heterogeneous conditions were reported (Tables 2.5 and 2.6). The use of other oxidising agents including chlorine, bromine, hypochlorous acid, perborates, permanganates and dichromates was reported. Sodium hypochlorite was found to be the most effective of those investigated.

TABLE 2.5   *Effect of treatment of chitosan with hydrogen peroxide under homogeneous conditions[8,9]*

H_2O_2 conc. (ppm)		0	2.4	4.8	9.2	14.4	20.0	28.0
Viscosity (poises)	after 1 hour	568	494	388	283	253	213	201
	after 4.3 hours	510	452	342	204	129	80	32.5

TABLE 2.6   *Effect of treatment of chitosan with hydrogen peroxide under heterogeneous conditions[8,9]*

H_2O_2 conc. (ppm)	0	25	50	100	200	400	800
Viscosity[a] (poises)	1976	760	282	151	46	17.8	2.7

[a] Viscosities determined using chitosan solutions (50 g dm^{-3}) prepared after oxidation of solid chitosan.

Despite a considerable amount of subsequent work on the deacetylation of chitin, no major error in the main conclusions outlined above has been found. Nud'ga et al.[55] studied the effect of the deacetylation conditions on the LVN and degree of deacetylation of the resultant chitosans. Chitin was refluxed in 49 wt-% NaOH solution, which gave a temperature of 140°C, at a liquor ratio of 10:1. The results confirmed that chain degradation increased with increase in time of treatment both in air and under nitrogen, but that the extent of degradation was less in the latter atmosphere. The use of an oxygen-free argon atmosphere was found to give a less degraded product than that obtained using a nitrogen atmosphere. However the extent of deacetylation, as measured by % N, appeared to be substantially independent of the time of treatment over the range 1–6 hours.

Fractionation of one of the samples by precipitation with acetone from a solution adjusted to pH 6.5 gave a total of 9 fractions whose % N values varied only slightly, showing that fractionation is by molecular weight and not by degree of deacetylation. The differential curve for the distribution of chitosan with respect to the LVN showed one maximum in the region of high LVN values, the sharpness of this peak indicating that the sample had a low polydispersity.

Wu and Bough[11] found that in general both the molecular weight and the % N-acetyl content showed initial rapid decreases, followed by continued decrease at much slower rates, and that the rates of these initial decreases were greater the higher the NaOH concentration used. When 35 wt-% NaOH was used the viscosity of the products started to rise from approximately zero after 12 hours treatment, and continued to rise up to a maximum after 20 hours treatment. A similar effect was noticeable with the $\bar{M}_w$ values except that the maximum was reached after 27 hours' treatment. These results suggest that when the higher concentrations of alkali are used, all the chitin is rapidly deacetylated to an extent sufficient to render it acid soluble and any continuation of the treatment can only reduce the molecular weight, and hence the viscosity, through chain degradation. When 35 wt-% NaOH is used, deacetylation proceeds more slowly so that in the initial stages only a small percentage of the material will have been deacetylated sufficiently to render it acid soluble, and it will be predominantly the lower-molecular-weight chains that will be soluble at the reduced extent of deacetylation. Once all the sample is soluble, the effect of chain cleavage on the viscosity becomes dominant.

In work reported by Kurita et al.[56] the temperature of reaction was controlled by the reflux temperature of the solution, so that it varied with variation in the NaOH concentration. Under these conditions it was found that while with 40 wt-% NaOH there is a very rapid initial deacetylation to an acid-soluble product, followed by a slow increase in the extent of deacetylation with time, with solutions having lower concentrations of up to 30 wt-% NaOH there appears to be an upper limit to the extent of deacetylation that may be obtained, and this is insufficient to produce an acid-soluble product (Figure 2.2). This upper limit of deacetylation increases with increase in the NaOH concentration but also with increase in the reaction temperature, and therefore it is not possible to differentiate between the effects of temperature and of NaOH concentration on the basis of these results.[56] However the results of Rigby,[8] who obtained an acid-soluble product using 5 wt-% NaOH at 150°C, suggest that it is the reduction in temperature at the lower alkali concentrations that is the important factor.

Typical deacetylation processes of other workers[12, 13, 49, 57, 58] involve the use of 40–50 wt-% NaOH solutions at 100–115°C for 0.5–6 hours. In the case of krill chitin, described as having a much more delicate structure than chitin from other crustacea, deacetylation was carried out using 50 wt-% NaOH at 80–96°C for 15–30 minutes.[59]

Although deacetylation of chitin by aqueous solutions of KOH was referred to in the early patents[8, 9] its use has received little attention from subsequent workers. Lusena and Rose[32] deacetylated lobster chitin by treatment with 55 wt-% KOH at 100°C for 0.5 hours. Decreasing the alkali concentration increased the time required to obtain an acid-soluble chitosan,

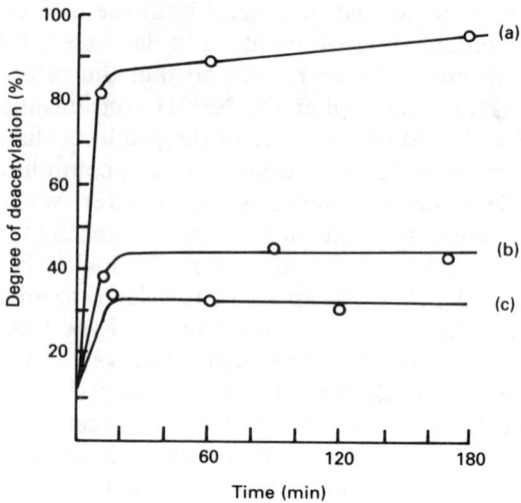

FIGURE 2.2   Effects of NaOH concentration and temperature on the heterogeneous
deacetylation of chitin: (a) 40 wt-% NaOH at 130°C; (b) 30 wt-% NaOH at
112°C; (c) 20 wt-% NaOH at 105°C[56]

and the products gave a less viscous solution than did those prepared
with 55 wt-% KOH and a shorter reaction time. Again, the use of a
nitrogen atmosphere was found to give a higher-molecular-weight product
than was obtained by deacetylation in air. However it was considered that
acid hydrolysis during demineralisation had potentially the greatest effect
on the molecular weight of the final chitosan product. Moorjani et al.[21]
attempted to compare the use of NaOH and KOH, and concluded that a
higher-molecular-weight product was obtained using the latter alkali.
However the comparison was made between samples deacetylated with
60 wt-% NaOH and with 60 wt-% KOH, which are 15 molal and 10.7 molal
solutions respectively. Since both the time and temperature were the same
for both deacetylations, it is only to be expected that the more concentrated alkali solution would give a lower-molecular-weight product.

One interesting observation of Lusena and Rose[32] was that the two 0.5-hour
deacetylation treatments separated by washing and air drying were as
effective as one continuous 15-hour deacetylation treatment, and the
product was more viscous. This was also found by later workers[15] to be the
case for deacetylation with NaOH solutions. In this work, deacetylation
was carried out with 47 wt-% NaOH at either 60°C or 110°C in a nitrogen
atmosphere, the washings between successive alkali treatments being
carried out at about 80°C. The results are given in Table 2.7.

The effectiveness of a multistage treatment interspersed with washing
can be seen from a comparison of L4 and L*1 and of H3 and H*1. A

TABLE 2.7 *Deacetylation of chitin by multistage alkali treatments*[15]

Sample[a]	Total reaction time (hours)	Deacetylation (%)	Insoluble material (%)	$\bar{M}_w \times 10^{-5}$
L1	2	57	40	—
L2	4	67	11	7.9
L3	6	87	Trace	8.5
L4	8	90	Trace	7.0
L*1	8	70	14	—
H1	1	78	Trace	6.0
H2	2	91	0	5.7
H3	3	96	0	5.7
H*1	4	82	Trace	6.3

[a] L and H denote reaction temperatures of 60°C and 110°C respectively, and the number after each letter indicates the number of successive treatments. The asterisk indicates a single continuous treatment.

possible explanation for the effect of the washing treatment can be found in studies on the swelling of cellulose with alkali[60] where it was shown that impregnation with a concentrated solution of NaOH, followed by dilution with water, causes a greater degree of swelling than is obtained by direct addition of NaOH solution of any concentration. Assuming similar behaviour in the case of chitin it can be argued that during washing, after one deacetylation treatment, the alkali concentration within the chitin particles would be gradually decreased down through the maximum swelling concentration, thereby increasing the accessibility of the chitin during the subsequent deacetylation treatment.

2.5.4. Alternative techniques for deacetylation of chitin

Broussignac[20] has described the use of an almost completely anhydrous deacetylation mixture comprising 2 parts KOH, 1 part monoethylene glycol and 1 part 96 vol.-% ethanol. Deacetylation was carried out by treating 1 part of chitin with approximately 13 parts of the above mixture at 120°C. The viscosity of the product decreased with increase in reaction time while the extent of deacetylation initially increased with time, then levelled off at about 85% in a similar manner to that observed with aqueous alkali treatments. Two advantages claimed for this system are that it is less corrosive to stainless steel than is aqueous alkali and that different viscosity grades can be produced in a reproducible manner simply by using different reaction times. However the use of H_2O_2 under either homogeneous or heterogeneous conditions was also described as a means of obtaining low viscosity products in a controlled way.[20]

Broussignac's method has been used in the deacetylation of krill and crab chitin,[14] the samples being treated under reflux for 20 hours. The chitosan obtained from krill chitin was tan coloured whereas that from crab chitin was white. However the yield of the former was much higher (90% compared with 62% from crab chitin) while the extent of deacetylation was slightly greater. The viscosities of 10 g dm^{-3} solutions of both materials were low (57–70 cps) as might be expected in view of the extended deacetylation treatment. No explanation was given for the use of such a prolonged treatment time.

Other variations on the standard aqueous alkali treatment have been directed towards reducing the quantity of alkali required in the treatment. In the conventional process the concentrated alkali acts both as the deacetylating reagent and the reaction medium, thus there is a large excess of alkali over what is actually consumed in the deacetylation reaction. As an example, using 40 wt-% NaOH and a liquor ratio of 10:1, there is a 2700% excess of NaOH over what would be consumed in the cleavage of all the amide groups present.

Fujita[61] treated 1 part of chitin with 1 part of 50 wt-% NaOH solution, then kneaded the mass to obtain uniform distribution of the NaOH before mixing with 10 parts liquid paraffin and stirring at 120°C for 2 hours. At the end of the reaction time the mixture was poured into 8 parts of cold water, filtered, washed with water and dried to give a 91% yield of product which was 92% deacetylated.

A development of this is the use of water-miscible solvents such as IPA, TBA or acetone as a diluent to ensure ease of stirring, and as a transfer medium to ensure uniform distribution of the aqueous alkali throughout the chitin mass.[62, 63] The chitin is slurried in the appropriate diluent then the NaOH solution is run in and the mixture heated to the reaction temperature. Although temperatures above the boiling point of the diluents may be achieved by carrying out the deacetylation step under pressure, much of the work reported was carried out at the reflux temperature of the mixture. In general, the extent of deacetylation was found to be less than that normally achieved using aqueous alkali alone but the products were acid soluble. Refluxing a mixture of 1 part chitin, 1.5 parts 50 wt-% NaOH and 12.5 parts acetone for 5 hours gave a product that was 62.4% deacetylated. A 10 g dm^{-3} solution in dilute acetic acid had a viscosity of 137 cps. Similar treatment using IPA and a reaction time of 3 hours gave a product that was 69% deacetylated. The viscosity of a 10 g dm^{-3} solution was 290 cps, which was increased to 440 cps on carrying out the deacetylation under nitrogen. In the latter case the product had a molecular weight in excess of 10^6 ($\bar{M}_v = 1.19 \times 10^6$). It was also found[63] that the addition of 1 part NaBH$_4$:10 parts chitin gave a 10-fold increase in the solution viscosity of the product, increasing it from 79 cps to 765 cps for a 10 g dm^{-3} solution in 1 vol.−% aqueous acetic acid. The efficiency of NaBH$_4$, which was assumed to function by preventing 'end-peeling' of sugar units from the reducing end, was considerably greater than that of thiophenol which functions as an oxygen scavenger.

Another technique to reduce the quantities of alkali required involves steeping chitin in 35–60 wt-% NaOH solution, removing the excess from the chitin so that 1 part chitin contains 0.5–2.5 parts NaOH solution, then heating the alkali-impregnated chitin at 45–110°C to bring about deacetylation.[64]

2.5.5 Deacetylation of fungal chitin

Although waste mycelia have been proposed as potential sources of chitosan, there are very few reports on the deacetylation of fungal chitin. As mentioned previously (section 2.1), such chitin is normally present in association with other polysaccharides which need to be removed in order to obtain chitin itself. However there are no reports in the literature of any attempts to isolate the chitin component prior to its deacetylation, so that this latter process is carried out on the total polysaccharide mass.

In a series of patents, Muzzarelli has disclosed the preparation of a chitosan–glucan complex from the mycelial biomass of filamentary fungi such as *Allomyces*, *Aspergillus* and *Mucor*, from yeasts and from other moulds and fungi. Lipids and proteins may be removed from the biomass at the same time as deacetylation is being carried out, or they may be removed prior to deacetylation by washing with a 0.625M NaOH solution.[65]

Infra-red spectra of the chitosan–glucan complex both before and after extraction with dilute acetic acid, and of the precipitate obtained from the supernatent on making alkaline, show that the complex can be resolved into two components, namely chitosan and glucan. In view of this, it is difficult to understand the claim that the chitosan–glucan complex is not simply a mixture of chitosan and glucan,[65c] except insofar as it is not a mixture of individual particles of chitosan and of glucan, rather it would appear to be a mixture at the molecular level, with no evidence of chemical links between the chitosan and glucan polymer chains.

More experimental details are given in two related papers by Muzzarelli *et al.*[66, 67] In the first of these, the IR spectra of *Mucor rouxii* after boiling in 40 wt-% NaOH for 30 minutes, and of *Streptomyces* after boiling for 4 hours, are given. The former is very similar to that for chitosan from crab chitin while the latter is very different because of the high concentrations of diatomaceous earth added as a filter aid in the initial isolation of the mycelia. Similar chitosan–glucan material has been produced[62] by treatment of fungal biomass with 40 wt-% NaOH solution at 85°C in the presence of a water-miscible organic diluent.

It is claimed[65–67] that the chitosan–glucan product may be used directly

TABLE 2.8   *Chitosan from a fermentation of* Absidia coerulea *at various harvest times*[68]

Cell harvest time (hours)	Dry cell weight ($g\ dm^{-3}$)	% yield from DCW		
		Hyphal material[a]	Chitosan[b]	Insolubles
24	14	26	16	2
36	18	30	27	1
48	19	30	11.5	20

[a] After a single 1 hour boil in 0.5M NaOH.
[b] Amount of pure chitosan extracted from hyphal mixture by treatment with 2 vol.-% acetic acid.

in place of chitosan itself in many of the application areas proposed for chitosan but, despite this, little interest to date has been shown in the commercial production of chitosan from mycelial waste.

A very interesting paper[68] describes the culture of the *Mucorales* organism *Absidia coerulea* to produce a biomass that, after a single treatment with 0.5M NaOH solution, is virtually pure chitosan, thereby avoiding the necessity of an alkaline deacetylation treatment. The composition of the product after deproteinisation by boiling in 0.5M NaOH depends on the stage at which the culture is harvested (Table 2.8), but products ranging from chitosan[0.06] to chitosan[0.2] were obtained. Analysis showed the chitosan from the culture harvested after 36 h to be 80% deacetylated and to have a viscosity of 390 cps for a 10 g dm^{-3} solution in 2 vol.$-$% acetic acid.

Shimahara *et al.* [69] screened 125 Mucoraceae strains for suitability for chitosan production and found considerable variation, both in the amount of chitosan produced per unit volume of culture and in its level of *N*-acetylation, as well as in the viscosity of solutions prepared from the products. The initial screening showed that strains possessing high chitosan productivity are concentrated in the genus *Absidia* and one strain, *Absidia butleri* HUT 1001, was selected for more detailed study. Yields of approximately 1 g dm^{-3} extractable chitosan were obtained and this was characterised as having 10–13% *N*-acetylation, with a molecular weight of 1.2×10^6. Chitosans from crab and prawn shell were characterised for comparison purposes and found to have molecular weights of 4×10^6 and 1.5×10^6 respectively, the former value being unusually high.

These results[68, 69] suggest that it may become both technically feasible and commercially viable to produce chitosan directly by fermentation of suitable organisms, thereby avoiding the need for a deacetylation step. Two factors that would aid the commercial viability are an increase in the yield per unit volume, and the development of organisms that combine high chitosan yields with the production of other useful products during the fermentation process.

2.5.6 Fully deacetylated chitosan

In general, alkaline deacetylation of chitin proceeds rapidly until the polymer is about 75–85% deacetylated, after which further treatment has only a very limited effect on the extent of deacetylation. The most probable explanation for this is that the morphology of the chitin is such that the remaining amide groups are inaccessible to the NaOH unless drastic conditions are used. If it is a morphological effect, solution and reprecipitation of the polymer after the initial deacetylation treatment should give a product in which the remaining amide groups are more accessible owing to the morphological changes induced. Although this is most probably the rationale behind the successful attempts at producing fully deacetylated chitosan[15, 63, 70, 71] in only one case has this argument been made.[71]

Mima et al.[15] dissolved samples that were 90–96% deacetylated after several deacetylation treatments interspersed with washing and drying – see Table 2.7, section 2.5.3 – and regenerated them by pouring in a fine stream into a large volume of 1M NaOH. The thread-like chitosan was then subjected to a further deacetylation treatment with 47 wt-% NaOH; either 2 hours at 60°C or 1 hour at 110°C. In the latter case the extent of deacetylation was increased from 96% to 99%, while the $\bar{M}_w$ value decreased from 5.7×10^5 to 5.0×10^5.

Domard and Rinaudo[70] modified a procedure, proposed by Kenne, Lindberg and co-workers[72, 73] for lipopolysaccharides, by the addition of thiophenol to act as an oxygen scavenger and so reduce chain degradation. It was also claimed[70] to have a catalytic effect, increasing deacetylation. The process involves precipitating chitosan from a solution in 83 vol.-% aqueous DMSO with excess 50 wt-% NaOH and heating for 1 hour at 100°C in a nitrogen atmosphere after the addition of thiophenol. After washing and drying the product, the whole procedure was repeated as many times as required. Two samples were treated in this way and the results are given in Table 2.9.

It was concluded that:

(a) the amount of NaOH should be approximately four times that of the total concentration of amine groups in the polymer;
(b) deacetylation treatments should be 1 hour at 100°C;
(c) multiple treatments are more effective than a single treatment of similar total time.[32]

The IR spectra of the final products in each case showed no evidence of the amide I band at 1655 cm^{-1}, while titration with HCl gave an equivalent weight for the amine groups of 161±1, in excellent agreement with the value expected for the fully deacetylated product.

Fully deacetylated chitin has also been produced in the author's laboratory and the use of the organic diluent process, as well as the standard aqueous

TABLE 2.9   *Effect of repeated deacetylation treatments on the extent of deacetylation and the molecular weight of chitosan*[70]

Sample	Step	Degree of N-acetylation[a]	$\bar{M}_v^b$
I	0	8	2.75×10^5
	1	1	1.97×10^5
	2	0	1.33×10^5
II	0	20	$5.0 \ \times 10^5$
	1	4	3.04×10^5
	2	1	1.63×10^5
	3	0	1.24×10^5

[a] Degree of *N*-acetylation was measured by IR spectroscopy.[74]
[b] Viscosities were measured using 0.2M acetic acid–0.1M sodium acetate as solvent and $\bar{M}_v$ values calculated from LVN values using the constants of Lee.[75]

TABLE 2.10   *Comparison of the aqueous NaOH and the TBA/aqueous NaOH systems for preparation of fully deacetylated chitosan by the multiple treatment method*[63]

Slurry composition	Sample number	Analysis	Number of deacetylation treatments			
			0	1	2	3
20 cm³ 50 wt-% NaOH: 1 g chitosan	1	% *N*-acetylation	22	11	0	0
		$\bar{M}_v^a/10^6$	2.01	1.79	1.79	1.63
	2	% *N*-acetylation	22	9	1	0
		$\bar{M}_v^a/10^6$	2.01	1.69	1.57	1.32
20 cm³ TBA: 1.25 cm³ 50 wt-% NaOH: 1 g chitosan	3	% *N*-acetylation	22	10	4	2.4
		$\bar{M}_v^a/10^6$	2.01	1.87	1.56	1.45
	4	% *N*-acetylation	22	7	4	2
		$\bar{M}_v^a/10^6$	2.01	1.69	1.67	1.29

[a] $\bar{M}_v$ were determined using the method of Roberts and Domszy.[76]

alkali process, were investigated.[63, 71] The chitosan was reprecipitated from solution in dilute acetic acid, then washed and dried, between successive alkali treatments. The results are given in Table 2.10.

The results show that complete deacetylation, together with retention of high molecular weight, can be achieved by a series of treatments with 50 wt-% NaOH solution at 75°C. The diluent slurry process, using TBA, is less effective both with regard to the extent of deacetylation and the retention of molecular weight.

2.5.7 Water-soluble chitosan*

Sannan *et al.*[77] observed that films cast from solutions of alkali chitin become water soluble if left to stand at room temperature for at least 48 hours before the alkali is washed out, and suggested that the changes in solubility might be due to a reduction in molecular weight and/or a change in the secondary structure. In further studies[78] the effect on the chitin of ageing of the alkali chitin solution at 25°C was examined by determining the flow time, the ease of precipitation on dilution with water, and the % *N*-acetyl content and aqueous solubility of the regenerated polymer.

Although increasing the temperature at which the alkali chitin solution was held increased the rate of deacetylation, the *N*-acetyl contents at which water solubility was obtained remained relatively constant in the range 45–55% *N*-acetyl (Figure 2.3).

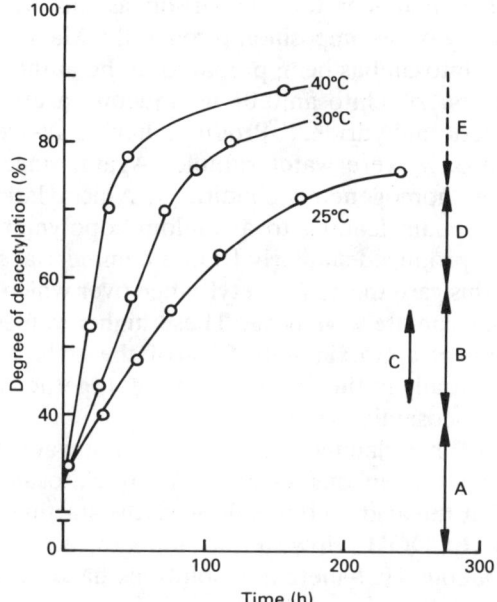

FIGURE 2.3　Effect of temperature on the homogeneous deacetylation of chitin and solubilities of the products. A and E: precipitated by dilution alone; D: precipitated during neutralisation; B: not precipitated on dilution or neutralisation; C: isolated samples redissolved in water[56]

---

* Although Sannan *et al.*[77] referred to 'water-soluble chitin' the material should correctly be called water-soluble chitosan since it is soluble in dilute acetic acid solutions (section 1.2).

The fact that water solubility occurs at an N-acetyl-D-glucosamine:D-glucosamine ratio of approximately 1:1 when prepared under homogeneous conditions, while chitin deacetylated to the same extent under heterogeneous conditions is insoluble even in dilute acid solutions, together with the differences in X-ray diffraction patterns – see section 2.5.8 – led the authors to propose that the former treatment gives a random copolymer of N-acetyl-D-glucosamine and D-glucosamine units and the latter treatment produces a block copolymer containing these two structural units.[56]

The kinetics of alkaline deacetylation of chitin under homogeneous conditions has been studied using solutions of alkali chitin.[79] The reaction was found to be pseudo-first order with an activation energy of 93.4 kJ mol^{-1}. An unsuccessful attempt was made to compare the homogeneous alkaline hydrolysis behaviour of chitin and of the monomeric N-acetyl-D-glucosamine, but alkaline solutions of the latter rapidly became brown in colour because of decomposition of either the starting material or the anticipated product, D-glucosamine. The methyl glycoside derivative would be a more suitable monomer for comparison as this would considerably reduce any possibility of decomposition through the Maillard reaction.

Water-soluble chitosan has been prepared in the author's laboratory by partial N-acetylation of chitosan[0.0] in aqueous acetic acid–methanol solution using acetic anhydride.[80] Products having N-acetyl contents of approximately 46–55% were water soluble. Again, since the reaction is taking place under homogeneous conditions, N-acetylation will occur at random along the chain leading to a random copolymer. Water-soluble chitosan has been produced similarly from a commercial sample of chitosan[0.14], but in this case the % N-acetyl range over which water solubility is obtained is approximately 54–60%. These higher values represent random N-acetylation of approximately 50% of the 86% of D-glucosamine residues present initially in the chitosan sample, together with the original 14% N-acetyl-D-glucosamine content.[80]

A recent patent[81] has claimed the preparation of water-soluble chitosan by this method of homogeneous N-acetylation of chitosan in acid solution, and it is stated that the acid used to solubilise the starting chitosan may be HCl, HNO$_3$ or CH$_3$COOH. However attempts to carry out N-acetylation of chitosan in aqueous HCl–methanol solutions have been unsuccessful[80] (Table 2.11).

The failure to achieve any significant N-acetylation, even in the presence of high concentrations of acetic anhydride, may be attributed to total protonation of the amine groups by the HCl, thereby preventing them from functioning as nucleophilic reagents. It is reasonable to expect that HNO$_3$ would have the same effect.

Another approach to obtaining random N-acetylation along the chain has been described recently.[82] In this, a solution of chitosan[0.1] in aqueous acetic acid–methanol was poured into a large volume of pyridine and

TABLE 2.11 *Attempted N-acetylation of chitosan[0.14] in aqueous HCl–methanol solutions.[80] In each case the solution contained 0.6 g chitosan[0.14] – equivalent to 3.1×10^{-3} mols $-NH_2$*

[Acid]	Acetic anhydride (mols)	% N-acetylation
0.1M CH$_3$COOH	5.9×10^{-3}	48
0.1M HCl	5.9×10^{-3}	14
0.066M HCl	5.9×10^{-3}	14
	21.0×10^{-3}	14
	47.0×10^{-3}	22
	233.5×10^{-3}	25

the highly swollen precipitate that formed was *N*-acetylated at room temperature using a large excess of acetic anhydride.

An improvement in this process, both in regard to the randomness of *N*-acetylation along the chain and to the efficiency of utilisation of acetic anhydride, is to precipitate the chitosan by pouring into a pyridine–acetic anhydride mixture.[82] Using this method, a 6-fold excess of acetic anhydride, based on the amine group concentration, gave chitosan[0.49], whereas a 20–25-fold excess was required to obtain the same level of *N*-acetylation using the first method. At this level, the products from both methods were water soluble, although that prepared by the second method had better solubility characteristics.

One surprising aspect of this reaction is the apparent absence of any *O*-acetylation, despite the large excesses of anhydride used and the presence of pyridine, which is frequently used as a catalyst in acetylation of sugars. However the IR spectra showed no evidence of ester formation, being identical to those of the corresponding samples prepared by homogeneous deacetylation of chitin.[78]

2.5.8 Changes in morphology during deacetylation

This was first studied by Clark and Smith[10] who concluded from X-ray diffraction studies that deacetylation of chitin – by treatment with saturated NaOH at 95°C for 6 hours – destroyed the regularity of packing between chains in the plane of the *c* axis. Nud'ga *et al.*[55] found that chitosan, as prepared by heterogeneous deacetylation, retained a high degree of order but that the peak positions were different from those for the starting chitin. However, chitosan films cast from solution gave diffraction diagrams characteristic of amorphous materials whether the chitosan was in the salt form or the free amine form.

Kurita *et al.*[56] reported that during heterogeneous deacetylation of chitin, the crystallinity as measured by X-ray diffraction decreased slightly up to a degree of deacetylation of 71% and then more rapidly, so that the

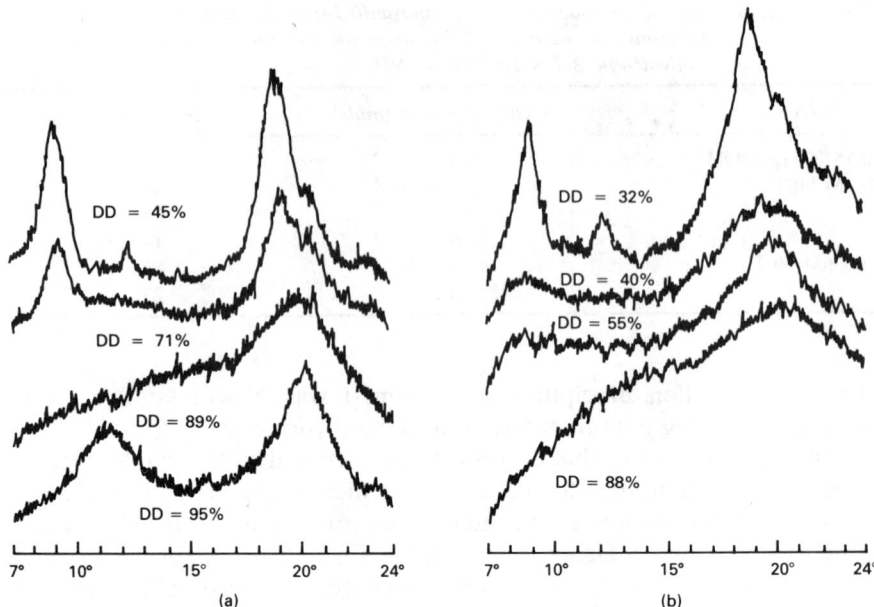

FIGURE 2.4   X-ray diagrams of (a) heterogeneously deacetylated chitin and
(b) homogeneously deacetylated chitin.[56] DD = degree of deacetylation

crystalline peaks had almost disappeared by 81% deacetylation while by
89% deacetylation the sample was completely amorphous. By 95% deacetylation, the material was again crystalline although the peak positions had
moved from $2\theta = 9.2°$ and $19.1°$ to $2\theta = 11.2°$ and $20.4°$. These new peaks
were considered to arise from the presence of crystalline chitosan, the
small amount (~5%) of N-acetylated D-glucosamine residues being insufficient to prevent the chains taking up the chitosan structure (Figure 2.4).
In contrast to this, when deacetylation was carried out under homogeneous
conditions it was found that although a sample having 32% deacetylation
was crystalline, its extent of crystallinity was only of the same order as that
for a heterogeneously prepared sample having 67% deacetylation. Furthermore, a sample having 40% deacetylation was almost completely
amorphous (Figure 2.4). These results were considered to support the
authors' conclusion that while deacetylation under homogeneous conditions takes place at random along the chain, giving rise to a random
copolymer of N-acetyl-D-glucosamine and D-glucosamine residues, heterogeneous deacetylation takes place preferentially in the amorphous
regions then continues more slowly from surface to centre of the crystalline
region, thereby giving rise to block copolymers of poly[β-(1→4)-2-acetamido-

2-deoxy-D-glucose] and poly[β-(1→4)-2-amino-2-deoxy-D-glucose] segments. This pattern of deacetylation under heterogeneous conditions,
coupled with the fact that there is no evidence of a crystalline chitosan
structure in the X-ray diffraction pattern of a sample having 71% deacetylated residues,[56] suggests that the crystalline chitosan diffraction pattern
obtained at 95% deacetylation arises from those chain segments originally
present in the crystalline region of the chitin and which are only deacetylated in the later stages of the treatment.

A number of changes in the IR spectrum of chitin during deacetylation
have been attributed to changes in the degree of crystallinity.[15] In the
2600–3800 cm⁻¹ region, as the extent of deacetylation increases from 91%
to 99%, three distinct, sharp peaks appear at 3455, 3365 and 3313 cm⁻¹
(Figure 2.5). These bands were described[15] as 'crystallisation-sensitive'
bands. Other changes were observed in the 400–1600 cm⁻¹ region. Attention was drawn to the fact that these changes in the spectrum appear rather
abruptly as the extent of deacetylation goes from 91% to 96%, and this
was equated with the transition in the average sequence length of
D-glucosamine residues of from 10 to 20, enabling crystallisation of blocks
of poly[β-(1→4)-2-amino-2-deoxy-D-glucose].

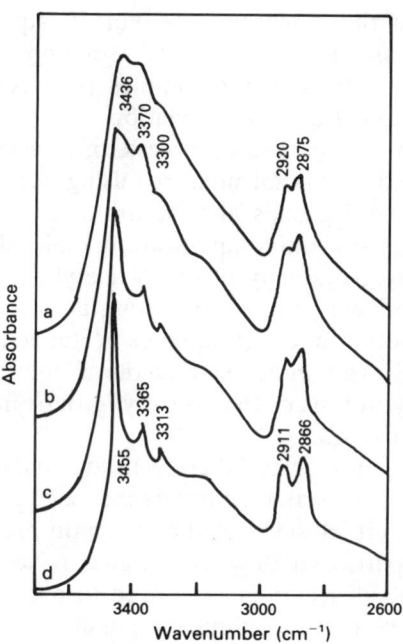

FIGURE 2.5   Infra-red spectra of (a) chitosan[0.22], (b) chitosan[0.09], (c) chitosan[0.04]
and (d) chitosan[0.01]

2.6   PREPARATION OF PURE CHITIN

As discussed in section 2.4, chitin as normally isolated is not a single
unique substance since it will contain varying amounts of residual protein
and will be partially deacetylated to an extent dependent on the source and
the conditions of purification. In view of this Kurita et al.[83] attempted to
prepare pure chitin – poly[ß-(1→4)-2-acetamido-2-deoxy-D-glucose] –
by N-acetylation of water-soluble chitosan. Conditions for N-acetylation
that were examined include: (a) acetic anhydride–pyridine; (b) acetic
acid–dicyclohexylcarbodiimide (DCC).

The acetic anhydride–pyridine system reacted rapidly with the gel
formed on pouring an aqueous solution of the polymer into pyridine.
Reaction times as short as 3 minutes at room temperature gave a product
whose IR spectrum showed the presence of ester groups, in addition to the
increase in intensity of the amide groups absorption bands. Hydrolysis of
the ester groups was carried out by treatment with saturated aqueous
$NaHCO_3$ at 20°C for 141 hours. Analysis of two samples by the method of
Elek and Harte[84] gave values for the extent of N-acetylation of 99% and
100%.

Treatment of water-soluble chitosan with acetic acid–DCC, using a
twenty-fold excess of the reagents in 60 vol.-% aqueous DMF, gave a
highly swollen gel whose IR spectrum, after isolation, showed no evidence
of ester formation. Analysis of the samples prepared in this way gave
values of 96%, 97% and 100% N-acetylation.

Hirano et al.[85] have developed a process for N-acetylation of chitosan
in aqueous acetic acid–methanol mixtures using acetic anhydride. Some
esterification of hydroxyl groups also occurs and these are removed by
treatment for 16 hours at room temperature in 0.5M alcoholic KOH. This
is claimed[85] not to hydrolyse any of the N-acetyl groups but it has been
demonstrated qualitatively[71] that chitin shows an increased adsorption of
anionic dye from solution in 0.1M aqueous acetic acid after treatment in
0.5M alcoholic KOH. This indicates that the alkaline treatment does, in
fact, cause limited hydrolysis of the N-acetyl groups in addition to hydrolysing the O-acetyl groups.

In view of the margin of error inherent in the analysis for the extent of
N-acetylation, it cannot be stated definitely that such products[83, 85] are pure
chitin, although they are closer to it than is chitin prepared directly from
crustacean shell. In particular they have higher N-acetyl contents and, in
view of the severe alkali treatment required to produce the chitosan for
subsequent re-N-acetylation, minimal or possibly zero residual protein.
They therefore represent the purest chitin obtained to date, with the
possible exception of that from diatoms.[6]

2.7 COMMERCIAL PREPARATION

The commercial-scale production of chitin and chitosan imposes additional constraints on the processing routes that may be used, the most important of these constraints being:

(a) the cost of the reagents and the need to recycle them where possible;
(b) the costs of disposal of waste liquors or solids for which no market can be found.

The use of HCl, the most commonly used reagent for the demineralisation step on a laboratory scale (section 2.2.3) has a number of disadvantages when used on an industrial scale. It is relatively expensive, its use results in large volumes of $CaCl_2$ solution that must be disposed of and, being a strong acid, it is likely to cause some hydrolysis of the chitin chains, thereby reducing the molecular weight of the purified material. The cheaper H_2SO_4 cannot be used since the $CaSO_4$ produced is too insoluble to be readily removed from the chitin matrix.

In view of these disadvantages the use of sulphurous acid has been proposed.[86] This must be used in excess of the amount of $CaCO_3$ present in the shell since, if stoichiometric amounts are used, $CaSO_3$ is formed. This has a very low aqueous solubility (~ 0.05 g dm^{-3}) so that removal from the chitin is difficult, but if excess H_2SO_3 is used the more soluble $Ca(HSO_3)_2$ is formed and this is readily removed. Half of the SO_2 may be recovered from the $Ca(HSO_3)_2$ solution by steam-stripping and the other half by treating the resultant precipitate of $CaSO_3$ with H_2SO_4. Thus assuming 100% recovery of the SO_2, the overall process is equivalent to demineralisation with H_2SO_4.

Another proposed process[87] involves carrying out the deacetylation step without a prior demineralisation treatment, using high NaOH concentrations and temperatures of 120–150°C to ensure that the equilibrium reaction

$$2NaOH + CaCO_3 \rightleftharpoons Ca(OH)_2 + Na_2CO_3$$

is driven over to the right-hand side. The chitosan produced contains substantial concentrations of $Ca(OH)_2$ and, after washing to remove the Na_2CO_3, this chitosan–$Ca(OH)_2$ mixture is subjected to countercurrent extraction with an aqueous solution of sucrose which removes the $Ca(OH)_2$ as the soluble calcium saccharate, from which it can be precipitated as $CaCO_3$ by treatment with CO_2.

Peniston and Johnson have also reported a deacetylation process for demineralised chitin that involves the use of considerably lower temperatures.[88] The process consists of mixing the chitin with the NaOH solution, heating to a temperature of 40–80°C, expelling any air from the container and letting the mixture stand, without stirring, for up to 160 hours. Although Peniston and Johnson claim the use of as little as 2 parts of NaOH solution (35–50 wt-%) per part of chitin they state that this gives insufficient liquor to wet the particles of chitin thoroughly, leading to limited swelling of the particles, reduced penetration by the NaOH into the crystalline regions, and a reduced rate of reaction. The use of higher liquor-to-solids ratios, to ensure thorough wetting of the chitin particles and provide a continuous liquid phase, is recommended by these workers.

Although these processes[86-88] have been patented it is not known whether they are being used industrially at the present time.

REFERENCES

1. G.G. Allan, J.R. Fox and N. Kong, in *Proceedings of 1st International Conference on Chitin/Chitosan (1977)*, R.A.A. Muzzarelli and E.R. Pariser (eds), MIT Sea Grant Program 78–7, 1978, p. 108.
2. M.V. Tracey, *Rev. Pure Appl. Chem.*, **7** (1957) 1.
3. A.C.P. Wutke, H. Gargantini and A.G. Gomes, *Bragantia*, **21** (1962) 795 (through *Chem. Abs.*, **61** (1962) 8854).
4. T.M. MacIntyre and M.H. Jenkins, *Sci. Agric.*, **32** (1952) 645.
5. J. Ruiz-Herrera, in ref. 1, p. 11.
6. M. Falk, D.G. Smith, J. McLachlan and A.G. McInnes, *Can. J. Chem.*, **44** (1966) 2269.
7. H.K. No, S.P. Meyers and K.S. Lee, *J. Agric. Food Chem.*, **37** (1989) 575.
8. G.W. Rigby, *US Patent 2,040,879* (1934).
9. Du Pont de Nemours and Co., *UK Patent 458,839* (1936).
10. G.L. Clark and A.F. Smith, *J. Phys. Chem.*, **40** (1936) 863.
11. A.C.M. Wu and W.A. Bough, in ref. 1, p. 88.
12. R.A.A. Muzzarelli, F. Tanfani, M. Emanuelli and S. Gentile, *J. Appl. Biochem.*, **2** (1980) 380.
13. P. Madhavan and K.G. Ramachandran Nair, *Fishery Technol.*, **11** (1974) 50.
14. C.G. Anderson, N. de Pablo and C.R. Romo, in ref. 1, p. 54.
15. S. Mima, M. Miya, R. Iwamoto and S. Yoshikawa, in *Chitin and Chitosan*, S. Hirano and S. Tokura (eds), The Japanese Society of Chitin and Chitosan, 1982, p. 21.
16. K. Shimahara, K. Ohkouchi and M. Ikeda, in ref. 15, p. 10.
17. R.H. Hackman, *Aust. J. Biol. Sci.*, **7** (1954) 168.
18. R.H. Hackman and M. Goldberg, *Carbohyd. Res.*, **38** (1974) 35.
19. R. Blumberg, C.L. Southall, N.J. Van Rensburg and O.B. Volckman, *J. Sci. Food Agric.*, **2** (1951) 571.
20. P. Broussignac, *Chim. Ind. Genie Chim.*, **99** (1968) 1241.
21. M.N. Moorjani, V. Achutha and D.I. Khasim, *J. Food Sci. Technol.*, **12** (1975) 187.
22. S.A. Karuppaswamy, in ref. 1, p. 437.
23. R.L. Whistler and J.N. BeMiller, *J. Org. Chem.*, **27** (1962) 1161.
24. S.T. Horowitz, S. Roseman and H.J. Blumenthal, *J. Amer. Chem. Soc.*, **79** (1957) 5046.

25. C.H. Giles, A.S.A. Hassan, M. Laidlaw and R.V.R. Subramanian, *J. Soc. Dyers Colourists*, **74** (1958) 645.
26. M. Takeda and E. Abe, *Norisho Suisan Koshusho Kenkyu Hokoku*, **11** (1962) 339.
27. A.B. Foster and R.H. Hackman, *Nature*, **180** (1957) 40.
28. M. Takeda and K. Katsuura, *Suisan Daigaku Kenkyu Hokoyu*, **13** (1964) 109.
29. K.L. Simpson, in ref. 1, p. 253.
30. D.L. Fox, *Comp. Biochem. Physiol.*, **44** (1973) 953.
31. R.A.A. Muzzarelli, in ref. 1, p. 25.
32. C.V. Lusena and R.C. Rose, *J. Fish. Res. Board Can.*, **10** (1953) 521.
33. G.A.F. Roberts and F.A. Wood, *Report to the Highlands and Islands Development Board*, 1983.
34. P.R. Austin, G.A. Reed and J.R. Deschamps, in ref. 15, p. 99.
35. R.T. Morrison and R.N. Boyd, *Organic Chemistry*, Allyn and Bacon, Boston, 3rd edn, 1975, p. 1106.
36. W.A. Szarek and D. Horton, *Amer. Chem. Soc. Symp. Ser.*, **87** (1979) 115.
37. R. Virudachalam and V. Rao, *Carbohyd. Res.*, **51** (1976) 135.
38. F. Mo and L.H. Jensen, *Acta Cryst.*, **34** (1978) 1562.
39. S.A. Barker, A.B. Foster, M. Stacey and J.M. Webber, *J. Chem. Soc.*, (1958) 2218.
40. P.R. Austin, *US Patent, 4,165,433* (1979).
41. F.A. Rutherford III and P.R. Austin, *University of Delaware Sea Grant Report, DEL-SG-13-78*, 1978.
42. P.R. Austin, C.J. Brine, S.N. Hirwe, G.A. Reed, H.A. Whelan and J.P. Zikakis, *University of Delaware Sea Grant Report*, DEL-SG-01-80, 1980.
43. P.R. Austin, *US Patent 4,309,534* (1982).
44. J.N. BeMiller, *Adv. Carbohyd. Chem.*, **22** (1967) 59.
45. S.A. Barker, E.J. Bourne, M. Stacey and D.H. Whiffen, *J. Chem. Soc.*, (1954) 171.
46. D. Gagnaire, J. Saint-Germain and M. Vincendon, *Makromol. Chem.*, **183** (1982) 593.
47. C. Bosso, J. Defaye, A. Domard and A. Gadelle, *Carbohyd. Res.*, **156** (1986) 57.
48. A. Domard and N. Cartier, *Int. J. Biol. Macromol.*, **11** (1989) 297.
49. D. Horton and D.R. Lineback, in *Methods in Carbohydrate Chemistry*, R.L. Whistler (ed.), Academic Press, New York, 1965, vol. 5, p. 403.
50. E. Winterstein, *Ber.*, **27** (1894) 3113.
51. F. Hoppe-Seyler, *Ber.*, **27** (1894) 3329.
52. O. von Fürth and M. Rosso, *Beit. Chem. Physiol. Pathol.*, **8** (1906) 163.
53. E. Löwy, *Biochem. Zeit.*, **23** (1909) 47.
54. R. Jeanloz and E. Forchielli, *Helv. Chim. Acta*, **33** (1950) 1690.
55. L.A. Nud'ga, E.A. Plisko and S.N. Danilov, *Zhur. Obsh. Khim.*, **41** (1971) 2555.
56. K. Kurita, T. Sannan and Y. Iwakura, *Makromol. Chem.*, **178** (1977) 3197.
57. M.L. Wolfrom, G.G. Maher and A. Chaney, *J. Org. Chem.*, **23** (1958) 1990.
58. M.L. Wolfrom and T.M. Shen Han, *J. Amer. Chem. Soc.*, **81** (1959) 1764.
59. M.M. Brzeski, in ref. 15, p. 15.
60. G. Saito, *Cellulosechemie*, **18** (1940) 106.
61. T. Fujita, *Japan Patent 13,599* (1970).
62. G.A.F. Roberts, *B.P. Application 8,719,704* (1987).
63. I. Batista and G.A.F. Roberts, *Makromol. Chem.*, **191** (1990) 429.

64. Lion Corporation, *Japan Patent 021,865* (1986).
65. R.A.A. Muzzarelli, (a) *Belgium Patent 876,990* (1979); (b) *German Patent 3,923,820* (1979); (c) *UK Patent 2,026,516* (1983).
66. R.A.A. Muzzarelli, F. Tanfani and M. Emanuelli, *J. Appl. Biochem.*, **3** (1981) 322.
67. R.A.A. Muzzarelli and F. Tanfani, in ref. 15, p. 183.
68. W.J. McGahren, G.A. Perkinson, J.A. Growich, R.A. Leese and G.A. Ellestad, *Process. Biochem.*, **19** (1984) 88.
69. K. Shimahara, Y. Takiguchi, T. Kobayashi, K. Uda and T. Sannan, in *Chitin and Chitosan*, G. Skjåk-Braek, T. Anthonsen and P. Sandford (eds), Elsevier, London, 1989, p. 171.
70. A. Domard and M. Rinaudo, *Int. J. Biol. Macromol.*, **5** (1983) 49.
71. B.D. Gummow, Ph.D. Thesis (CNAA), Nottingham Polytechnic, UK, 1984.
72. C. Erbing, K. Granath, L. Kenne and B. Lindberg, *Carbohyd. Res.*, **47** (1976) C5.
73. L. Kenne and B. Lindberg, in *Methods in Carbohydrate Chemistry*, R.L. Whistler (ed.), Academic Press, New York, 1980, vol. 8, p. 403.
74. M. Miya, R. Iwamoto, S. Yoshikawa and S. Mima, *Int. J. Biol. Macromol.*, **2** (1980) 323.
75. V.F. Lee, *University Microfilms (Ann Arbor) 74/29446*, 1974.
76. G.A.F. Roberts and J.G. Domszy, *Int. J. Biol. Macromol.*, **4** (1982) 374.
77. T. Sannan, K. Kurita and Y. Iwakura, *Makromol. Chem.*, **176** (1975) 1191.
78. T. Sannan, K. Kurita and Y. Iwakura, *Makromol. Chem.*, **177** (1976) 3589.
79. T. Sannan, K. Kurita and Y. Iwakura, *Polymer J.*, **9** (1977) 649.
80. G.G. Maghami and G.A.F. Roberts, unpublished work.
81. Lion Corporation, *Japan Patent 142,710* (1985).
82 K. Kurita, Y. Koyama, S. Nishimura and M. Kamiya, *Chem. Lett.*, (1989) 1597.
83. K. Kurita, T. Sannan and Y. Iwakura, *Makromol. Chem.*, **178** (1977) 2595.
84. A. Elek and R.A. Harte, *Ind. Eng. Chem.*, *Anal. Ed.*, **8** (1936) 267.
85. S. Hirano, Y. Ohe and H. Ono, *Carbohyd. Res.*, **47** (1976) 315.
86. Q.P. Peniston and E.L. Johnson, *US Patent 4,066,735* (1978).
87. Q.P. Peniston and E.L. Johnson, *US Patent 3,862,122* (1975).
88. Q.P. Peniston and E.L. Johnson, *US Patent 4,195,175* (1980).

3

Analysis of Chitin and Chitosan

3.1 INTRODUCTION

When considering the analysis of chitin and chitosan it must be remembered that there is no definitive 'standard' material for either polymer. Thus in many instances analysis is carried out in order to characterise the material rather than to determine whether or not it conforms to a definite structure. In the case of commercial products, much of the analysis is again a matter of characterisation since in many applications it is not critical whether, for example, the sample is chitosan[0.20] or chitosan[0.10]. However in a number of important applications the intended end-use will impose certain critical specifications such as the absence of toxic heavy metals or of residual protein, which might be allergenic, in chitin and chitosan intended for medical applications. Proposed standards for chitosan for use in pharmaceutical and medical applications have been published.[1] These cover general characteristics, chemical and microbiological purity levels, physiological properties and biological activity.

The characteristics most frequently assessed are the degree of N-acetylation, molecular weight, residual protein, moisture content, ash content, lipid content, heavy metal content and colour, their relative importance depending on the intended use.

3.2 DEGREE OF N-ACETYLATION

3.2.1 Introduction

The degree of N-acetylation and the free amine group content are, of course, inversely related and the former may be obtained directly by determining the amide group concentration, or indirectly by determining the amine group concentration. Which approach is the more suitable is partly controlled by the relative abundance of the two functional groups and also by any special requirements of the technique such as a need for sample solubility. In addition there are a few methods that do not depend on the direct determination of either functional groups.

Finally it must be remembered that not only the total level of N-acetylation but also the pattern of substitution of the minor component – D-glucosamine residues in chitin, N-acetyl-D-glucosamine residues in chitosan – can have a considerable effect on the solubility, extent of swelling in water and susceptibility to biodegradation. Different patterns of substitution, that is random or block, have primarily been deduced from physical behaviour and properties[2] rather than actually determined. However Hirano et al.[3] have attempted to study the arrangement of N-acetylated and deacetylated residues by periodate oxidation followed by sodium borohydride reduction and acid hydrolysis (Figure 3.1). The N-acetylated fragments produced were analysed by chromatography. It is most likely that developments in this area will come from enzymatic degradation studies.

3.2.2 Determination of the N-acetyl content

Hydrolytic techniques

These involve hydrolysis of the N-acetyl groups under either acid or alkaline conditions, followed by determination of the acetic acid produced. Such methods have the advantage that there is no requirement for the sample to be soluble.

The liberation of acetic acid from chitin on alkaline hydrolysis and on acid hydrolysis was originally reported by Braconnot[4] and Ledderhose[5] respectively, while the stoichiometry between D-glucosamine and acetic acid was established as early[6] as 1912 (section 1.3.2). Indeed, determination of the acetic acid liberated on hydrolysis in HCl has been developed as a technique for the quantitative estimation of the chitin present in a sample,[7, 8] although it is necessary in this case to assume that any chitin present is fully N-acetylated and this is not necessarily the case.

One of the earliest methods was that of Elek and Harte[9] which, while not developed specifically for chitin, has been used to analyse both chitin[10, 11] and chitosan.[11] The method employs a 25 wt-% solution of p-toluene sulphonic acid as the hydrolytic medium. Other workers have used 80 wt-% H_2SO_4[12, 13] or 50 wt-% NaOH;[14] where NaOH is used, the reaction mixture is acidified prior to the distillation step. Other workers[15] have used the method of Lemieux and Purves[16] which involves the oxidation of terminal CH_3- and related groups with hot, concentrated CrO_3, followed by determination of the acetic acid produced.

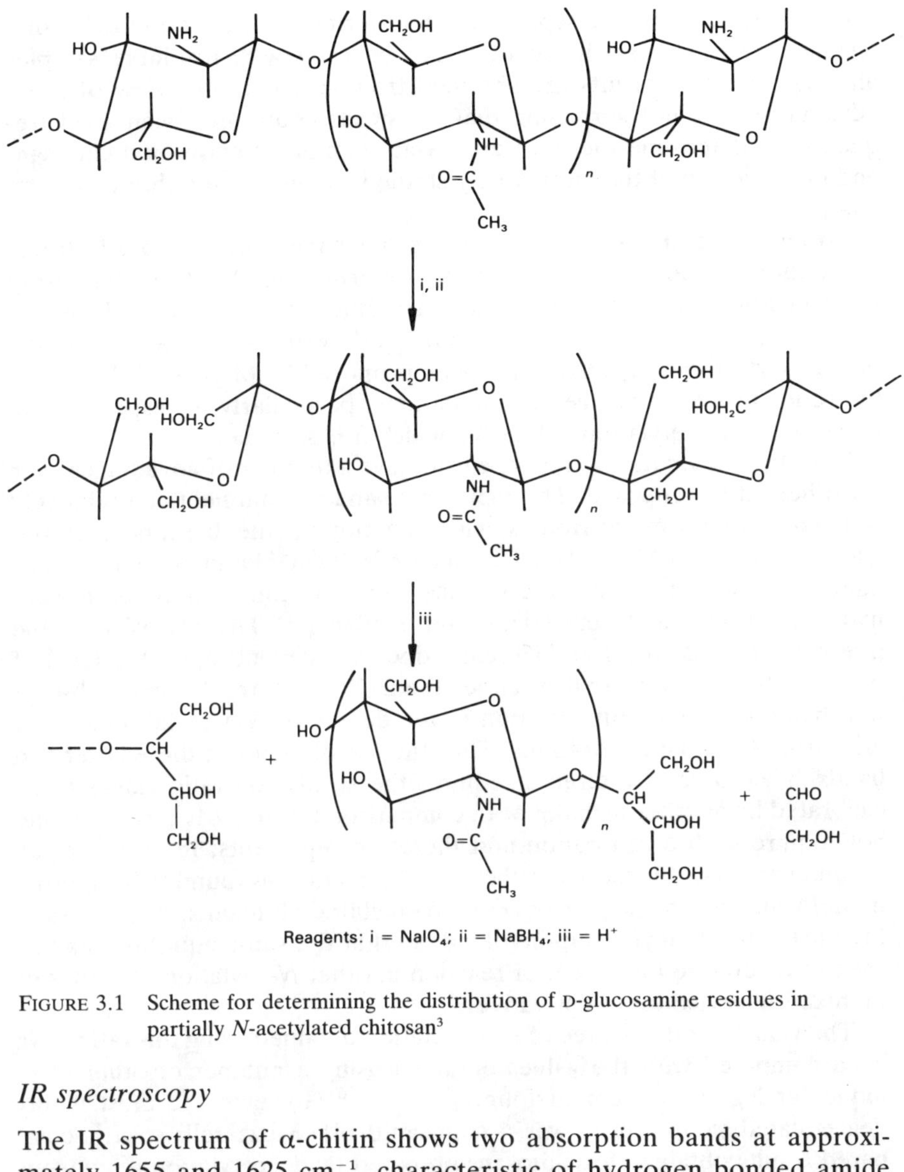

Reagents: i = NaIO$_4$; ii = NaBH$_4$; iii = H$^+$

FIGURE 3.1   Scheme for determining the distribution of D-glucosamine residues in partially N-acetylated chitosan[3]

IR spectroscopy

The IR spectrum of α-chitin shows two absorption bands at approximately 1655 and 1625 cm^{-1}, characteristic of hydrogen bonded amide groups[17] and although Darmon and Rudall noted the disappearance of these bands during deacetylation,[17] their use for determining the degree of N-acetylation was not proposed until some 25 years later.[18]

The use of IR spectroscopy has a number of advantages; it is relatively rapid, makes use of instrumentation found in most laboratories and, provided use is made of an internal reference peak to correct for variation in the amount of material in the beam, the purity of the sample does not need to be determined separately. Furthermore, use of the KBr disc technique means that the method can be used with insoluble samples although the best results are obtained from cast films. In view of these advantages it is not surprising that IR spectroscopy has been used frequently to determine the degree of *N*-acetylation of chitin and chitosan, and that it is one of the most widely studied techniques for chitin/chitosan analysis.

At least three IR absorption band ratios have been proposed, differing either in the band selected for determining the *N*-acetyl group concentration or in that selected as the internal reference. These are A_{1655}/A_{3450},[18-20] A_{1550}/A_{2878}[11] and A_{1655}/A_{2867},[21] while Aiba has reported[22] that a fourth ratio, A_{1554}/A_{897}, has been proposed by Miya *et al.*[23] Each of the ratios proposed has certain limitations, particularly in respect of the range of % *N*-acetylation values for which it is suitable.

The A_{1655}/A_{3450} ratio would appear to have a number of advantages over the other ratios proposed. The 3450 cm^{-1} band is prominent and relatively isolated, two useful attributes when employing the baseline method, whereas the use of either the 2878 cm^{-1} or 2867 cm^{-1} band as the reference band is complicated[24] by interference from the much more prominent hydroxyl group band centred around 3450 cm^{-1}. The intensity of the internal reference band at 3450 cm^{-1} does not depend upon the level of *N*-acetylation whereas that of either the 2878 cm^{-1} or 2867 cm^{-1} absorption band will vary with variation in the degree of *N*-acetylation as they arise from C–H band vibrations. Thus the use of either of these latter two bands as an internal reference requires the absorption ratio values to be calibrated by another technique. In comparison the A_{1655}/A_{3450} ratio should not require such a calibration and therefore represents, in principle, an absolute method. Furthermore the A_{1655}/A_{3450} ratio was found to be approximately constant for a range of fully *N*-acylated chitosans, regardless of the particular *N*-acyl group involved, so that the same equation may be used to determine the extent of reaction in other *N*-acylation reactions or in mixed *N*-acylchitosan derivatives.[19]

The values for the degree of *N*-acetylation obtained using this ratio have been compared with the values obtained using a number of other techniques and good agreement found.[19, 20, 25, 26] However the most highly deacetylated sample used was a chitosan[0.14]. Muzzarelli *et al.*[27] compared results obtained for chitosans prepared by deacetylation in aqueous NaOH and by the method of Broussignac,[28] using the A_{1550}/A_{2878} ratio, and found that different calibration plots were required depending on the method used for deacetylation. Aiba[22] compared the results obtained with both the A_{1550}/A_{2878} and the A_{1655}/A_{2878} ratio with those obtained by a number of other techniques (colloid titration, elemental analysis, UV spectroscopy and GLC). The results showed poor agreement except between the A_{1655}/A_{2867} IR spectroscopic results and those from colloid

TABLE 3.1   *Comparison of the % N-acetyl values determined for a range of chitosan samples by various methods[22]*

Sample	Elemental analysis	Colloid titration[29]	IR		UV[22]	GPC[22]
			A_{1655}/A_{2867}[21]	A_{1550}/A_{2878}[11]		
1	0.1	1.5	0.8	0.3	0.6	2.9
2	—	1.3	2.4	6.5	10.4	6.7
3	—	4.6	5.4	5.2	4.1	7.2
4	—	7.5	6.4	6.6	3.4	8.4
5	—	12.8	14.1	10.6	18.9	12.8
6	—	18.9	23.4	15.7	38.8	19.3
7	14.4	22.3	21.6	13.9	23.3	22.1
8	29.2	27.7	26.2	15.8	25.5	28.5
9	24.3	45.4	—	16.7	30.2	30.0
10	58.0	55.5	—	41.1	55.0	52.6

titration[29] where the agreement was moderate (Table 3.1). This latter correlation is not surprising since the A_{1655}/A_{2867} band ratio was originally calibrated[21] using the colloid titration method.

Domard and Rinaudo[24] examined chitosan samples having low N-acetyl contents and found that the original method of Miya et al.[21] was to be preferred over the A_{1655}/A_{3450} ratio for such samples. In a later, more detailed study Domard[30] examined 16 chitosan samples and compared the N-acetyl values obtained using each of the IR absorption band ratios with those obtained using circular dichroism (CD) measurements (discussed later in this section). The samples examined covered the range chitosan[0.0]–chitosan[0.25], as measured by CD, and good agreement was found between these values and those obtained using the A_{1655}/A_{2867} ratio, but not the values obtained using the other two ratios (Table 3.2). It was pointed out[30] that both of these latter ratios make use of an incorrect baseline in determining the absorbance of the amide I[18–20] and amide II[11] bands, making it impossible to determine the N-acetyl content accurately at low levels – see Figure 3.2 for the respective baselines employed. The A_{1550}/A_{2878} ratio gave the worst correlation and Domard concurred with the conclusion of Miya et al.[21] that the amide I band is better than the amide II band for analysing samples of low N-acetyl content. As had been previously pointed out,[19, 20] there is a considerable change in the 1600–1500 cm^{-1} region with change in the extent of N-acetylation over the range chitosan[0.0]–chitosan[0.30]. This was erroneously attributed[19, 20] to changes in hydrogen bonding whereas the change is in fact due to the variation in relative intensity of the amide II band at 1555 cm^{-1} and the –NH$_2$ band centred at 1590 cm^{-1}. At low levels of N-acetylation the latter band predominates, swamping the amide II band and making accurate determination of its absorbance difficult.[30]

TABLE 3.2   *Comparison of the % N-acetyl values determined for a range of chitosan samples by IR spectroscopy and by CD measurements[30]*

Sample	IR spectroscopy			CD spectroscopy[30]
	A_{1655}/A_{3450} [19]	A_{1550}/A_{2878} [11]	A_{1655}/A_{2867} [21]	
1	6.0	4.5	0.6	0.0
2	9.4	6.0	4.0	3.5
3	11.2	8.6	8.3	7.4
4	12.7	11.2	10.6	9.4
5	13.9	11.2	11.3	10.9
6	16.5	14.0	13.6	10.9
7	15.4	13.7	14.3	11.5
8	15.4	13.7	14.6	11.8
9	16.3	14.2	15.0	14.4
10	18.3	16.0	17.6	14.6
11	18.5	17.8	16.8	14.9
12	19.6	18.8	18.6	17.5
13	20.1	20.6	19.4	18.8
14	22.5	22.2	22.3	21.4
15	22.7	25.2	23.8	23.0
16	24.5	29.2	25.0	24.7

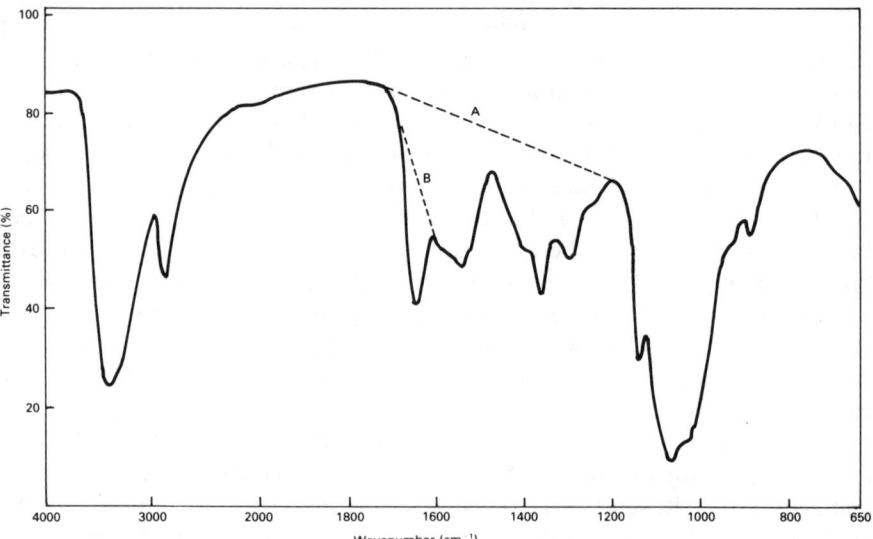

FIGURE 3.2   Baselines used for amide group absorbance: A – used for A_{1655}/A_{3450} [18-20] and A_{1550}/A_{2878} [11] ratios; B – used for A_{1655}/A_{2867} [21] and A_{1655}/A_{3450} [31] ratios

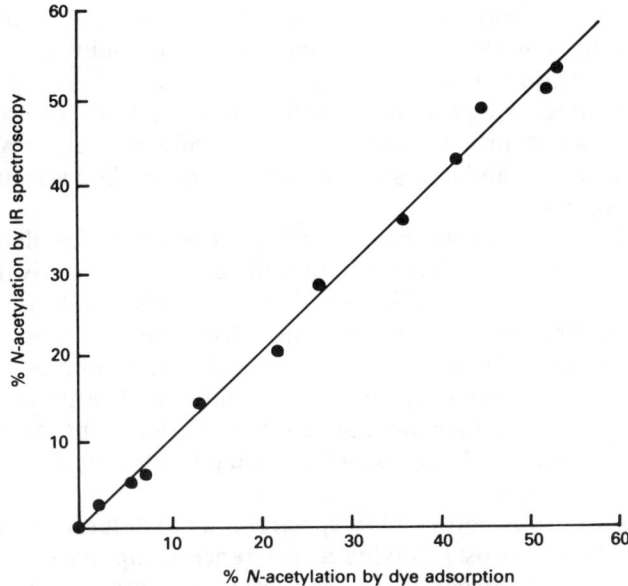

FIGURE 3.3 Correlation between % N-acetylation values obtained by dye adsorption and by IR spectoscopy[31] using the A_{1655}/A_{3450} ratio and the baseline of Miya et al.[21]

More recently a modification of the A_{1655}/A_{3450} has been proposed[31] in which the amide I absorption band is measured using the baseline proposed by Miya et al.[21] In this way the advantages of using this ratio are retained while the error pointed out by Domard[30] is corrected. The N-acetyl content is calculated from the equation.

$$\% \ N\text{-acetylation} = (A_{1655}/A_{3450}) \times 115$$

which was determined from analysis of a number of fully N-acetylated samples. Results for a series of samples covering the approximate range of chitosan[0.0]–chitosan[0.55] gave a good correlation ($r = 0.99$) between the N-acetyl values obtained using this equation and those calculated from dye adsorption measurements (section 3.2.3), indicating its suitability for analysis of chitin/chitosan samples over a wide range of N-acetylation values down to completely deacetylated material (Figure 3.3).

UV spectroscopy

Castle et al.[32] appear to have been the first to attempt to use UV spectroscopy quantitatively with chitin or chitosan. UV spectra of chitin dissolved in DMAc–LiCl and related solvent systems were recorded but although an absorption band attributable to the *N*-acetyl group appeared at 274–278 nm, its intensity was dependent on the age of the solution. This was explained as being due to the setting up of complex equilibria, up to 5 days being required to reach equilibrium. Furthermore the relationship between concentration and absorbance was not linear in all cases, again attributed to equilibrium effects, and it was concluded that the method was unsuitable for quantitative work.

Muzzarelli and Rocchetti have reported a method for determining the degree of *N*-acetylation of chitosan using first derivative UV spectroscopy[33] and *N*-acetyl-D-glucosamine solutions for calibration. It was noted that the presence of D-glucosamine affects the calibration but that correction factors can be derived. A series of chitosan samples which had been previously characterised by other methods (unspecified) were analysed using the technique and the results were said to be in agreement but to be more accurate. Unfortunately no quantitative results were given for the chitosan analysis.

Aiba[22] has used conventional UV spectroscopy to determine the degree of *N*-acetylation of chitosan, taking as reference compounds *N*-acetyl-D-glucosamine, di-*N*-acetylchitobiose and tetra-*N*-acetylchitotetrose. The respective molar absorption coefficients at 220 nm were found to be 177, 173 and 150, and in view of the much lower value for the tetramer this was discarded and the average for the monomer and dimer, 175, was used. This compares with a value of less than 1 for the molar absorption coefficient of D-glucosamine at the same wavelength. The results obtained by this method were found to be more in agreement with the results obtained by colloid titration than with those from IR spectroscopic methods.

Circular dichroism

N-Acetyl-D-glucosamine and its oligomers show two CD bands due to the CH$_3$CONH– group; the first arises from the $\pi-\pi^*$ transition and occurs near 190 nm and the second, which occurs near 210 nm, may be assigned to the $n-\pi^*$ transition.[34, 35] Domard has shown[36] that this latter transition becomes independent of the DP above, and including, the dimer and that in the solid state it shows a bathochromic shift relative to its position in the spectrum of an aqueous solution. This bathochromic shift is also observed in the CD spectrum of a film of chitin.[36, 37] Since the CD spectrum of chitosan[0.0] gives only one band, which is centred around 185 nm and disappears on protonation of the amine group,[38] an acidic solution of a partially acetylated chitosan will only show the CD bands due to the *N*-acetyl group and Domard has made use of this to measure the degree of *N*-acetylation of chitosan.[30] The analysis is straightforward, the extent of *N*-acetylation being calculated from the equation

$$\% \ N\text{-acetyl} = 161 \times H_s \times 100/[161 \times H_s + 203(H_{100} - H_s)]$$

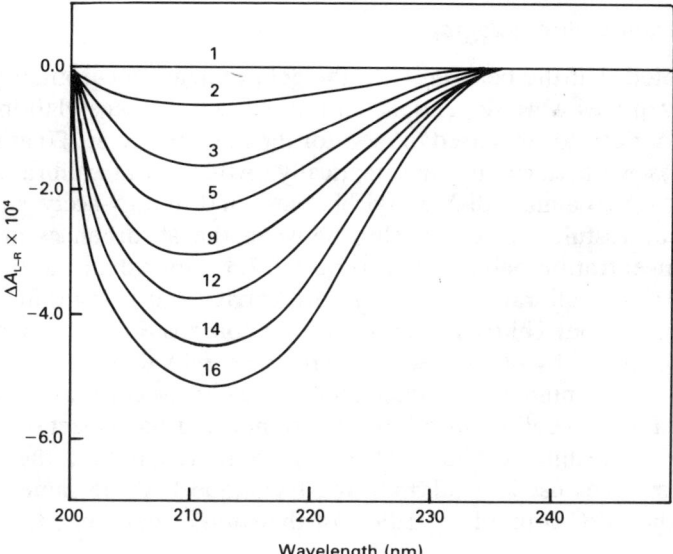

FIGURE 3.4   CD spectra of chitosan samples[30] covering the range chitosan[0.0]–chitosan[0.25]. The numbers on the curves correspond to the sample numbers in Table 3.2

where H_s is the height of the 211 nm peak for the sample and H_{100} is the height of this peak for the same concentration, in g dm^{-3}, of a fully N-acetylated sample. In view of the constancy of both the wavelength and the ellipticity per monomer unit, di-N-acetylchitobiose was used to establish H_{100}. The CD method gave results in very good agreement with those from the IR method of Miya et al.[21] (see above). Figure 3.4 shows the CD spectra of a number of chitosan samples.

NMR spectroscopy

Determination of the degree of N-acetylation by NMR spectroscopy was first carried out by Hirano and Yamguchi[39] in their work on N-acetylchitosan gels. The samples, dissolved in DCOOH, were analysed from the ratio of N-acetyl methyl protons: (methine and methylene) protons in the NMR spectrum.

The method is most accurate at high levels of N-acetylation but problems might be expected in the case of α-chitin which is not soluble in HCOOH and hence, presumably, insoluble in DCOOD. However the use of hexafluoroacetone which is a solvent for α-chitin[40] might enable this technique to be extended to α-chitin, while it has been claimed[41] that the extent of deacetylation of chitin may be readily determined using solid state ^{13}C CP/MAS NMR spectroscopy.

Gel permeation chromatography

Aiba[22] noted that the peak areas in the gel permeation chromatograms of chitosan samples were dependent on the degree of N-acetylation. In this work a UV detector was used to monitor the absorbance at 220 nm and the peak areas were calculated automatically. Attempts at calibration using N-acetyl-D-glucosamine, di-N-acetylchitobiose and tetra-N-acetylchitotetrose were unsuccessful, the relationship between the absorbances at a given molar concentration being found to be 1:1.7:3 rather than the theoretical ratio of 1:2:4. Calibration was therefore carried out using different concentrations of four chitosan samples whose compositions had been previously determined by other methods. The N-acetyl values obtained by this technique for a number of other chitosan samples correlated best with those obtained by colloid titration. One rather unusual aspect of this work was that, according to the relevant figure in the paper, the different chitosan samples used to calibrate the technique had the same retention time on the GPC column regardless of their molecular weight.

3.2.3 Determination of the amine group content

Acid–base titration methods

In the method first proposed by Broussignac,[28] chitosan is dissolved in a known excess of acid and the solution is then titrated potentiometrically with NaOH. This gives a titration curve having two inflection points, the difference between the two along the abscissa corresponding to the amount of acid required to protonate the amine groups (Figure 3.5). The amine group concentration is determined using the equation

$$\% \text{ NH}_2 = \frac{16.1(y-x)f}{w}$$

where f = molarity of the NaOH solution and w = weight, in grams, of the sample.

This technique has been used by a number of workers, but its precision has been questioned[24] by Domard and Rinaudo because of the tendency towards precipitation of the chitosan in the neutralisation pH range. These workers reported an improved technique in which the chitosan is progressively titrated with HCl while the conductivity is monitored to determine the stoichiometry of the interaction. This procedure gave an equivalent weight of 161 ± 1 for the amine group of a sample of chitosan[0.0].[24] Prior to this, Moore examined a back titration method in which a sample is steeped in a known excess of 0.5M HCl, in which chitosan is insoluble, followed by titration of aliquots of the supernatent liquor.[42] The results

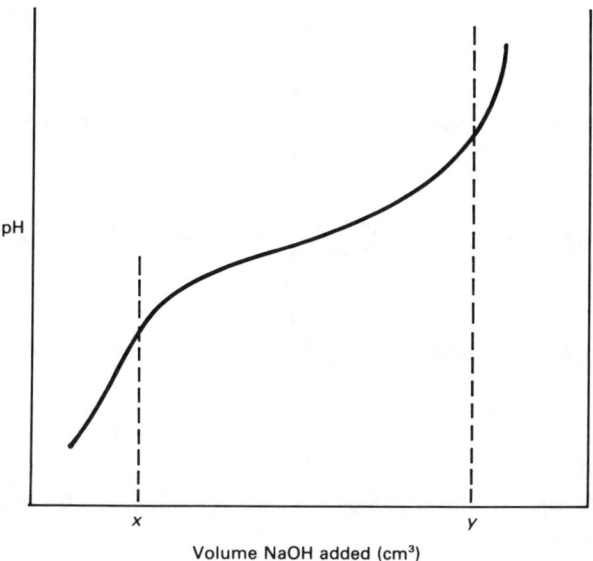

pH

Volume NaOH added (cm³)

FIGURE 3.5 Titration curve of chitosan dissolved in excess acetic acid showing the two points used in Broussignac's relationship[28]

were thought to give too low a value for the free amine group concentration because of the Donnan membrane effect[43] which would result in the external solution having a higher $H^{\oplus}$ concentration than the internal solution, leading to too low an estimate of the quantity of acid adsorbed.

This problem does not arise in the case of titration of water-soluble chitosan salts, and Hayes and Davies[44] titrated solutions of chitosan hydrochloride, prepared by precipitation with concentrated HCl from solutions of chitosan in aqueous acetic acid. The solutions were titrated either with standard alkali, followed potentiometrically or by using phenolphthalein as indicator, or with $AgNO_3$ using 2,7-dichlorofluoroscein as indicator. All three titrimetric methods gave results in very close agreement for a given sample of chitosan.

Chitosan hydrobromide salts have been used in preference to the hydrochloride salts.[20] Use of the former salts had been rejected by Hayes and Davies who stated that they decomposed and became coloured on standing.[44] However precipitation of chitosan hydrobromide by addition of concentrated HBr to a solution of chitosan in dilute HBr gave a product that did not discolour on standing provided it had been washed adequately with methanol prior to drying.[20] The results obtained by titration of these salts in solution were in agreement with those obtained by IR spectroscopy and by the salicylaldehyde method[20] (discussed later in this section).

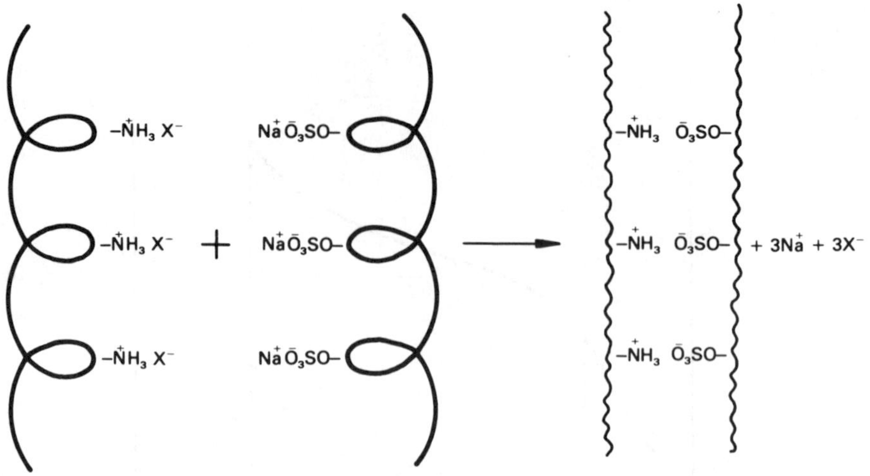

FIGURE 3.6   Principle of charged group stoichiometry involved in colloid titration[29]

Colloid titration

The colloid titration technique was developed by Terayama[29] as a method for analysing polyelectrolytes in aqueous solution. It is based on the reaction between cationic and anionic polyelectrolytes where neutralisation of electrical charge proceeds stoichiometrically (Figure 3.6). Thus if the concentration of ionic groups in one of the polyelectrolytes is known, the concentration in the other may be calculated from the relative volumes at the end point.

If a cationic polyelectrolyte is titrated with a solution of an anionic polyelectrolyte, of known composition, the end point may be detected by the addition of a suitable dye such as Methylene Blue (3.1) (Colour Index Basic Blue 9; C.I. No. 52015) or Toluidine Blue (3.2) (C.I. Basic Blue 17; C.I. No. 52040) – see Figure 3.7 – which undergo a metameric colour change in the presence of free anionic polyelectrolyte. Although Terayama did not actually use the colloid titration technique to determine the amine group content of chitosan he indicated that this method could be used for this analysis, stating that the greater the degree of deacetylation of chitosan the greater is its capacity for combining with the anionic polyelectrolyte, and it has been used for this purpose by other workers.[10, 21, 22]

Metachromatic titration

Metachromatic titration is similar to colloid titration insofar as it involves polyelectrolyte-induced metachromasy in a suitable dye, but it differs in that there is no second, oppositely charged, polyelectrolyte present.

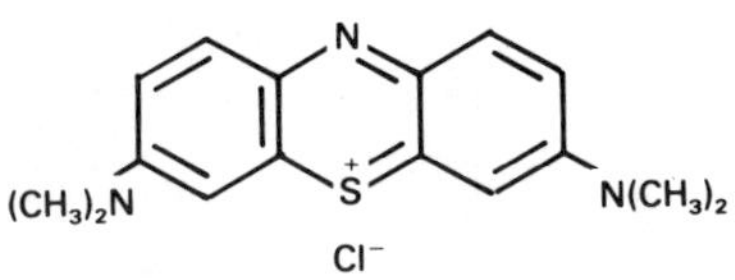

3.1

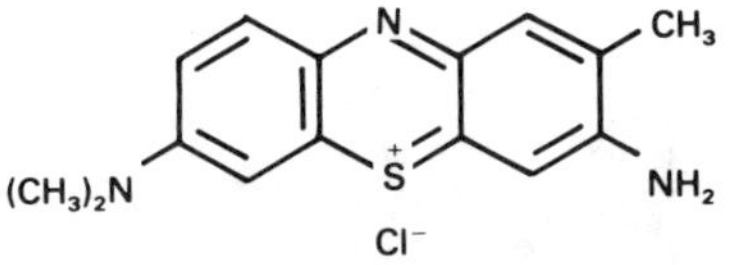

3.2

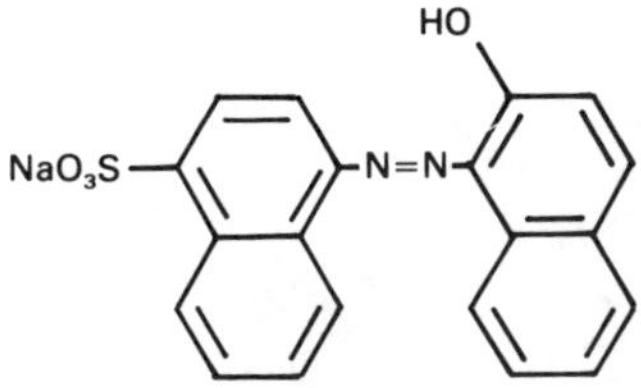

3.3

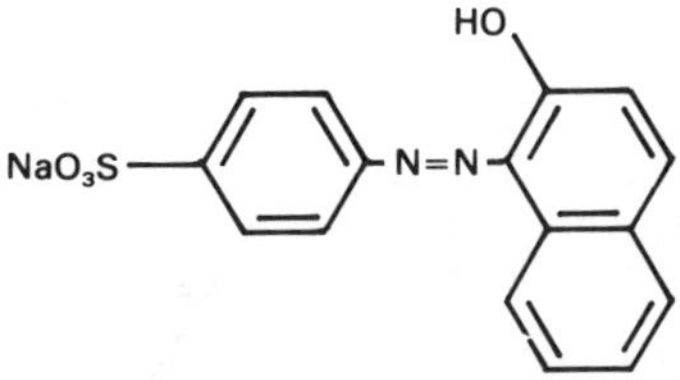

3.4

FIGURE 3.7   Structures of representative dyes used in colloid titration and metachromatic titration techniques[25, 29, 45]

The absorbance at λ_{max} of a dye showing polyelectrolyte-induced meta-chromasy will decrease with increase in the concentration of added polyelectrolyte until a minimum value is obtained, after which further additions of polyelectrolyte will have no further effect on the absorbance. The intersection in a plot of Absorbance *versus* Volume of polyelectrolyte solution added represents the point at which the mixture contains an equivalent number of dye ions and charged groups on the polyelectrolyte (Figure 3.8).

Gummow and Roberts[25] observed chitosan-induced metachromasy in a solution of 4-(2-hydroxy-1-naphthylazo)-1-naphthalene sulphonic acid, sodium salt (**3.3**) (C.I. Acid Red 88; C.I. No. 15620) and used this behaviour to analyse a series of chitosans by metachromatic titration. The results obtained were in good agreement with those obtained by IR spectroscopy. In subsequent work it was found that Orange II (**3.4**) (C.I. Acid Orange 7; C.I. No. 15510) was a more suitable dye for use in metachromatic titrations[45] despite not giving such a visually apparent metameric change (section 6.2.5).

Periodate oxidation

Jeanloz and Forchielli showed that $NaIO_4$ oxidises the α-aminoalcohol group of the deacetylated residues in chitosan while the *N*-acetylated

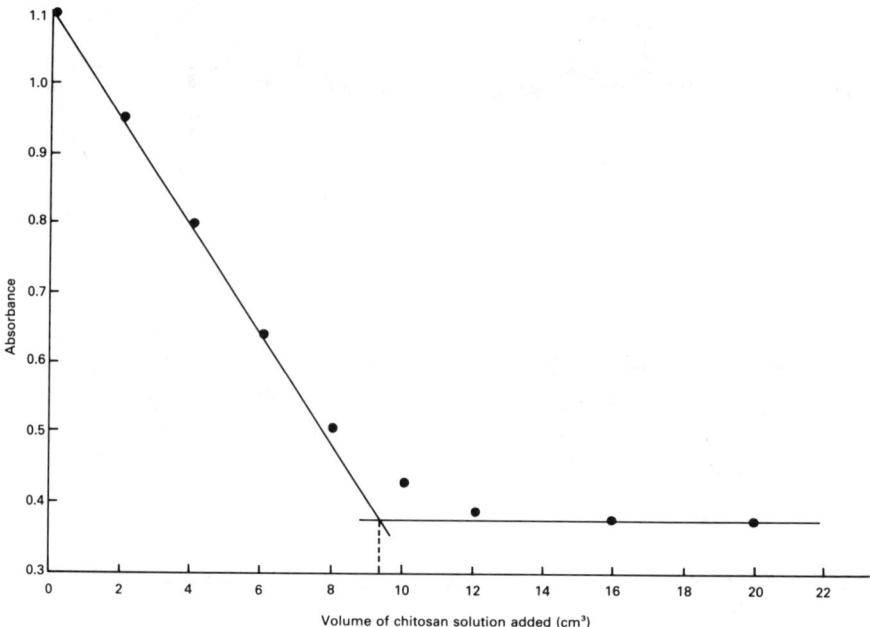

FIGURE 3.8   Typical metachromatic titration plot[25]

residues are resistant, and used the reaction to determine the position of
the β-glycosidic link between the residues in a chitosan sample having a
known degree of N-acetylation.[46] Subsequently Moore and Roberts used
the reaction to calculate the extent of deacetylation of chitin and chitosan
samples from measurements of the amount of periodate consumed.[19] The
values obtained for the amine group content were, in general, greater than
those calculated on the basis of the N-acetyl content determined by IR
spectroscopy. This was attributed to alkaline degradation of dialdehyde
units by the NH_3 liberated during oxidation of the α-aminoalcohol groups,
leading to water-soluble fragments that are themselves subject to periodate
oxidation so that the periodate consumption is in excess of that required
solely for oxidation of the α-aminoalcohol groups.

Residual salicylaldehyde analysis

The free amine groups of chitosan react readily with salicylaldehyde to give
the yellow Schiff's base N-salicylidenechitosan, the reaction being complete after 16 hours at room temperature using methanol as the reaction
medium and a 3:1 mole ratio of salicylaldehyde: amine group.[47] This
reaction was used by Domszy and Roberts[20] to determine the amine group
content of chitosan, the basis of the technique being the reaction of
chitosan with excess salicylaldehyde followed by spectroscopic determi-

nation of the salicylaldehyde remaining at the end of the reaction period. Suppression of hemiacetal and acetal formation by reaction between methanol and salicylaldehyde was achieved by using an 80 vol.-% solution of methanol in 1 vol.-% aqueous acetic acid as the reaction medium. The results obtained agreed with those from IR spectroscopy and from titration of the chitosan hydrobromide salt.[20] Subsequently the technique has been found to give comparable results to those obtained from picric acid adsorption measurements[48] (see later in this section). Another related method involves determining the amount of salicylaldehyde released from a known weight of N-salicylidenechitosan on hydrolysis in HBr.[49]

Reaction with 2,4-dinitrofluorobenzene

The use of 2,4-dinitrofluorobenzene (DNFB), a logical extension of its use in protein chemistry, was first proposed by Araki and Ito[50, 51] in their studies on enzymatic deacetylation of chitin. However the results obtained differed significantly from those obtained by determination of the radioactive acetic acid produced during deacetylation of chitins prepared by N-acetylation of chitosan with labelled acetic anhydride. In some cases the value obtained by the DNFB technique was less than 50% of the value obtained by the labelled acetic acid method.

The DNFB method was subsequently studied in some detail[52] and it was found that the conditions used in the hydrolysis step must be carefully controlled to avoid degrading the chromophore. However even when appropriate precautions were taken it was found that complete reaction of the available amine groups was not achieved under either heterogeneous or homogeneous conditions, the fraction of the available amine groups reacting decreasing as their concentration increased. It was concluded that the method is not suitable as an analytical procedure except perhaps for very highly N-acetylated samples.[52] This failure of the amine groups to react completely with DNFB was subsequently confirmed by Aiba.[22]

Dye adsorption

Maghami and Roberts[26] have shown that at equilibrium there is a 1:1 stoichiometry for the interaction of sulphonic acid groups on anionic dyes with protonated amine groups on chitosan. The results obtained using equilibrium dye adsorption measurements were compared with those obtained by IR spectroscopy[19] and found to correlate reasonably well, although there was some scatter on both sides of the line representing exact correspondence between the two techniques.

Picric acid adsorption

This method[48] is related to the dye adsorption method above (indeed at one time picric acid was an important dye for silk) but there are two

TABLE 3.3   *Comparison of the % N-acetyl values determined for a range of chitin and chitosan samples by the picric acid adsorption and residual salicylaldehyde methods[48]*

Sample	% N-acetyl	
	Picric acid adsorption[48]	*Residual salicylaldehyde determination[20]*
1	98.8	No free amine groups detected
2	98.8	98.8
3	96.5	96.1
4	94.5	92.7
5	91.4	92.3
6	91.2	91.4
7	83.6	85.4
8	62.6	70.3

major differences. Firstly the amount of picric acid adsorbed was not determined by measuring the residual concentration of picric acid in solution but by subsequent desorption of the adsorbed picric acid by treating the 'dyed' chitosan with a solution of DIPEA in methanol and measuring the concentration of the DIPEA–picric acid salt in the eluent. Secondly, it is claimed to be applicable to very small quantities of chitosan, 5×10^{-3} g, being suitable. The results obtained for samples in the range chitin[0.98]–chitosan[0.28] agreed well with those obtained by the residual salicylaldehyde method[20] and, for soluble samples, by the titration method of Broussignac[28] (Table 3.3).

3.2.4   Determination based on overall composition

Elemental analysis

Pure chitin[1.0] (**1.1**) has a nitrogen content of 6.89% while pure chitosan[0.0] (**1.4**) has 8.69% N. Thus in theory it is possible to determine the degree of *N*-acetylation from a knowledge of the nitrogen content of the sample, the relative change between the two extremes being greater for nitrogen than for carbon (Table 3.4). However because of the problems caused by the presence of moisture, which is difficult to eliminate, and the possible presence of some inorganic materials, the use of the N/C ratio is to be preferred. It is also crucial that no residual protein remains on the sample prior to analysis since the average %N and %C in proteins is considerably different from those of either chitin or chitosan.

Gas chromatography

Muzzarelli *et al.*[27] examined the use of chitin and chitosan as the column packing for the GC separation of amines and alcohols. They

TABLE 3.4 *Relationship between the % N-acetyl value of a sample of chitin or chitosan and the % N or % C content*

	N-acetyl (%)					
	0	*20*	*40*	*60*	*80*	*100*
N content (%)	8.69	8.26	7.87	7.52	7.19	6.89
C content (%)	44.71	45.33	45.89	46.39	46.89	47.29
N/C ratio	0.194	0.182	0.171	0.162	0.153	0.146

reported that the retention time for methanol was proportional to the degree of *N*-acetylation of the packing, decreasing with decrease in the degree of *N*-acetylation in a rectilinear manner. This was attributed to formation of a chitin/methanol complex, as previously proposed by Austin.[53] The feasibility of using methanol retention times as a measure of the degree of *N*-acetylation is doubtful since differences in particle size between samples could give rise to differences in retention times even if all the samples had the same % *N*-acetyl content.

Pyrolysis techniques

A number of papers have been published on attempts to use pyrolysis–MS, pyrolysis–GC or pyrolysis–GC/MS to characterise chitin and chitosan. Hayes and Davies[44] were the first to study chitin and chitosan by pyrolysis–MS. They demonstrated the greater thermal stability of chitin and suggested that since the fragmentation patterns for the two polymers were different, the technique could provide information on the degree of *N*-acetylation. Subsequently Mattai and Hayes[54] found that the peak ratios 80/60, 67/60 and 80/42 increased with increase in the amine group content. The peaks at *m*/*z* 80 and 67 were assigned to fragments originating from D-glucosamine, and those at *m*/*z* 60 and 42 to fragments originating from *N*-acetyl-D-glucosamine. Of the three ratios, the 80/60 ratio showed the best correlation with amine group content and this was attributed to the fact that only single fragments contribute to these two peaks whereas the peaks at *m*/*z* 67 and 42 are both made up of contributions from two ions. The authors concluded that an advantage of the technique is that the origin of the chitin is unimportant for mass spectral correlations of peak ratios, but that a disadvantage is that it is not a suitable technique for distinguishing between chitosan samples having similar amine group contents.

Lal and Hayes[55] and Davies *et al.*[56] investigated the use of pyrolysis–GC and obtained a correlation between the amine group content and the ratio of the areas of pyrolysis peaks derived from the D-glucosamine and *N*-acetyl-D-glucosamine residues. Davies and Hayes[57] state that the best correlation is obtained for the ratio of the peaks having retention times of 13.69 min and 9.75 min. The former is claimed to be a major peak in the pyrogram of chitosan, but absent from that of chitin, while the latter is stated to be a major peak in the pyrogram of chitin and to be absent from that of chitosan. Davies and Hayes,[57] using different operating conditions, reported a good correlation between the amine group content and the ratio of the peaks having retention times of 2.2 min (1,2-diaminopropane peak) and 9.5 min (acetamide peak).

Finally, Alonso et al.[58] have reported that in the differential thermal gravimetric analysis of chitin and chitosan there is a correlation (correlation coefficient = 0.98) between the percentage weight loss at 320°C and the degree of N-acetylation of the sample. This correlation is valid for samples down to chitin[0.80]. For more highly deacetylated samples there is a correlation (correlation coefficient = 0.998) between the percentage weight loss at 280°C and the degree of N-acetylation. The different temperatures required for correlation at different levels of N-acetylation reflect the greater thermal stability of chitin.[44]

3.3 MOLECULAR WEIGHT

3.3.1 Introduction

The determination of a molecular weight average for either chitin or chitosan poses a number of problems. Chief among these in the case of chitosan are the possible presence of microgels due to uneven treatment during the deacetylation step, and the tendency of the polymer chains to aggregate on standing in solution,[24, 59] while the limited choice of solvents and the nature of the solutions formed are problems associated with working with chitin.[60] Some of the studies that have been carried out on chitin have made use of water-soluble derivatives but, while such an approach has the benefit of enabling aqueous solutions to be used, it must be remembered that the preparation process itself will almost certainly cause a reduction in the molecular weight. It must also be remembered that the origin and preparation history of a sample will affect its molecular weight so that attempts to compare molecular weight values obtained by different workers using different samples, as has on occasion been done, are of little or no worth.

It has been suggested by Rinaudo and Domard[61] that the characteristic DP be used, rather than the molecular weight average, because of uncertainties introduced by variation in the degree of N-acetylation or by the nature of the counter ion, particularly if the chitosan is in the form of its acid salt.

Although a number of techniques have been applied to the problem only light scattering, GPC, and dilute solution viscosity have been used to any extent. Osmometry[24] and end group analysis[62-64] have each had only limited use as has ultracentrifugation.[65-67] The application of the hypoiodate method[68] for determining reducing end groups of chitin sulphate samples[63, 64] yielded peculiar results, in that the $\bar{M}_n$ values obtained by this method were larger than the $\bar{M}_w$ values obtained from light scattering measurements, and it was suggested[63] that this discrepancy between $\bar{M}_n$ and $\bar{M}_w$ values can be attributed to shielding of the reducing end groups, thereby preventing reaction of all the reducing end groups present in the sample.

3.3.2 Light scattering spectrophotometry

Light scattering spectrophotometry has been used with chitosan by a number of workers[24, 69-73] but the measurements are prone to give an overestimate of the molecular weight because of aggregation of the chains on standing. Association of chains was first proposed by Van Duin and Hermans[69] to explain the apparent increase in $\bar{M}_w$ with decrease in total ionic strength. Since the effect of changing ionic strengths should operate in the opposite direction if association is due to hydrogen bonding, they suggested that it was due to the presence of small amounts of divalent ions. In support of this they observed that use of deionised water gave a reduction in the measured $\bar{M}_w$ value compared with that obtained using non-deionised water.

Direct evidence of aggregation with solutions of chitosan[0.0] has been obtained from the variation in the scattered intensity of light as a function of the time elapsed since preparation of the solution.[24] It was also shown that preheating the solution for 2 h at 50°C reduced the rate at which this increase in scattering intensity occurs. In this case the mechanism of aggregation was considered to be that typical of crystallisation processes. More recently the use of laser light scattering spectrophotometry has been reported by Domard and Rinaudo[72] and Muzzarelli et al.[73] No evidence of aggregation with variation in electrolyte concentration was observed[73] over the range 0.1–0.5M NaOAc for a sample of chitosan[0.20].

Hackman and Goldberg[74] examined solutions of chitin, carboxymethyl chitin and hydroxyethyl chitin, by light scattering spectrophotometry. The solvent used for chitin was aqueous LiSCN at 95°C – the LiSCN was a saturated solution at room temperature – while aqueous NaCl solutions (0.01–2.5M) were the solvents used for the chitin derivatives. Aggregation was observed in the case of the carboxymethyl chitin, the extent of association being reduced on increase in the NaCl concentration from 0.5M to 2.5M. In the light of the observations of Van Duin and Hermans,[69] this effect of electrolyte concentration on the extent of aggregation would suggest that the association is due to the presence of a small concentration of some multivalent ion.

The molecular weight of chitin has also been determined in the DMAc–LiCl system.[60] A number of commercial samples were examined but solutions containing more than 5.0 g dm^{-3} could not be examined because of their very high viscosity. Another problem is that this system has a low dn/dc value, causing the possible error to be quite high.

In carrying out light scattering studies, a very accurate determination of the concentration dependence of the refractive index is most important. Various values have been obtained for chitosan and these are given in Table 3.5. The values vary from 0.166 to 0.180 cm^3 g^{-1} when measured at 546 nm, and whether this variation is due to differences in solvent or in the structure of the chitosan samples used is not known; no study of the effect of extent of N-acetylation on dn/dc has been carried out. However there is a much greater difference between the two values obtained at approximately 633 nm, one being 25% less than the other. It seems unlikely that so large a difference can be due to the difference in solvent systems in the two cases.

3.3.3   Gel permeation chromatography

Gel permeation chromatography (GPC) is a frequently used technique for determining the molecular weights and molecular weight distributions (MWD) of polymers, and it is not surprising that it has been employed in studies on chitosan[75–77] and that a procedure claimed to give optimum conditions for GPC of chitosan has been published.[78]

The cationic nature of chitosan in acid solution causes difficulties in using GPC for molecular weight determinations of chitosan. For the separation process to operate correctly it is necessary that there is no interaction between the gel and the polymeric solute, so in the case of polyelectrolytes any electrostatic effects, whether of repulsion or attraction, must be eliminated. Conventional GPC gels carry a negative surface charge so that the protonated chitosan chain would be expected to be strongly adsorbed onto the gel through electrostatic attraction. This problem was examined in some detail by Domard and Rinaudo[72] who used a cationic gel prepared from porous silica beads grafted with quaternary ammonium groups, together with a low-molecular-weight electrolyte (NH$_4$OAc) in the solvent to suppress any electrostatic repulsion between the cationic gel and protonated chitosan. The optimum concentration of NH$_4$OAc was found to be 0.05M and with this concentration the universal calibration curve of Log([η].MW) *versus* V_e (the elution volume) is valid for a variety of neutral and cationic polymers. Any decrease or increase in the NH$_4$OAc concentration was found to lead to deviation from this curve owing to incomplete suppression of electrostatic repulsion in the former case and to charge reversal on the chitosan, leading to its adsorption on the gel, in the latter case.

TABLE 3.5 Values of dn/dc for chitin and chitosan in different solvent systems

Polymer	Solvent	λ (nm)	dn/dc (cm³ g⁻¹)	Reference
Chitosan[0.12].HCl	H_2O	[a]	0.166	69
Chitosan[0.13][b]	1.85M HCOOH–0.5M NaOOCH	546	0.174	70
Chitosan[0.0].HCl	0.2M NaCl	546	0.170	24
Chitosan[0.4]	0.1M HOAc–0.2M NaCl	[a]	0.180	71
Chitosan[0.0].HOAc	0.05M NH_4OAc	633	0.119	72
Chitosan[0.13]–[0.21]	0.08M HOAc–0.5M NaOAc	632.8	0.160	73
Chitin[0.90]	5.55M LiSCN	546	0.164[c]	74
Chitin	DMAc–LiCl (50 g dm⁻³)	546	0.10	60

[a] Wavelength used not specified but most probably 546 nm.
[b] Degree of N-acetylation calculated from the elemental analysis figures given in the text.
[c] Determined using hepta-N-acetylchitoheptose as a model compound.

This study[72] demonstrates that the results obtained previously for chitosan, using conventional gels,[75-77] are incorrect, as must be the proposed procedure for obtaining optimum conditions.[78] The use of a conventional gel may explain the peculiar molecular weight distribution results obtained by Wu and Bough in their study of the alkaline deacetylation of chitin.[76] They reported that the ratio $\bar{M}_w/\bar{M}_n$ remained constant, at approximately 4.8, for deacetylation times of 0.5 to 5 h, despite a reduction in molecular weight of about 50% over the same period of time. Such constancy is surprising since random scission of the polymer chain would be expected[79] to move the MWD towards the 'most probable distribution', thereby shifting the $\bar{M}_w/\bar{M}_n$ ratio towards 2.0.

3.3.4 Viscometry

Viscometry is one of the simplest and most rapid methods for determining the molecular weights of polymers, despite having the disadvantage of not being an absolute method since it requires the determination of constants through correlation of LVN values with molecular weight values measured by an absolute method. The most commonly used equation relating LVN values to molecular weights is the Mark–Houwink equation

$$[\eta] = K \bar{M}_v^a \qquad (3.1)$$

where $[\eta]$ is the LVN and K and a are constants that are independent of molecular weight over a considerable range of molecular weights. They are dependent on polymer, solvent, temperature and, in the case of polyelectrolytes, the nature and concentration of the added low-molecular-weight electrolyte. The constants are normally evaluated from a plot of $\text{Log}[\eta]$ versus $\text{Log } \bar{M}_w$ for a series of carefully prepared fractions having very narrow MWD values.

Several sets of values for K and a have been proposed for chitosan in the literature and these are given in Table 3.6. The units used for K in Table 3.6 are $\text{cm}^3 \text{ g}^{-1}$ and the literature values have been converted to these units from dl g^{-1} where necessary. The solvent systems used and the molecular weight ranges over which the constants were determined in each case are also given.

Lee[65] used ultracentrifugation to determine the molecular weights ($\bar{M}_n$ and $\bar{M}_w$) of three samples of chitosan[0.21], the two lower-molecular-weight samples being prepared from the third sample by shear degradation. This was done to ensure that the three samples all had the same degree of N-acetylation, thereby eliminating any effects that might otherwise arise from variations in charge density along the chains. It has subsequently been shown[45] however that the amine group content, and hence the linear charge density in acid solution, has no effect on the LVN over the approximate range of chitosan[0.0]–chitosan[0.4]. The three

TABLE 3.6 Values for the Mark–Houwink equation constants K and a for chitosan

Solvent system	K (cm³ g⁻¹)	a	Molecular weight range (10⁵)	Reference
1. 0.2M HOAc–0.1M NaCl–4M urea	8.93×10^{-2}	0.71	1.13–4.92	65
2. 0.167M HOAc–0.47M NaCl	111.5	0.147	0.13–1.7	66
3. 0.1M HOAc–0.2M NaCl	1.81×10^{-3}	0.93	0.9–11.4	62
4. 0.33M HOAc–0.3M NaCl	3.41×10^{-3}	1.02	0.13–1.35	67
5. 0.33M HOAc–0.2M NaOAc– 0.67M Cl₂CH₂COOH	1.28×10^{-2}	0.85	0.61–1.60	80

samples, although having the same linear charge density, had appreciably different MWD values ($\bar{M}_w/\bar{M}_n = 3.51$, 1.76 and 1.07) whereas samples having similar, narrow MWD values are required for determining K and a. This, together with the fact that only three samples were used in the determination, reduces the accuracy of the values obtained – $K = 8.93 \times 10^{-2}$ cm^3 g^{-1}, $a = 0.71$.

The value for a reported by Berkovich et al.[66] is very unusual since for a linear polymer having a random coil conformation a should lie between 0.5 in a θ solvent and about 1.2 for a highly expanded coil conformation in a good solvent. A value of less than 0.5 is typical of globular proteins where the polymer chain adopts a very tightly packed conformation, but in such cases the LVN is low, usually less than 5 cm^3 g^{-1}, and is more or less independent of the molecular weight. Neither of these two characteristics is observed with solutions of chitosan.

Although K and a are normally determined using samples calibrated in terms of $\bar{M}_w$, other approaches are possible. Roberts and Domszy[62] made use of a technique first proposed by Charlesby[81] for determining a and subsequently extended to cover determination of K as well by Sharples and Major who applied the method to cellulose acetate.[82] The method considers the properties of the molecular weight distribution resulting from random degradation of polymer chains and for this the Mark–Houwink equation is expressed as

$$\bar{M}_v = K' \, [\eta]^{1/a} \tag{3.2}$$

where $K' = 1/(K)^{1/a}$.

Consideration of the MWD resulting from random degradation leads to

$$\bar{M}_v = \bar{M}_n [\Gamma(2 + a)]^{1/a} \tag{3.3}$$

where Γ = gamma function. Combining equations (3.2) and (3.3) gives

$$\bar{M}_n = K'[\eta]^{1/a} \, / \, [\Gamma(2 + a)]^{1/a} \tag{3.4}$$

or expressing it in logarithmic form

$$\log \bar{M}_n = (1/a) \log[\eta] + \log K' - (1/a) \log[\Gamma(2 + a)]$$

Hence a plot of Log $\bar{M}_n$ versus $\log[\eta]$ should give a straight line of slope $(1/a)$ and intercept equal to $\log K' - (1/a) \log[\Gamma(2 + a)]$.

Random degradation of a sample of chitosan[0.24] was achieved by homogeneous hydrolysis in 1.67M acetic acid and the $\bar{M}_n$ values were determined by UV/visible spectroscopic analysis of the phenylosazone derivatives prepared by reaction of the reducing terminal unit in each chain with phenylhydrazine. The values of K and a obtained by this method are similar to those reported[83, 84] for related β-(1→4)-linked ionic polysaccharides (Table 3.7).

The most recent attempt to obtain values for K and a is that of Pogodina

TABLE 3.7   *Values for the Mark–Houwink equation constants for ionic β-(1→4)-linked polysaccharides*

Polysaccharide	[NaCl] (M)	K ($cm^3 g^{-1}$)	a	Reference
Chitosan acetate	0.2	1.81×10^{-3}	0.93	62
Sodium carboxymethylcellulose	0.1	1.23×10^{-4}	0.91	83
Sodium cellulose sulphate	0.5	7.91×10^{-2}	0.93	84
Chitosan acetate	0.02	3.04×10^{-5}	1.26	62
Sodium carboxymethylcellulose	0.01	6.46×10^{-6}	1.2	83

et al.[67] Four samples of chitosan[0.2] having different LVN values were fractionated by precipitation from solution in aqueous acetic acid, using acetone as the precipitant, to give a total of 32 fractions. Each fraction was subjected to ultracentrifugation and their molecular weights (M_{SD}) calculated from the measured diffusion and sedimentation coefficients using the Svedberg equation. The values of K and a were then determined in the usual way from the double logarithmic plot. One interesting aspect of this work is that the four original samples all had quite narrow MWD values, the $\bar{M}_w/\bar{M}_n$ ratios being between 1.33 and 1.08. Despite the very narrow distribution of chain lengths implied by this latter figure, that particular sample was separated into 5 fractions having molecular weights ranging from 6.8×10^4 to 1.7×10^5.

Gamzazade *et al.*,[80] like Berkovich *et al.*[66] and Pogodina *et al.*,[67] used ultracentrifugation to obtain M_{SD} values for chitosan samples ranging from chitosan[0.08] to chitosan[0.47]. Two solvent systems were used, 0.33M HOAc–0.2M NaOAc and 0.67M $Cl_2CHCOOH$. The results for both solvent systems fell on the same line when plotted as log[η] *versus* log M_{SD}, the graph giving a = 0.85 and K = 1.38×10^{-2} $cm^3 g^{-1}$.

One test to evaluate the different pairs of constants is to apply them to LVN values and compare the $\bar{M}_v$ values obtained with $\bar{M}_n$ and $\bar{M}_w$ values where these are known. Such an evaluation is complicated by the fact that each pair of constants has been determined using a different solvent system, but is simplified by the apparent insensitivity of the LVN to the degree of *N*-acetylation over the approximate range of chitosan[0.0]–chitosan[0.4].[45] The most fully characterised chitosan in terms of molecular weight that has been reported in the literature is a chitosan[0.0] sample prepared and studied by Domard and Rinaudo.[24, 59, 72] The $\overline{DP}_n$ value has been reported[59] to be ≃ 960 (GPC) or 1100 (osmometry) and the $\overline{DP}_w$ to be ≃ 3360 (light scattering) or 3370 (GPC). In addition, the LVN has been determined in a number of different solvent systems.[24] Applying the different pairs of constants to the LVN measured in the solvent system closest to that used in determining those particular constants gives the $\overline{DP}_v$ values listed in Table 3.8. It can be seen that only pair 3 gives a $\overline{DP}_v$ value

TABLE 3.8 $\overline{DP}_v$ *values calculated from the literature LVN values.*[24] *The most appropriate LVN has been used with each pair of constants*

Mark–Houwink equation constants[a]	1	2	3	4	5
$\overline{DP}_v$ calculated from LVN values	654	18	2600	420	840

[a] Numbers for the pairs of constants are taken from Table 3.6.

between the $\overline{DP}_n$ and $\overline{DP}_w$ values as is required by theory ($\overline{DP}_n < \overline{DP}_v \leqslant \overline{DP}_w$), and it must be concluded that the most accurate constants currently available for the Mark–Houwink equation for chitosan are $K = 1.81 \times 10^{-3}$ $cm^3 \, g^{-1}$ and $a = 0.93$. This conclusion is supported by the experience of Rinaudo and Domard who have reported[61] that the $\overline{M}_v$ values obtained using these constants are the only ones to be always located between $\overline{M}_n$ and $\overline{M}_w$.

The LVN values reported for the same chitosan[0.0] sample in different solvent systems[24] illustrates the correctness of the suggestion[61] that the characteristic DP should be used rather than the characteristic molecular weight average. The LVN values obtained with chloride, bromide and acetate counterions give $\overline{M}_v$ values of 5.2×10^5, 6.4×10^5 and 5.8×10^5 respectively, but $\overline{DP}_v$ values of 2640, 2645 and 2625.

Very little work has been reported on viscosity measurements on chitin. Lee[65] applied the constants obtained for chitosan, using 0.2M HOAc–0.1M NaCl–4M urea as solvent, to the LVN for chitin measured in anhydrous formic acid. Apart from the rather drastic change in solvent used, the application of constants determined for a polyelectrolyte to a non-ionic polymer is questionable.

Recently K and a values have been determined for chitin in the DMAc–LiCl solvent system.[60] The values obtained are $K = 2.4 \times 10^{-1} \, cm^3 \, g^{-1}$, $a = 0.69$. Two different concentrations of LiCl were used, 50 g dm^{-3} and 80 g dm^{-3}, the constants being obtained using the former concentration of LiCl. Although actual values were not given, it was stated that the LVN values were much higher in the 80 g dm^{-3} LiCl solution.

3.4   OTHER ANALYSES

3.4.1   Residual protein

Procedures that have been used involve hydrolysis of the residual protein followed by determination of either the amino acids produced or the loss in weight of the sample. The protein hydrolysis step may be carried out using acid or alkali but if the former is used, weight-loss measurements cannot be made since the sample will be completely hydrolysed. Furthermore the amino acid concentration cannot be determined using ninhydrin since

D-glucosamine, which is produced on acid hydrolysis of chitin, also gives a colour reaction with this reagent.

In a weight-loss method developed by Takiguchi et al.[85] (the author is indebted to Professor K. Kurita for providing a translation of this paper) the chitin sample of known degree of N-acetylation is treated with 10M NaOH under reflux to hydrolyse the protein. After neutralising and filtering the reaction mixture, the filtrate is analysed for its amino acid content using a ninhydrin–hydrindantin solution, and the filtered solid washed, dried, weighed and the degree of N-acetylation measured. The weight of protein in the original sample is given by

$$P = W - \{W' + W'[(16116 + 42.04x)/(16116 + 42.04y) - 1]\}$$

where W = initial weight of sample/g
$\quad W'$ = final weight of sample after deproteinisation/g
$\quad x$ = initial N-acetyl content/%
$\quad y$ = final N-acetyl content/%.

The separate determination of amino acid concentration acts as a valuable check on the protein content value, as measured by the loss in weight, and Shimahara and Takiguchi subsequently[86] published a method based solely on the reaction of the amino acids in the neutralised hydrolysate. Again 10M NaOH is used in the hydrolysis step which is carried out at 121°C for 60 minutes. After neutralising, and filtering off the deproteinised chitin particles, the filtrate is treated under standardised conditions with the ninhydrin–hydrindantin solution and the absorbance at 564 nm measured. The protein content is given by

$$\text{Protein}/\% = 2.37(A_{564}/W)$$

where W = the weight of chitin/g. Although not stated in this otherwise detailed description of the method,[86] it is probable that the protein used as a standard for calculating the factor in the equation was bovine Achilles tendon collagen, as described in an earlier outline report of the method.[87]

Acid hydrolysis has been used by several workers. The procedure involves hydrolysis at 100–110°C in sealed tubes using either 6M HCl[88] or 12M HCl[89] for 16–24 h, followed by amino acid analysis on an automatic analyser. Although requiring considerably more in terms of laboratory resources, this approach enables the composition of the residual protein to be determined in addition to the total amount.

3.4.2 Colour

There is no standard test for the evaluation of colour for chitin and chitosan, but many solutions of chitosan exhibit a very definite yellow colour while chitin frequently appears to have a pronounced yellow-to-brown appearance. One assessment of chitosan[90] used an empirical relationship based on the fact that, being yellowish in colour, a chitosan solution will have its lowest transmission at the short wavelength end of the visible spectrum. The relationship used was

$$\text{Colour (arbitrary units)} = (\%T_{560,C} - \%T_{400,C})/(\%T_{560,S} - \%T_{400,S})$$

where $\%T_{560}$ and $\%T_{400}$ are the percentage transmissions at 560 and 400 nm and the subscripts C and S refer to the chitosan solution (10 g dm^{-3} in 0.5 vol.-% acetic acid) and the solvent respectively. Solid samples could be assessed from reflectance measurements on the powered material, one suitable instrument being the ICS Micro Match which is an abridged spectrophotometer that measures the reflectance characteristics at 20 nm intervals over the range 400–700 nm. The results may be obtained in the form of a Yellowness Index defined[91] as

$$\text{Yellowness Index} = (131.6X - 116.4Z)/Y$$

where X, Y and Z are the tristimulus values of the sample.

3.4.3   Percentage insoluble material

Again there is no standard procedure, but determination is most simply carried out by dissolving in a suitable solvent, 0.1–0.2M acetic acid for chitosan and DMAc–LiCl or related complex organic solvent for chitin, followed by filtering off, rinsing, and drying any undissolved material which may then be weighed directly.

3.4.4   Heavy metal content

Chitosan, and to a much lesser extent chitin, readily forms complexes with transition metal ions and other heavy metal ions (section 5.2). Hence care must be taken in the various preparation steps to avoid contact with such ions and the use of demineralised water is advisable. It has been claimed[92] that if this is done the common metal content can be kept below 5 ppm, with the highest levels being those for nickel and copper, indicating that most of the trace metals present in commercial chitosans come from contaminants, equipment, and water, and are not present in the original *in vivo* chitin.

At the levels likely to be encountered, the most appropriate method of analysis is atomic absorption spectroscopy following solution of the material in concentrated mineral acid.

3.4.5   Ash, lipids, moisture and nitrogen content

The ash, lipids, and nitrogen contents are best determined using the standard procedures (18.025, 18.046 and 18.026 respectively) recommended for the analysis of fish and other marine products by the AOAC.[93] For determination of the moisture content the British Standard for textiles,[94] which specifies heating at $105\pm3°C$ for 4–16 h, is appropriate.

REFERENCES

1. J. Knapczyk, L. Krówczyński, J. Krzek, M. Brzeski, E. Nürnberg, D. Schenk and H. Struszczyk, in *Chitin and Chitosan*, G. Skjåk-Braek, T. Anthonsen and P. Sandford (eds), Elsevier, London, 1989, p. 657.
2. K. Kurita, T. Sannan and Y. Iwakura, *Makromol. Chem.*, **178** (1977) 3197.
3. S. Hirano, S. Tsuneyasu and Y. Kondo, *Agric. Biol. Chem.*, **45** (1981) 1335.
4. H. Braconnot, *Ann. Chim. (Paris)*, **79** (1811) 265.
5. G. Ledderhose, *Ber.*, **9** (1876) 1200.
6. H. Brach and O. von Fürth, *Biochem. Zeit.*, **38** (1912) 468.
7. B. Radhakrishnamurthy, E.R. Dalferes and G.S. Berenson, *Anal. Biochem.*, **26** (1978) 61.
8. Z. Holan, J. Votruba and V. Vlasáková, *J. Chromatog.*, **190** (1980) 67.
9. A. Elek and R.A. Harte, *Ind. Eng. Chem., Anal. Ed.*, **8** (1936) 267.
10. T. Sannan, K. Kurita and Y. Iwakura, *Makromol. Chem.*, **177** (1976) 3589.
11. T. Sannan, K. Kurita, K. Ogura and Y. Iwakura, *Polymer*, **19** (1978) 458.
12. C.K. Kandaswamy, in *Proceedings 1st International Conference on Chitin/Chitosan (1977)*, R.A.A. Muzzarelli and E.R. Pariser (eds), MIT Sea Grant Report MITSG 78–7, 1978, p. 517.
13. N. Gowri, G. Sundara Rajulu and M. Aruchami, in *Chitin and Chitosan*, S. Hirano and S. Tokura (eds), The Japanese Society of Chitin and Chitosan, Tottori, 1982, p. 77.
14. F.A. Rutherford and P.R. Austin, in ref. 12, p. 182.
15. L.J. Filar and M.G. Wirick, in ref. 12, p. 169.
16. R.U. Lemieux and C.B. Purves, *Can. J. Res.*, **25B** (1947) 485.
17. S.E. Darmon and K.M. Rudall, *Disc. Faraday Soc.*, **9** (1950) 251.
18. G.K. Moore and G.A.F. Roberts, in ref. 12, p. 421.
19. G.K. Moore and G.A.F. Roberts, *Int. J. Biol. Macromol.*, **2** (1980) 115.
20. J. G. Domszy and G.A.F. Roberts, *Makromol. Chem.*, **186** (1985) 1671.
21. M. Miya, R. Iwamoto, K. Ogura and Y. Iwakura, *Int. J. Biol. Macromol.*, **2** (1980) 323.
22. S. Aiba, *Int. J. Biol. Macromol.*, **8** (1986) 173.
23. M. Miya, R. Iwamoto, K. Ohta and S. Mima, *Kobunshi Ronbunshu*, **42** (1985) 181.
24. A. Domard and M. Rinaudo, *Int. J. Biol. Macromol.*, **5** (1983) 49.
25. B.D. Gummow and G.A.F. Roberts, *Makromol. Chem.*, **186** (1985) 1239.
26. G.G. Maghami and G.A.F. Roberts, *Makromol. Chem.*, **189** (1988) 2239.
27. R.A.A. Muzzarelli, F. Tanfani, G. Scarpini and G. Laterza, *J. Biochem. Biophys. Methods*, **2** (1980) 299.
28. P. Broussignac, *Chem. Ind. Genie Chim.*, **99** (1968) 1241.
29. H. Terayama, *J. Pol. Sci.*, **8** (1952) 243.
30. A.D. Domard, *Int. J. Biol. Macromol.*, **9** (1987) 333.
31. A. Baxter, M. Dillon, K.D.A. Taylor and G.A.F. Roberts, *Int. J. Biol. Chem.*, submitted for publication.
32. J.E. Castle, J.R. Deschamps and K. Tice, in *Chitin, Chitosan and Related Enzymes*, J.P. Zikakis (ed.), Academic Press, London, 1984, p. 273.

33. R.A.A. Muzzarelli and R. Rocchetti, in *Chitin in Nature and Technology*, R.A.A. Muzzarelli, C. Jeuniaux and G.W. Gooday (eds), Plenum Press, New York, 1986, p. 385.
34. A. Stone, *Biopolymers*, **7** (1969) 173.
35. E. Kabat, K. Lloyd and S. Beychock, *Biochemistry*, **8** (1969) 747.
36. A. Domard, *Int. J. Biol. Macromol.*, **8** (1986) 243.
37. L.A. Buffington and E.S. Stevens, *J. Amer. Chem. Soc.*, **101** (1979) 5159.
38. A. Domard, *Int. J. Biol. Macromol.*, **9** (1987) 98.
39. S. Hirano and R. Yamaguchi, *Biopolymers*, **15** (1976) 1685.
40. R.C. Capozza, *German Patent 2,505,305* (1975).
41. H. Saito, R. Tabeta and S. Hirano, in ref. 13, p. 71.
42. G.K. Moore, Ph.D. Thesis (CNAA), Trent Polytechnic, UK, 1978.
43. F.G. Donnan, *Chem. Rev.*, **1** (1924) 73.
44. E.R. Hayes and D.H. Davies, in ref. 12, p. 406.
45. G.G. Maghami and G.A.F. Roberts, *Makromol. Chem.*, **189** (1988) 195.
46. R. Jeanloz and E. Forchielli, *Helv. Chim. Acta*, **33** (1950) 1690.
47. G.K. Moore and G.A.F. Roberts, *Int. J. Biol. Macromol.*, **3** (1981) 337.
48. W.A. Neugebauer, E. Neugebauer and R. Brzezinski, *Carbohyd. Res.*, **189** (1989) 363.
49. J.G. Domszy, Ph.D. Thesis (CNAA), Trent Polytechnic, UK, 1983.
50. Y. Araki and E. Ito, *Biochem. Biophys. Res. Commun.*, **56** (1974) 669.
51. Y. Araki and E. Ito, *Eur. J. Biochem.*, **55** (1975) 71.
52. J.G. Domszy and G.A.F. Roberts, *Int. J. Biol. Macromol.*, **7** (1985) 45.
53. P.R. Austin, *US Patent 4,063,016* (1977).
54. J. Mattai and E.R. Hayes, *J. Anal. Appl. Pyrolysis*, **3** (1982) 327.
55. G.S. Lal and E.R. Hayes, *J. Anal. Appl. Pyrolysis*, **6** (1984) 183.
56. D.H. Davies, E.R. Hayes and G.S. Lal, in ref. 33, p. 365.
57. D.H. Davies and E.R. Hayes, in *Methods in Enzymology*, W.A. Wood and S.T. Kellogg (eds), Academic Press, London, 1988, vol. 161, p. 442.
58. G. Alonso, C. Peniche-Covas and J.M. Nieto, *J. Therm. Anal.*, **28** (1983) 189.
59. A. Domard, and M. Rinaudo, in ref. 3, p. 315.
60. M. Terbojevich, C. Carraro, A. Cosani and E. Marsano, *Carbohyd. Res.*, **180** (1988) 73.
61. M. Rinaudo and A. Domard, in ref. 1, p. 71.
62. G.A.F. Roberts and J.G. Domszy, *Int. J. Biol. Macromol.*, **4** (1982) 374.
63. K. Nagasawa, Y. Tohira, Y. Inoue and N. Tanoura, *Carbohydr. Res.*, **18** (1971) 95.
64. K. Nagasawa and N. Tanoura, *Chem. Pharm. Bull.*, **20** (1972) 157.
65. V.F. Lee, *University Microfilms (Ann Arbor) 74/29446*, 1974.
66. L.A. Berkovich, G.I. Timofeyeva, M.G. Tsyurupa and V.A. Davankov, *Vysokomol. Soedin., Ser. A*, **22** (1980) 1834.
67. N.V. Pogodina, G.M. Pavlov, S.V. Bushin, A.B. Mel'nikov, Ye. B. Lysenko, L.A. Nud'ga, V.N. Marcheva, G.N. Marchenko and V.N. Tsvetkov, *Polym. Sci. U.S.S.R.*, **28** (1986) 251.
68. K.H. Meyer, R.P. Piroué and M.E. Odier, *Helv. Chim. Acta*, **35** (1952) 574.
69. P.J. Van Duin and J.J. Hermans, *J. Pol. Sci.*, **36** (1959) 295.
70. R.A.A. Muzzarelli, A. Ferrero and M. Pizzoli, *Talanta*, **19** (1972) 1222.
71. M. Terbojevich, A. Cosani, M. Scandola and A. Fornasa, in ref. 33, p. 349.
72. A. Domard and M. Rinaudo, *Polym. Comm.*, **25** (1984) 55.
73. R.A.A. Muzzarelli, C. Lough and M. Emanuelli, *Carbohyd. Res.*, **164** (1987) 433.
74. R.H. Hackman and M. Goldberg, *Carbohyd. Res.*, **38** (1974) 35.

75. A.C.M. Wu, W.A. Bough, E.C. Conrad and K.E. Alden, *J. Chromatog.*, **128** (1976) 87.
76. A.C.M. Wu and W.A. Bough, in ref. 12, p. 88.
77. S. Mima, M. Miya, R. Iwamoto and S. Yoshikawa, in ref. 13, p. 21.
78. A.C.M. Wu, in ref. 57, p. 447.
79. P.J. Flory, *Principles of Polymer Chemistry*, Cornell University Press, Ithaca, 1953, p. 321.
80. A.I. Gamzazade, V.M. Shlimak, A.M. Sklyar, E.V. Shtykova, S.A. Pavlova and S.V. Rogozhin, *Acta Polym.*, **36** (1985) 420.
81. A. Charlesby, *J. Pol. Sci.*, **15** (1955) 263.
82. A Sharples and H.M. Major, *J. Pol. Sci.*, **27** (1958) 433.
83. G. Sitaramaih and D.A.I. Goring, *J. Pol. Sci.*, **58** (1962) 1107.
84. K. Kamida and K. Okajima, *Pol. J.*, **13** (1981) 163.
85. Y. Takiguchi, K. Ohkouchi, O. Okada and K. Shimahara, *Seikai Daigaku Kogaku Hokoku* **1** (1984) 2451.
86. K. Shimahara and Y. Takiguchi, in ref. 57, p. 417.
87. K. Shimahara, K. Ohkouchi and M. Ikeda, in ref. 13, p. 10.
88. O.L. Oke, S.O. Talabi, and I.B. Umoch, in ref. 12, p. 327.
89. C.J. Brine, in ref. 13, p. 105.
90. G.A.F. Roberts and F.A. Wood, *Report to the Highlands and Islands Development Board*, 1983.
91. American Society for Testing Materials, *ASTM D1925*.
92. R.A.A. Muzzarelli, in *The Polysaccharides*, G.O. Aspinall (ed.), Academic Press, New York, 1985, vol. 3, p. 417.
93. *Official Methods of Analysis of the Association of Official Analytical Chemists*, W. Horwitz (ed.), AOAC, Washington, 13th edn, 1980.
94. *British Standard 4407*, 1989.

4

Derivatives of Chitin and Chitosan

4.1 INORGANIC ESTERS AND RELATED DERIVATIVES

4.1.1 Nitrates

Compared with the extensive research that has been carried out on both the production and the properties of cellulose nitrate,[1] because of its commercial importance as an explosive and as a plastic, very little research has been carried out on the nitration of chitin and chitosan.

Von Fürth and Scholl[2] found that chitin can be nitrated by fuming nitric acid, the product being partially soluble in acetic acid. Their work was repeated some 30 years later by Schorigin and Hait[3] who concluded that the HNO_3–H_2SO_4 nitrating mixtures used for preparing cellulose nitrate are unsuitable for use with chitin, owing to the extensive chain degradation brought about by H_2SO_4. The use of concentrated HNO_3 (d, 1.5) on its own was recommended instead, nitration occurring in 1–2 h after which the chitin nitrate was isolated by pouring into a large volume of water. Analysis showed the maximum DS to be ~1.5 and complete denitration could be achieved by treating with NaSH at room temperature. The product was in general quite stable, although it ignited at ~163°C and burned very vigorously. A considerable portion was soluble in formic acid, the % N contents of both the soluble and the insoluble fractions being similar. This fact led Schorigin and Hait to suggest that the difference in solubility is due to the effect of different molecular weights but, since the elemental analysis figures show the material to be less than completely nitrated an alternative explanation is that the two fractions have different patterns of substitution within the same overall DS value.

116

At about the same time Clark and Smith[4] also prepared chitin nitrate by dissolving chitin in fuming nitric acid and precipitating in water after reaction times of 1–2 h. These latter workers found, contrary to their expectations, that it was not possible to estimate the DS by determining the increase in weight of the sample, the values obtained by this method being less than those obtained from determination of the % N content. Based on the observation that the solubility of the chitin nitrate in concentrated HCl was similar to that of the original chitin, it was concluded that very little deacetylation takes place during the nitration step, a view supported by its lack of solubility in dilute aqueous acids. They did not give any analytical details for their chitin nitrate but Hackman,[5] using a similar process, obtained a product having a DS of 1.25.

Clark and Smith also carried out heterogeneous nitration of chitin,[4] using a 5:1 mixture of fuming HNO_3:concentrated HNO_3, to obtain an oriented sample for X-ray diffraction studies, the results of which indicated an orthorhombic cell with $a = 9.0$ Å, $b = 10.3$ Å and $c = 23.0$ Å. The increase in the c axis was stated to be close to that expected for substitution of nitrate groups for hydroxyl groups along the chain.

Wolfrom et al.[6] examined the preparation of chitosan nitrate and its nitrate and perchlorate salts. Two routes were investigated; in the first, chitosan[0.15] was dissolved in absolute HNO_3, while in the second it was suspended in a 1:1:1.3 mixture of glacial acetic acid:acetic anhydride:absolute nitric acid for 5.5 h at < 5°C. The products from both processes were very similar, being the nitric acid salt of chitosan nitrate (**4.1**) having a $DS_{(ONO_2)}$ value of 1.65. These were converted to the chitosan nitrate (**4.2**),

4.1   **4.2**

with the O-nitrate content unchanged, by addition of dilute alkali to a solution of the salt in 50 vol.-% aqueous acetone. Care must be taken to avoid the solution becoming alkaline since the analogous cellulose nitrate is very readily degraded by alkali. The chitosan nitrate (**4.2**) was then converted to the perchlorate salt by suspending it in glacial acetic acid containing perchloric acid. The perchlorate salt, unlike the nitrate, was found to be unstable at room temperature, decomposing slowly in most cases but with detonation in one instance.

The most comprehensive study to date is that of Hirano and Yano[7] who

TABLE 4.1   *Substitution values of* N-*acylchitosan* O-*nitrate derivatives*[7]

| | $DS_{(ONO_2)}$ | |
N-*acyl group*	N-*acylated chitosan* O-*nitrate*	O-*nitrated* N-*acyl-chitosans*
Acetyl	1.4	1.9
Propionyl	1.5	2.0
Butyryl	1.5	1.7
Hexanoyl	1.5	2.0
Octanoyl	1.5	2.0
Decanoyl	1.5	2.0
Dodecanoyl	1.6	1.9
Tetradecanoyl	1.7	1.9
Hexadecanoyl	1.7	1.9
Benzoyl	1.8	3.0

have reported the preparation and properties of O-nitrated chitosan and of the O-nitrate derivatives of a number of N-acylchitosans. These latter products were prepared by two routes:

(a) nitration of chitin and its N-acylchitosan analogues using fuming nitric acid–acetic anhydride mixtures;
(b) nitration of chitosan[0.0], followed by conversion to the free base[6] and N-acylation of the O-nitrated chitosan[0.0].

The $DS_{(ONO_2)}$ was 1.7–2.0 for the products obtained by nitration of N-acylchitosans and 1.5–1.8 for those obtained by N-acylation of O-nitrated chitosan[0.0] (Table 4.1). The values of 1.7–2.0 for the DS of samples prepared by route (a) are in agreement with the results obtained by Marchenko *et al.*[8] who reported DS values close to the theoretical maximum of 2.0 for chitin nitrate samples. The high DS values obtained by these latter workers are somewhat surprising since they used aqueous HNO_3 as the nitrating reagent and it is well known, from studies carried out on the nitration of cellulose, that the presence of water reduces the extent of nitration.[1]

Nitration[7] of N-benzoylchitosan gave a product having a DS value of 3.0, and although it was assumed that the additional NO_2 group was a $C–NO_2$ group located on the aromatic ring, this could not be confirmed since the IR absorption bands at 750 and 720 cm^{-1} are similar to those in the spectrum of the N-benzoyl derivative of O-nitrated chitosan.

O-Nitrated chitosan ($DS_{(ONO_2)}$ = 1.5–1.8) was flammable at 151–156°C but no ignition point was observed for the N-acylated derivatives, all of which decomposed at temperatures in excess of 240°C. This stability contrasts with the instability of cellulose nitrate and is presumably due to the reduction in the proportion of high-energy bonds. The solubilities of the nitrated derivatives of the various N-acylchitosans differ considerably from that of cellulose nitrate. Nitrated N-benzoyl- and the N-acylchitosans up to N-butyrylchitosan were found to be soluble in DMSO, while up to the N-hexanoylchitosan were soluble in DMF. All products up to O-nitrated N-hexadecanoylchitosan were soluble in methanesulphonic acid but none was soluble in either acetone or ether–ethanol mixtures, both of which are common solvents for cellulose nitrate.[1]

4.1.2 Phosphates

Karrer et al.[9] were the first to attempt the phosphorylation of either chitin or chitosan. They treated chitosan with 15 parts pyridine and 5 parts phosphorous oxychloride at 40°C for 5 h, but although they claimed that products having 24% P were obtained, Hackman[5] was unable to obtain phosphorylated derivatives of chitin using their method. More recently interest in phosphate esters of chitin and chitosan has increased, mainly because of their metal ion binding capabilities (section 5.2.6) and two preparative techniques have been developed. The first is based on a method for preparing cellulose phosphates, originally through heating with mixtures of phosphoric acid and urea by an impregnation-baking sequence[10] but subsequently modified by the use of an inert liquid as the reaction medium.[11] Both DMF[12, 13] and toluene[14] have been used as the inert reaction medium for preparation of chitin and chitosan phosphates.

The second method, which has been developed by Nishi et al.,[15–18] involves low temperature (0–5°C) reaction of phosphorus pentoxide with chitin or chitosan dissolved in methanesulphonic acid. A phosphorylated chitin having a DS of 1.6 was obtained using 2 equivalents of P_2O_5, but the efficiency of the reaction decreases sharply at higher levels and the DS was only increased to about 1.75 on doubling the amount of P_2O_5 used.[16, 17] Analysis by ^{13}C NMR spectroscopy indicated that the C(3)OH group was at least as reactive as the C(6)OH group. Although the DS obtained increases steadily with time of reaction, the molecular weight of the product decreases concomitantly and the preferred reaction conditions were stated to be[17] 1–2 h at 5–8°C.

Since the preparation process described in the literature[16, 17] does not involve a neutralisation step the products obtained from chitosan, or from a chitin having an appreciable concentration of amine groups, may have quite complex structures. The phosphate ester group will be present in the free acid form while many of the amine groups will be in the form of the methanesulphonate salt. Zwitter-ion type structures may also be formed between adjacent phosphate ester groups and protonated amine groups, with elimination of a molecule of methanesulphonic acid (Figure 4.1), if the affinity of the phosphate ester anion for the protonated amine group is greater than that of the methanesulphonate ion. This will introduce

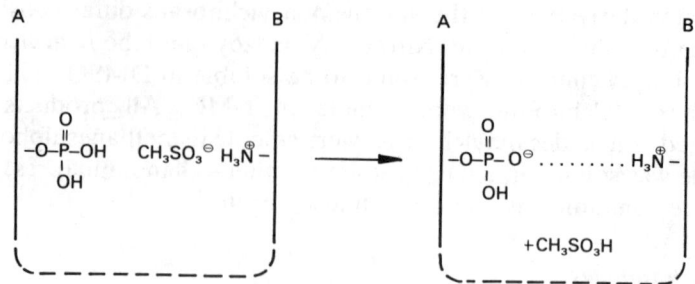

FIGURE 4.1   Formation of salt linkage in chitosan phosphate. Broken line indicates that
chain sections A and B may be sections of the same chain or of two different
chains

inter- or intrachain ionic bonds – salt linkages – the likelihood of formation
increasing with increase in either the DS of the product or the extent of
deacetylation of the starting chitin or chitosan.

Such salt linkages have been invoked to explain the dependence of the
solubility on the DS in the case of chitosan phosphates.[17] Although chitin
phosphates are water-soluble at any DS level above a minimum value,
chitosan phosphates are insoluble in water at high DS values and this has
been attributed to the formation of such interchain salt linkages, thereby
producing a polyelectrolyte complex (section 6.1.4). It would be of interest
to evaluate the solubility of the fully neutralised product sodium chitosan
phosphate.

Aqueous solutions of both chitin phosphate and chitosan phosphate
show the viscosity behaviour typical of polyelectrolytes, the viscosity
number increasing rapidly with decreasing polymer concentration in the
absence of added low-molecular-weight electrolyte, but giving a rectilinear
decrease in viscosity number with decrease in polymer concentration in the
presence of low concentrations of electrolyte. However the LVN value for
a chitin phosphate sample was considerably lower than that for a chitosan
phosphate sample prepared under similar conditions, the values being
approximately 15 cm^3 g^{-1} and 100 cm^3 g^{-1} respectively for a chitin phosphate having $DS = 1.0$ and a chitosan [0.55] phosphate having $DS = 0.4$,
when measured in 0.05M NaCl solution. Furthermore, while the viscosities
of chitin phosphate solutions were found to increase with increase in pH
because of increasing dissociation of the phosphate group, those of
chitosan[0.55] phosphate solutions decrease. This latter decrease has been
attributed to a decrease in the extent of protonation of the amine groups as
the methanesulphonate salt is neutralised. Nishi et al. concluded[17] that the
difference in LVN values between the chitin phosphate and chitosan
phosphate samples was due to the inherent low pH of the system rather
than to differences in molecular weight. They calculated the $\bar{M}_v$ value for

chitin phosphate using the Mark–Houwink equation constants for heparin[19] and obtained a value of approximately 1×10^4. However, application of the constants for chitosan[20] to the LVN for chitosan[0.55] phosphate gives $\bar{M}_v = 1.26 \times 10^5$ or 1.49×10^5 depending on whether the constants for 0.2M NaCl or for 0.02M NaCl are used, the NaCl concentration used in determining the LVN for chitosan[0.55] phosphate being 0.05M.

The much larger LVN value for chitosan[0.55] phosphate may be an artefact arising from interchain ionic interactions between -O-P(O) (OH)O$^{\ominus}$ and $-\overset{\oplus}{N}H_3$ groups, as suggested by Nishi et al.,[17] or it may be a true difference arising from the much greater stability of chitosan towards acid hydrolysis compared with that of chitin, so that the former would be expected to undergo much less chain scission during reaction in methanesulphonic acid. Again it would be of interest to compare the LVN values for neutral products, sodium chitin phosphate and sodium chitosan phosphate.

The same workers have also measured the titration curves for these two products and for D-glucose 6-phosphate. All three show two pK_a values for the phosphate group, those for the polymeric phosphates being lower than those for the monomeric phosphate. A comparison of the values for the chitin and chitosan phosphates shows that the lower pK_a values for each polymer are similar (3.65 and 3.7 respectively, 4.2 for D-glucose 6-phosphate) but there is a much larger difference between the two higher pK_a values (8.2 and 8.5 respectively, 8.65 for D-glucose 6-phosphate). Furthermore, the titration curve for chitosan[0.55] phosphate shows an additional, broad inflection point in the pH 6–7 region due to the presence of the amine groups.

Insoluble derivatives of chitin phosphate and chitosan phosphate were prepared by gradual addition of adipoyl chloride to the reaction mixture. The crosslinked products precipitated out of the reaction mixture and could be easily collected and purified.[18]

4.1.3 Sulphates

This has been studied more extensively than any of the other inorganic esterification reactions mainly, but not exclusively, in a search for heparin substitutes.

Non-specific sulphation reactions

The earliest attempt at sulphation was that of Karrer et al.[9] who sulphated chitin using a mixture of 3 parts chitin:30 parts pyridine:7 parts chlorosulphonic acid at 0°C. The product was isolated as its sodium salt and had 14.4% S, which is equivalent to a DS of 1.7. However its physiological activity was very low, presumably because the sulphate groups were predominantly, if not completely, O-sulphate ester groups. Once the suggestion[21] that the anticoagulant behaviour of heparin depends on the presence of the N-sulphate groups* had been confirmed,[22, 23] attention switched to sulphation of chitosan, two papers being published simultaneously.[23, 24] Doczi et al.[23] demonstrated that the contribution of N-sulphate groups to anticoagulant activity is far greater than that of O-sulphate groups but gave no description of their preparation route. At the same time Wolfrom et al.[24] described the preparation of chitosan sulphate by the heterogeneous reaction between pyridine-swollen chitosan and chlorosulphonic acid in pyridine, the reaction being carried out at 100°C for 1 h. The water-soluble product was isolated in the sodium salt form and found to contain two N-sulphate groups and one O-sulphate group per anhydrochitobiose unit. The anticoagulant activity was 56 International Units per mg, heparin has an anticoagulant activity ≥ 100 IU mg^{-1}, and the toxicity was approximately twice that of heparin. This increased toxicity was attributed to the high molecular weight of the chitosan sulphate.

Hackman[5], using similar reaction conditions to those used by Wolfrom et al.,[24] obtained chitin sulphate having a DS of 1.1. Enzymatic hydrolysis of the sodium chitin O-sulphate yielded D-glucosamine in slight excess over that obtained on hydrolysis of the starting chitin, indicating that a limited amount of de-N-acetylation occurs during sulphation under these conditions. The use of chlorosulphonic acid–pyridine mixtures to sulphate chitin was also examined by Cushing et al.[25, 26] and rejected on the grounds of lack of reproducibility and the poor colour of the product. Instead they used ClSO$_3$H in an inert solvent, 1,2-dichloroethane, at 25°C. Products having 13.2–15% S ($DS = 1.45$–1.8) and $\bar{M}_w$ values of 1.2×10^4–1.7×10^4 were obtained. The sodium chitin O-sulphate could be bleached with H$_2$O$_2$ without any apparent degradation. The anticoagulant activity of freshly prepared material was 22–34 IU mg^{-1} but there was a drop on storage in the solid state, while in unbuffered solution there was a gradual fall in pH, owing to hydrolysis of the sulphate groups, leading to a decrease both in the anticoagulant activity and in the solution viscosity. Omission of the dialysis step in the initial purification process gave products having improved stability. The sulphation process caused considerable chain cleavage which was considered beneficial since it reduces the toxicity problem associated with higher-molecular-weight materials.[24]

A patent issued the same year as the paper of Cushing et al.[25] claims the

---

*The product formed on sulphation of the amine groups of chitosan is more correctly designated as an N-substituted sulphamic acid but the normal convention, which is used here for convenience, is to distinguish between the two by use of the terms O-suphate and N-sulphate.

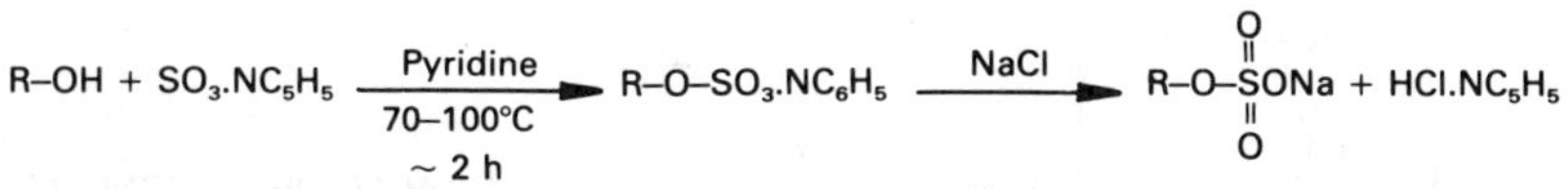

$$R\text{–}OH + SO_3.NC_5H_5 \xrightarrow[\sim 2\,h]{\substack{\text{Pyridine} \\ 70\text{–}100°C}} R\text{–}O\text{–}SO_3.NC_6H_5 \xrightarrow{\text{NaCl}} R\text{–}O\overset{\overset{O}{\|}}{\underset{\underset{O}{\|}}{S}}ONa + HCl.NC_5H_5$$

FIGURE 4.2   Proposed mechanism of formation of chitin O-sulphate by reaction of chitin
with an SO_3–pyridine complex[27]

use of SO_3 complexed with pyridine, dioxane, N,N-dimethylaniline or
2,2′-dichlorodiethylether for sulphating chitin.[27] The SO_3–pyridine complex was the preferred one since pyridine could also be used as the reaction
medium. The reaction mechanism proposed is shown in Figure 4.2.

Although Cushing's patent[26] includes chitosan in a list of polysaccharides
suitable for modification by sulphation, no examples were given. However
within a couple of years patents were issued dealing with the sulphation of
chitosan[28] and of N–formylchitosan.[29, 30] The first of these patents claimed
the use of SO_3–SO_2 mixtures at their reflux temperature ($\sim -10°C$).
Products having 9–20.6% S ($DS = 0.75$–3.0) were obtained. The other two
patents[29, 30] describe the preparation of N-formylchitosan sulphates either
by sulphation of N-formylchitosan, prepared by treating chitosan with
concentrated HCOOH at $\sim 100°C$ in the presence of pyridine, or by
N-formylation of chitosan sulphate. Sulphation was carried out at 20°C
using either $ClSO_3H$–$CHCl_3$ or $ClSO_3H$–$HCONH_2$ mixtures. The use of
H_2O_2 to bring about a reduction in the molecular weight was also claimed.

Wolfrom and Shen Han[31] reprecipitated chitosan[0.1] and subjected it
to solvent exchange through the series water→ethanol→absolute
ethanol→diethyl ether→pyridine. The final product was suspended in
pyridine and treated with a $ClSO_3H$–pyridine mixture for 1 h at 100°C to
give, after neutralisation and purification, a product containing two sulphate groups per anhydro-D-glucosamine unit. The presence of acid-
sensitive N-sulphate groups and the absence of free amine groups was
demonstrated, indicating that the product was a chitosan N-sulphate-O-
sulphate derivative – chitosan N,O-disulphate. Wolfrom and Shen Han
also reported a homogeneous sulphation process by treatment of chitosan,
reprecipitated and solvent-exchanged through the same series except that
the pyridine was replaced by DMF, with an SO_3–DMF complex in an
excess of DMF, the reaction taking place at room temperature. The
proposed reaction mechanism involves formulation of the SO_3–DMF complex as a dipolar ion which is susceptible to nucleophilic attack by either
the amine or hydroxyl groups of chitosan (Figure 4.3). Again the product
contained one N-sulphate and one O-sulphate groups per anhydro-D-
glucosamine residue.

The chitosan sulphate obtained using $ClSO_3H$–pyridine had an anticoagulant activity of 56 IU mg^{-1} but its acute LD_{50} was 380 mg kg^{-1}, more
than twice the toxicity of heparin ($LD_{50} = 750$ mg kg^{-1}), whereas that

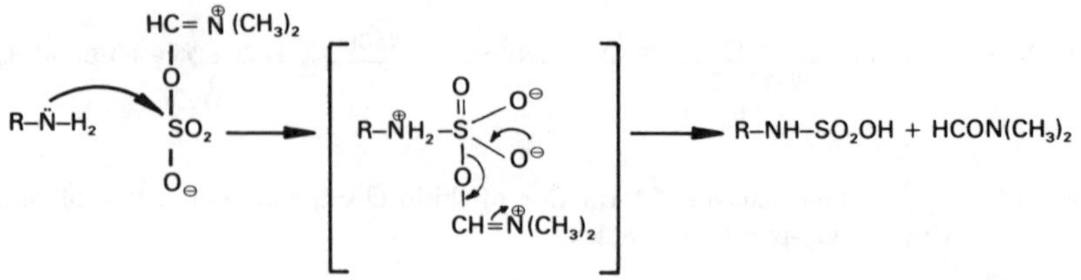

FIGURE 4.3   Proposed mechanism of formation of chitosan N-sulphate and O-sulphate by reaction of chitosan with SO_3–DMF[31]

prepared using the SO_3–DMF complex had an anticoagulant activity of 50 IU mg^{-1} and an acute LD_{50} of 775 mg kg^{-1}. The major difference between the two products was their molecular weights, being 4.56×10^5 and 1.86×10^5 respectively ($DP = 1280$ and 530). The lower toxicity of the second sample was attributed to its lower molecular weight. A patent issued to Wolfrom the previous year also deals with sulphation of chitosan with $ClSO_3H$–pyridine and with SO_3–DMF in DMF but gives no additional information.[32]

Nagasawa and co-workers have published a series of papers[33-35] dealing with the reaction between H_2SO_4 and a number of polysaccharides including chitin[33, 34] and chitosan.[33, 35] Treatment of chitin with 96 wt-% H_2SO_4 for 2 h at $-5°C$ gave a product which could be separated into a dialysable and a non-dialysable fraction, both fractions having a $DS \sim 1.2$. Similar treatment of a chitosan[0.08] sample yielded only a non-dialysable product having a DS of ~2. The hydrolysis constant for the acid-labile sulphate of chitosan sulphate in 0.1M HCl at 99.5°C was found to be 1.17×10^{-3} s^{-1}, very similar to the value of 1.03×10^{-3} s^{-1} reported previously for heparin.[36] This was taken as evidence of the presence of N-sulphate groups in the product. It was shown that neither extending the reaction time from 2 to 10 h, nor varying the reaction temperature between 0 and 30°C, had any noticeable effect on the level of substitution of the product, but that reducing the H_2SO_4 concentration from 96 to 80 wt-% reduced the DS from 2 to 0.7.

Selective 6-O-sulphation reactions

More recently two methods for the selective 6-O-sulphation of chitosan have been developed. The first[37] involves the use of a 2:1 mixture of 95 wt-% H_2SO_4 and 98 wt-% $ClSO_3H$ at 0–4°C. Reaction times of about 1 h are used and the product is isolated by precipitation. Chitosan[0.20] was used and characterisation of the product by conductiometric titration, IR and ^{13}C NMR spectroscopy showed the DS to be 0.95–1.0 and that sulphation had taken place at the C(6)OH group. The spectrum in D_2O

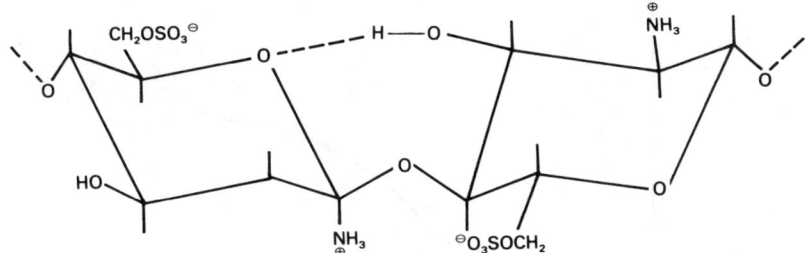

FIGURE 4.4 Proposed Zwitter-ionic interaction between C(6′)OSO$_3^\ominus$ and C(2)$\overset{\oplus}{N}$H$_3$ leading to increased rigidity of chitosan 6-O-sulphate chains at low pH values[37]

solution was compared with that for a low-molecular-weight chitosan (DP ~30) and whereas the signals for C(1)–C(5) of chitosan 6-O-sulphate were shifted by relatively small amounts compared with their position in the spectrum for the chitosan oligomer, the C(6) signal is shifted significantly downfield (−6.2 ppm). The more or less complete absence of any C(1) signal attributable to reducing end-groups indicates that the sulphation reaction did not involve concomitant hydrolysis of the polymer chain.

The authors argued that in alkaline solution chitosan 6-O-sulphate should behave like a typical polyanion and that the local conformation of its chains should be similar to that of cellulose or chitin, with ribbon-like segments stabilised by C(3′)OH . . . O(5) intramolecular hydrogen bonds. However at low pH values there is the possibility of forming Zwitter-ionic structures having intramolecular electrostatic bonds between C(6′)OSO$_3^-$ . . . $\overset{\oplus}{N}$H$_3$ (2)C and these should lead to an increase in chain rigidity (Figure 4.4).

The second method[38, 39] involves protecting the amine group by complexing it with Cu(II) ions, followed by treatment of a DMF suspension of the Cu(II)–chitosan complex with SO$_3$–pyridine at 25–75°C for 8–48 h. Elimination of Cu(II) ions from the Cu(II)–chitosan 6-O-sulphate complex was achieved by percolating an aqueous solution of the complex through a selective cation exchange resin. At 25°C the reaction shows an induction period of about 16 h followed by a very rapid increase in reaction rate and a levelling off after about 24 h (Figure 4.5). At a mole ratio of SO$_3$–pyridine:Cu(II)–chitosan of 6:1, a reaction time of 16 h at 25°C gives a product having a DS of approximately 1. The selectivity of the reaction was confirmed by IR and ^{13}C NMR spectroscopy. Higher DS values than 1 could be achieved by increasing the temperature, a product having DS = 1.8 being obtained after 16 h reaction at 75°C using a 6:1 mole ratio of reactants to substrate. The additional sulphation occurred partially at C(3)OH but mainly at C(2)NH$_2$, as shown by ^{13}C NMR spectroscopy,

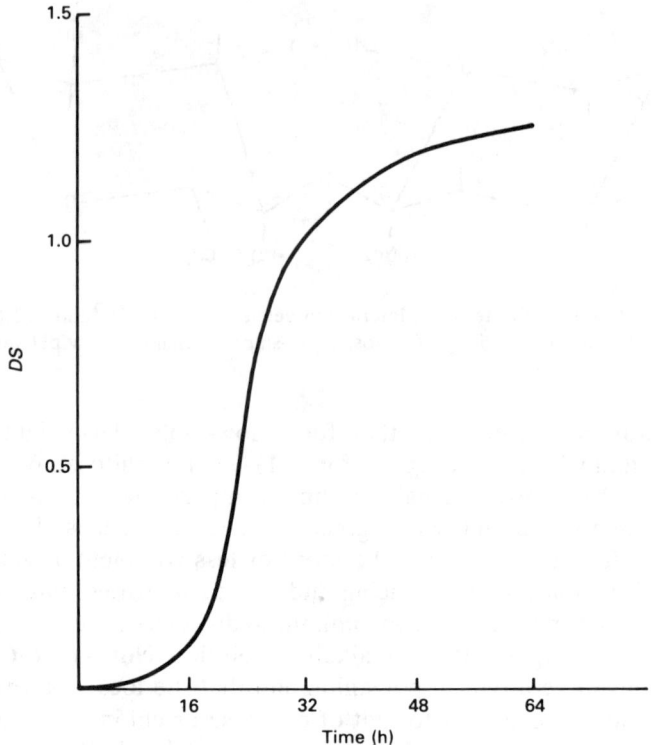

FIGURE 4.5   Rate of formation of chitosan 6-O-sulphate by treatment[39] of chitosan–Cu(II)
complex with SO₃–pyridine complex at 25°C. Mole ratio of SO₃:Py = 6:1

while the molecular weight (light scattering) of a product obtained after 24
h reaction time at 25°C was 1.3×10^6 with an LVN of 270 cm³ g⁻¹
measured in 0.2M NaCl solution.

A very elegant study of chitosan 6-O-sulphate by Stivala et al.[40] using
small-angle X-ray scattering led to the conclusion that the sample, as
prepared by Naggi et al.,[37] is quite highly branched. The study determined
the solution parameters of chitosan 6-O-sulphate and compared them with
those of heparin;[41] the most important parameters are shown in Table 4.2.

The total length L, both that calculated on the basis of the molecular
weight, monomer molecular weight and length per monomer unit, and that
determined experimentally, are very similar for heparin but differ considerably for chitosan 6-O-sulphate. In the latter case the calculated value
of L is more than twice the experimentally measured value. Similarly the
experimental and calculated mass per unit length (M_u) of heparin are very
similar while those for chitosan 6-O-sulphate differ, the experimentally
determined value being more than twice the calculated value, the latter
being the value calculated on the basis of a linear, unbranched structure.

TABLE 4.2   *Solution parameters of chitosan 6-O-sulphate and heparin*[40]

	Chitosan 6-O-sulphate	Heparin
$\bar{M}_w$	3.16×10^4	1.29×10^4
L(exp.)	271 Å	237 Å
L(calc.)	620 Å	260 Å
Persistence length	20.3 Å	21.1 Å
M_u(exp.)	117 Å	54.5 Å
M_u(calc.)	51 Å	52.7 Å
$[\eta]$	80 cm^3 g^{-1}	417 cm^3 g^{-1}

Finally, since $\bar{M}_w$ for the chitosan 6-O-sulphate sample is more than twice that for the heparin sample, the LVN of the former would be expected to be greater than the LVN of the heparin whereas in fact the latter is considerably greater. These three factors all support a branched structure for the chitosan 6-O-sulphate investigated.

Branching was explained as arising from chain degradation through acid hydrolysis, and linking of the oligomer chains to the backbone through sulphate diester groups formed between a backbone C(6)OH and an oligomer C(6)OH. Evidence of sulphate diester groups was obtained by IR spectroscopy, a weak band at 1390 cm^{-1} being observed in addition to the much stronger band at about 1230 cm^{-1} characteristic of the sulphate monoester group. Consideration of the M_u (calc.) value, together with the reported potentiometric results,[37] enabled a structure to be proposed in which one oligomer chain having a *DP* ~15 is attached for every ten monomer units in the backbone chain (a *DS* of ~0.1).

The preparation of chitin sulphate under homogeneous conditions has recently been reported.[42] Chitin dissolved in DMAc–LiCl (50 g dm^{-3}) was treated with SO$_3$–pyridine complex, in one instance after subjecting the chitin–DMAc–LiCl solution to ultrasonic treatment. In both cases ^{13}C NMR spectroscopy showed selective substitution of the C(6)OH groups, the *DS* values being 0.3 and 0.9.

Sulphation of chitin and chitosan derivatives

Other workers have studied the preparation and properties of sulphate derivatives of modified chitins and chitosans. The first work of this type was that of Horton and Just[43] who prepared the N-sulphate derivative of poly[β(1→4)-2-amino-2-deoxy-D-glucopyranuronic acid] (**4.3**). The synthesis route, which is outlined in Figure 4.6, involved formation of the perchloric acid salt to protect the amine group, followed by oxidation of the C(6)OH with CrO$_3$–glacial acetic acid, sulphation by the procedure of Wolfrom and Shen Han[31], and neutralisation. The product had an anticoagulant activity of 25.8 IU mg^{-1} and an LD_{50} of 237 mg kg^{-1}. The low

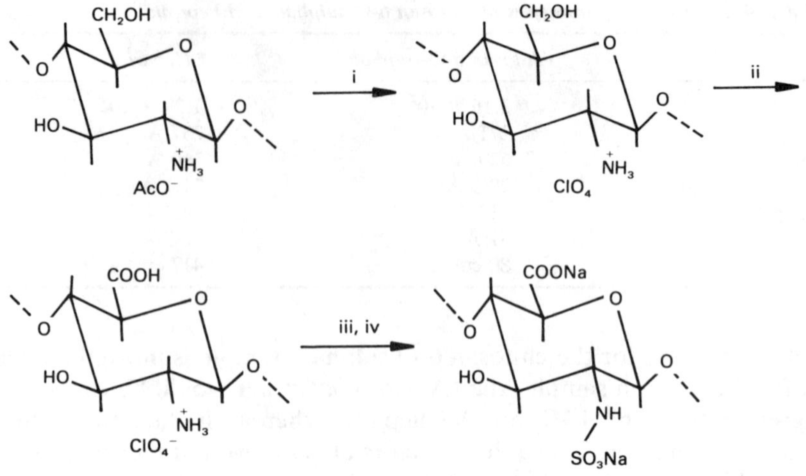

Reagents: i = HClO$_4$; ii = CrO$_3$–CH$_3$COOH; iii = ClSO$_3$H–pyridine
iv = NaOH

FIGURE 4.6   Route for the preparation of the N-sulphate of poly[β(1→4)-
2-amino-2-deoxy-D–glucopyranuronic acid][43]

anticoagulant activity and high toxicity relative to heparin was attributed to
its molecular weight, 4.3×10^5, being very high compared with that of
therapeutic grade heparin which is $\sim 1.3 \times 10^4$.

Okiei et al.[44] prepared chitin O-sulphate and carboxymethyl chitin
O-sulphate by reaction of chitin or carboxymethyl chitin respectively, with
ClSO$_3$H–pyridine. The products (Table 4.3) were examined for their
inhibitory effect on thrombin activity and it was found that the carboxymethyl chitin O-sulphate (4.4c) had similar thrombin inhibition activity to
that of heparin while chitin O-sulphate (4.4d) was much less effective.

Hirano et al.[45] prepared sulphate derivatives of chitosan and N-acylated
chitosans (Table 4.4). The N-acetylchitosan di-O-sulphate (chitin di-O-
sulphate, 4.5c) was approximately twice as active as heparin as an anticoagulant but had much less lipoprotein lipase activity, while the chitosan
N-sulphate-O-sulphates (4.5a, b) were 10–60% more active as anticoa-
gulants than heparin and had 2–3 times the lipoprotein lipase activity.

Finally Muzzarelli et al.[46, 47] have prepared N-carboxymethyl chitosan
sulphates and examined their behaviour as blood coagulants. The starting
N-carboxymethyl chitosan was prepared by the reductive alkylation
technique[48] (section 4.3.3) and this was then sulphated with either a 1:1
H$_2$SO$_4$:ClSO$_3$H mixture[37], SO$_3$–DMF in DMF[31], or ClSO$_3$H–pyridine, to
give the mono-, di- and trisulphate derivatives respectively (Table 4.5). It
was found that sulphation of the C(3)OH group, which is required in the
preparation of the N-carboxymethyl chitosan 3,6-O-N-trisulphate, necessi-

TABLE 4.3 *Structures of sulphated chitin derivatives evaluated for thrombin inhibition activity*[44]

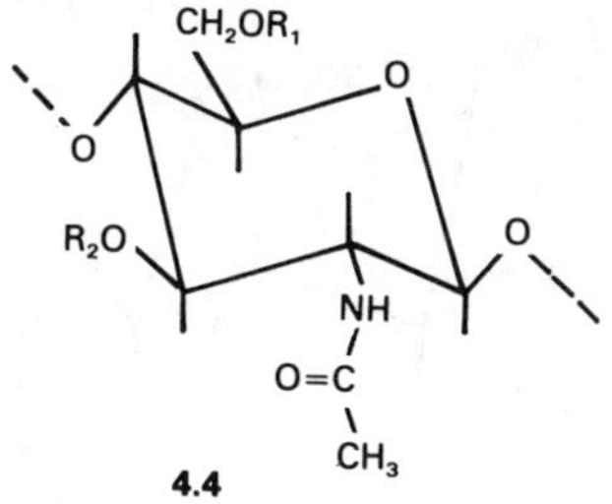

4.4

	R_1	R_2	Molecular weight
a	0.8 CH$_2$COONa 0.2 H	H	6.3×10^4
b	0.8 CH$_2$COONa 0.14 SO$_3$Na 0.06 H	H	1.8×10^4
c	0.56 CH$_2$COONa 0.44 SO$_3$Na	0.37 SO$_3$Na 0.63 H	2.4×10^4
d	0.69 SO$_3$Na 0.31 H	H	2.0×10^4

TABLE 4.4 *Structures of sulphated derivatives of chitosan and N-acylchitosans evaluated for anticoagulant activity*[45]

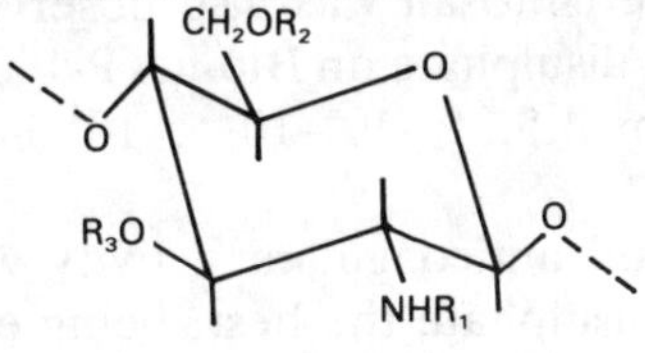

4.5

	R_1	R_2	R_3
a	SO$_3$Na	SO$_3$Na	0.63 SO$_3$Na 0.37 H
b	0.92 SO$_3$Na 0.08 CH$_3$CO	0.73 SO$_3$Na 0.27 H	H
c	CH$_3$CO	SO$_3$Na	SO$_3$Na
d	CH$_3$(CH$_2$)$_4$CO	SO$_3$Na	0.8 SO$_3$Na 0.2 H
e	H	0.74 SO$_3$Na 0.26 H	H
f	SO$_3$Na	0.54 CH$_2$COONa 0.36 H	H

TABLE 4.5   *Structures of sulphated N-carboxymethyl chitosans evaluated for anticoagulant activity*[47]

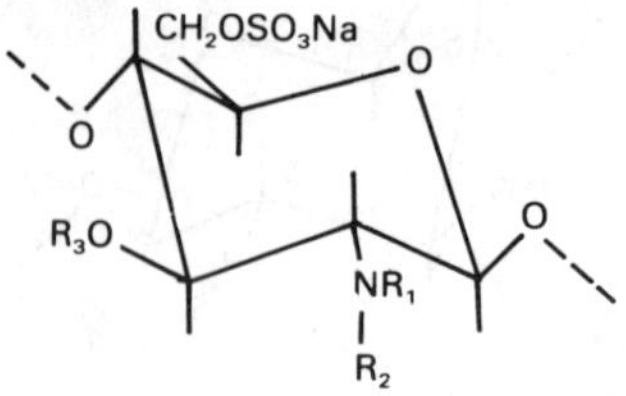

4.6

	R_1	R_2	R_3	Reagent
a	0.5 CH$_2$COONa 0.5 CH$_3$CO	H	H	H$_2$SO$_4$–ClSO$_3$H
b	0.58 CH$_2$COONa 0.42 CH$_3$CO	H	0.8 SO$_3$Na 0.2 H	SO$_3$–DMF in pyridine
c	0.50 CH$_2$COONa 0.50 CH$_3$CO	0.5 SO$_3$Na 0.5 H	SO$_3$Na	ClSO$_3$H–pyridine

tated prior hydrolysis of the *N*-carboxymethyl chitosan by boiling in 6M HCl for at least 10 minutes (since the starting chitosan in this case was a chitosan[0.50], the maximum *DS* for sulphate groups is actually 2.5 and not 3.0). This increase in the extent of substitution with decrease in the molecular weight of the chitosan was also observed on fractionating the *N*-carboxymethyl chitin disulphate on Bio-gel P-100, when fractions having molecular weights from 4.5. × 10^4–10 × 10^4 had sulphur contents of 12.4–9.5%.

All showed very good anticoagulant activity with the low-molecular-weight fraction of the disulphate the best, being equal in effectiveness to heparin. The better activity of the *N*-carboxymethyl chitosan disulphate was considered to be due to the presence of a high proportion of *N*-acetyl groups, in agreement with the results of Hirano *et al.*[45] None of the samples exhibited adverse effects on red blood cells.

A naturally occurring chitin sulphate–chitin 6-*O*-sulphate, *DS* = 1, molecular weight (GPC) = 4 × 10^4, has been isolated from the test of tunicates *H. roretzi* and *S. plicata* and its interaction with lectins studied.[49]

4.1.4   Xanthates

Formation of alkali chitin

Preparation of chitin xanthate requires the initial formation of alkali chitin followed by reaction between the alkali chitin and carbon disulphide:

$$\text{Chit–OH} + \text{NaOH} \longrightarrow \text{Chit–O}^-\text{Na}^+ + \text{H}_2\text{O}$$

$$\text{Chit–O}^-\text{Na}^+ + \text{CS}_2 \longrightarrow \text{Chit–O–}\overset{\overset{\text{S}}{\|}}{\text{C}}\text{–S}^-\text{Na}^+$$

Thus the formation of alkali chitin is of considerable importance in the preparation of chitin xanthate and of a number of chitin ethers (section 4.3.1).

Three patents granted to Thor[50–52] cover both the preparation of alkali chitin and the xanthation step. Thor found that the amount of NaOH bound/N-acetylglucosamine residue increases with increase in the concentration of the steeping solution and with decrease in the steeping temperature. A plot of his tabulated results shows (Figure 4.7) that the Absorption *versus* Concentration curve is S-shaped and similar to those reported for various cellulose substrates,[53] although the concentration at which the

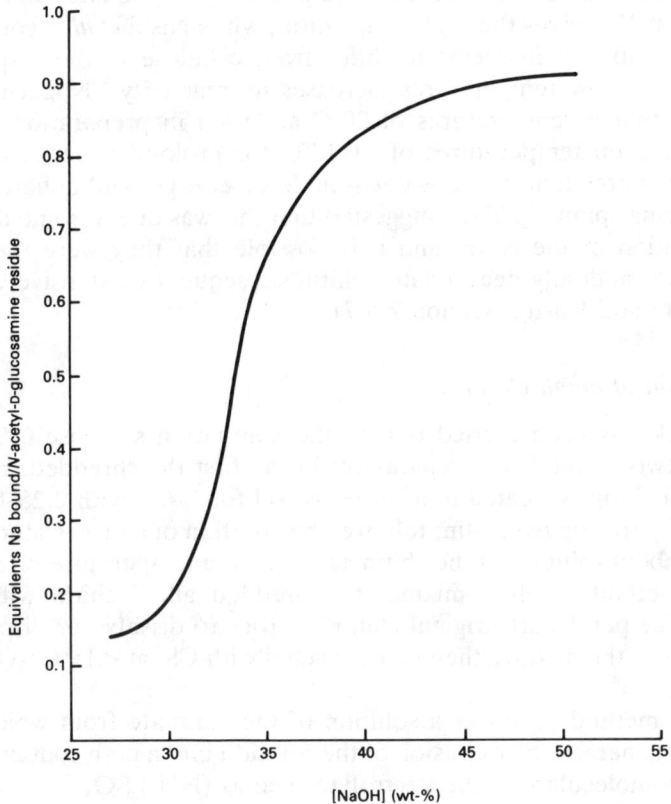

FIGURE 4.7   Relationship between NaOH combined with chitin and the concentration of the NaOH steeping solution at 25°C[50–52]

steepest rise occurs, 30–35 wt-% NaOH, is higher than for cellulose which shows this rapid increase in NaOH absorption between 8–16 wt-% NaOH depending on the particular cellulose substrate. Since Thor worked with chitin from shrimp, that is α-chitin, this decreased readiness to swell may be attributed to the greater extent of interchain hydrogen bonding in α-chitin compared with either cellulose I or cellulose II (section 1.4.2).

Thor[50-52] emphasised the advantage of working at low temperature stating that it is 'generally desirable to work at somewhat below the maximum permissible temperature,' but it was Danilov and Plisko[54] who appear to have been the first actually to freeze the alkali chitin mixture, their process involving subjecting the alkali chitin to up to three cycles of freezing and thawing and the removal of any residual aqueous liquor after each freezing treatment. This technique of freezing alkali chitin has been used by a number of workers[55-60] including in some instances freezing under vacuum.[56, 58] Tokura et al.[60] claim that high alkali concentrations, addition of a surfactant such as sodium dodecylsulphonate, and freezing for at least 10 h gives the best alkali chitin, while Nishi et al.[56] commented that 'generally chitin seems to differ from cellulose in the respect that treatment at low temperatures increases its reactivity.' Noguchi et al.[55] reported that if temperatures of 20°C are used in preparation of alkali chitin, or even temperatures of 11–13°C for prolonged periods, the re-generated fibres tend to be water-soluble or else gel and adhere to each other during spinning. They suggested that this was due to some degree of deacetylation of the chitin and it is possible that they were the first to obtain the randomly deacetylated chitin subsequently extensively studied by Sannan and Kurita (section 2.5.7).

Xanthation of alkali chitin

Less work has been carried out on the xanthation step itself. Thor[50-52] claimed two methods of preparation; in the first the shredded, relatively dry alkali chitin is treated in a closed vessel for 1–6 h with 0.25–0.5 parts CS_2 per 1 part original chitin, followed by addition of an ice–water mixture to bring about solution of the chitin xanthate at a temperature <15°C. The second method involves mixing the shredded alkali chitin with about 10 parts ice per 1 part original chitin in order to dissolve or disperse the alkali chitin, the mixture then being reacted with CS_2 at <15°C to form the xanthate.

Either method produces a solution of the xanthate from which chitin may be regenerated by extrusion of the solution into a bath containing acid and a low-molecular-weight electrolyte such as $(NH_4)_2SO_4$.

$$\text{Chit–O–}\overset{\overset{\displaystyle S}{\|}}{\text{C}}\text{–S}^-\text{Na}^+ + \text{HX} \longrightarrow \text{Chit–OH} + CS_2 + \text{NaX}$$

Chitin xanthate is not normally used in its own right but mainly as an intermediate in the preparation of regenerated chitin, although its use as an adhesive has been reported.[61] It has been used in preparation of chitin fibres,[50–52, 62, 63] chitin films[50–52, 62] and chitin sponges.[62] Tokura et al.[57, 63] have stated that the properties of chitin fibres produced from chitin xanthate are inferior to those spun from solutions of chitin in 99% HCOOH but this may be due to the use of non-optimum conditions, particularly in the coagulation/regeneration steps.

4.2   ORGANIC AMIDES, ESTERS AND RELATED DERIVATIVES

4.2.1   N-Acyl derivatives of chitosan

N-Acylation using carboxylic acids

N-Acylation may be brought about in a number of ways, not all of which are necessarily applicable to the preparation of all N-acyl derivatives.

The simplest procedure is reaction between a carboxylic acid and chitosan, and the preparation of N-formylchitosan by heating a solution of chitosan in 100% HCOOH at 90°C, with the gradual addition of pyridine, has been claimed,[29] while Aiba[64] has followed the course of N-acetylation of chitosan in 20 vol.-% acetic acid. This latter method was much slower, the extent of N-acetylation being approximately 50% after 300 h at 80°C.

Alternatively, heating the as-cast films prepared from solutions of chitosan in the appropriate dilute aqueous acid has been shown to bring about some degree of N-acylation. Thus a film cast from a mixture of equal proportions of chitosan and phthalic acid in 0.33M acetic acid became insoluble in water, 0.33M acetic acid or dilute NH_4OH on heating at 50°C for 48 h, and similar behaviour was found with the as-cast film prepared from a solution of chitosan in maleic acid.[65] Insolubilisation of chitosan acetate films on heating at temperatures >60°C has also been observed.[66]

N-Acylations involving carbodiimide-mediated reactions have also been carried out. Yaku and Yamashita[67] acetylated chitosan film with acetic acid and dicyclohexylcarbodiimide (DCC) using aqueous DMF as the reaction medium, but did not achieve complete acetylation when the reaction medium contained more than 40 vol.-% H_2O. Kurita et al.[68] carried out homogeneous reactions using water-soluble chitosan[0.5] and a 20-fold excess of both acetic acid and DCC, in aqueous DMF. Again the extent of N-acetylation achieved was dependent on the water content of the reaction medium; with either 55 or 60 vol.-% H_2O the products remained in solution in the reaction mixture and on isolation were found to be soluble in dilute HCl, indicating incomplete N-acetylation, but with 40 vol.-% H_2O the product precipitated out as a highly swollen gel that on working up was

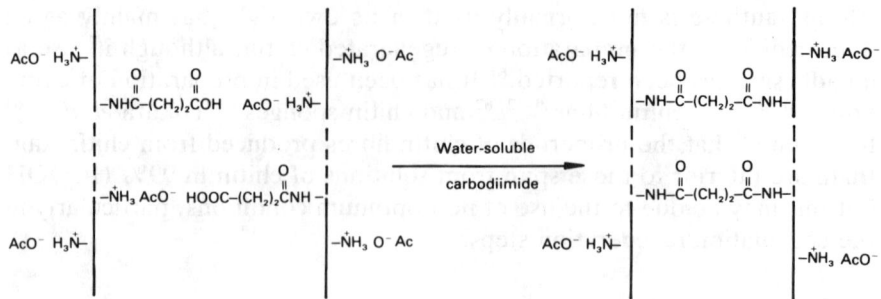

FIGURE 4.8   Crosslinking of partially *N*-succinylated chitosan through carbodiimide-mediated amide formation[69]

found to be chitin[1.0]. The IR spectrum showed no evidence of any *O*-acetylation having occurred. The use of a water-soluble carbodiimide, 1-ethyl-3-(3-dimethylaminopropyl)-carbodiimide, and no DMF was also examined but negligible *N*-acetylation was achieved, a result that was taken as support for the contention that the reaction is hampered by too high a water content.

Yamaguchi et al.[69] used a water-soluble carbodiimide to induce gelation in solutions of partially *N*-succinylated chitosan and hydroxyethyl chitosan through crosslink formation by reaction between pendant carboxylic acid groups and amine groups on the chitosan chain (Figure 4.8). Roberts and Taylor[70] used carbodiimide-mediated reactions to couple chitosan to pre-formed crosslinked polymer beads of poly(methacrylic acid) resin and carboxymethyl Sephadex. Three techniques were investigated: direct coupling by treatment of the beads with a water-soluble carbodiimide, 1-cyclohexyl-3-[2-(*N*-methylmorpholino)ethyl]carbodiimide, followed by the addition of a solution of chitosan; coupling through active ester formation[71] by treatment of the beads with 1-hydroxybenzotriazole and the carbodiimide; and coupling by the Sundaram technique[72] using 2-ethoxy-1-ethoxycarbonyl-1,2-dihydroquinoline. Both chitosan[0.14] in 0.1M acetic acid and water-soluble chitosan[0.58] in distilled water were used with each technique and, in the case of direct coupling, the effect of incorporating a spacer was also examined. The results showed that chitosan was coupled more efficiently under neutral conditions and that the amount bound to the beads increased with increase in the concentration of added NaCl. It was concluded that the effect of acid is to increase both the chain dimensions and interchain repulsion through protonation of the amine groups, thereby increasing the extent of screening of adjacent potential sites by chains already attached to the surface.

N-*Acylation using acyl chlorides*

Acid chlorides, because of their high reactivity and hence their lack of discrimination, have found little application in the selective *N*-acylation of chitosan. Kurita *et al.*[68] attempted to *N*-acetylate water-soluble chitosan[0.50] by an interfacial reaction technique but only a very limited amount of *N*-acetylation was obtained. This was attributed to the steric arrangement of the chitosan molecules at the liquid interface limiting the accessibility of the amine groups to the acetyl chloride molecules. In a later paper[73] acetyl chloride was used to *N*-acetylate a highly swollen chitosan precipitate prepared by addition of a chitosan solution in aqueous acetic acid–methanol to a large excess of pyridine, the acetyl chloride being added to the stirred mixture. The efficiency of the reaction was very low, at a mole ratio of 307:1 for $CH_3COCl:-NH_2$ the starting chitosan[0.12] was converted to chitosan[0.34]. The efficiency was improved by solvent exchange of the precipitate five times with pyridine, followed by addition of the acetyl chloride in THF. Under these conditions a mole ratio of 25:1 gave chitin[0.84]. In both series the acetylation reaction, although carried out at 0°C, was not specific for the amine group, the presence of a weak band at 1730 cm^{-1} in the IR spectra indicating some esterification of the hydroxyl groups. The ester groups were removed by treatment with 1M methanolic KOH at room temperature for 4 h.

N-*Acylation using acyl anhydrides*

The most common reagents for *N*-acylation of chitosan are, without doubt, the acyl anhydrides and these have been used under both heterogeneous and homogeneous conditions, principally the latter. One detailed study has been made of heterogeneous *N*-acylation.[74, 75] Three systems were examined: (a) acetic anhydride–glacial acetic acid–HClO$_4$; (b) acetic anhydride at room temperature for 120 h followed by refluxing in acetic anhydride for 2 h; (c) acetic anhydride–methanol at room temperature. Of these, the last method was found to be the most efficacious.

The effect of solvent on the ease of *N*-acetylation of films of chitosan[0.22] was examined with a number of solvents covering the solubility parameter range of 7.4–19.2 Hildebrands and only methanol and formamide gave any appreciable *N*-acetylation. *N*-Acetylation was also carried out in a series of binary solvent mixtures; methanol–ethanol, methanol–formamide, methanol–propanol and ethanol–formamide, and the extent of reaction determined after 30 minutes. The results suggest that although there is a considerable increase in the rate of *N*-acetylation in the solubility parameter value range $12.75 < \delta < 14.75$ Hildebrands, with a maximum at $13.1 < \delta < 13.5$ Hildebrands, the crucial factor is not the

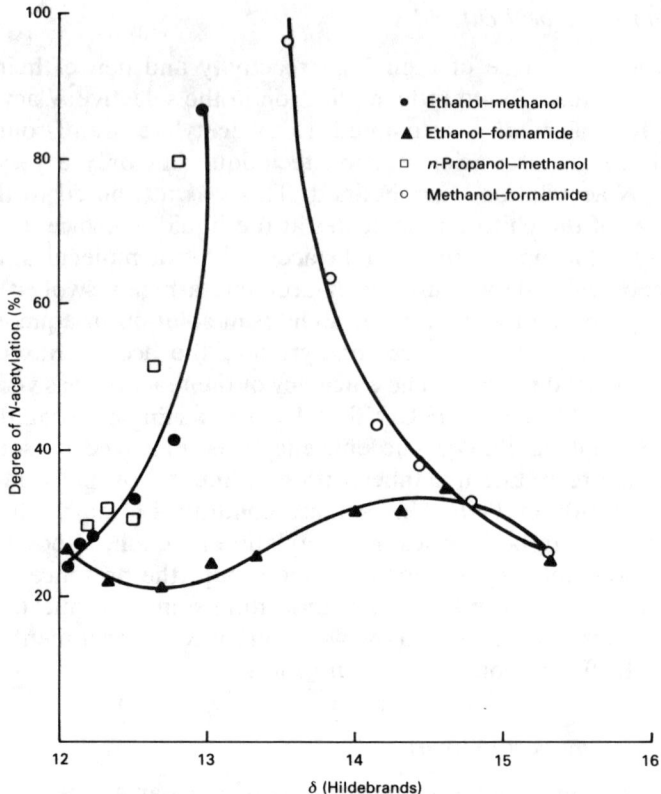

FIGURE 4.9   Degree of *N*-acetylation, after 30 minutes' reaction, as a function of the solubility parameter value (δ) of the reaction mixture. Films not presteeped[74, 75]

δ value of the reaction medium but rather the presence of methanol (Figure 4.9).

However on presteeping films in the reaction media for 24 h prior to the addition of acetic anhydride, and measuring the extent of *N*-acetylation after 5 minutes, it was found that reaction media having the same δ value give the same degree of *N*-acetylation, and that those having $13 < \delta < 15.5$ Hildebrands allow complete *N*-acetylation within 5 minutes. This shows that the degree of swelling of chitosan, and hence the accessibility of the amine groups, depends primarily on the δ value of the reaction medium provided sufficient time is allowed for the solvent molecules to penetrate and swell the chitosan. Methanol appears to be unique among the solvents examined in that only a very short time is required for its diffusion into chitosan so that rapid N-acetylation can take place without the need for a presteeping treatment.

The rate of N-acylation of chitosan[0.22] with different acyl anhydrides in methanol – acetic, propionic, butyric, hexanoic and benzoic anhydrides – was also examined.[74] The results for butyric anhydride, typical for the aliphatic acyl anhydrides, are shown in Figure 4.10. The induction period decreases with increase in temperature and with decrease in the molecular size of the anhydride, typical of a diffusion-controlled reaction, and could be completely eliminated by a pretreatment in methanol. The Arrhenius energies of activation decreased with increase in molecular size of the anhydride, from 95–90 kJ mol^{-1} for the series acetic, propionic and butyric, to ~75 kJ mol^{-1} for hexanoic anhydride.

Heterogeneous N-acylation using highly swollen precipitated chitosan has been extensively studied by Kurita's group. In their first paper on this technique,[68] water-soluble chitosan[0.5] was precipitated out by addition of its aqueous solution to a large volume of pyridine. The precipitate, in the form of a highly swollen gel, was solvent-exchanged several times with

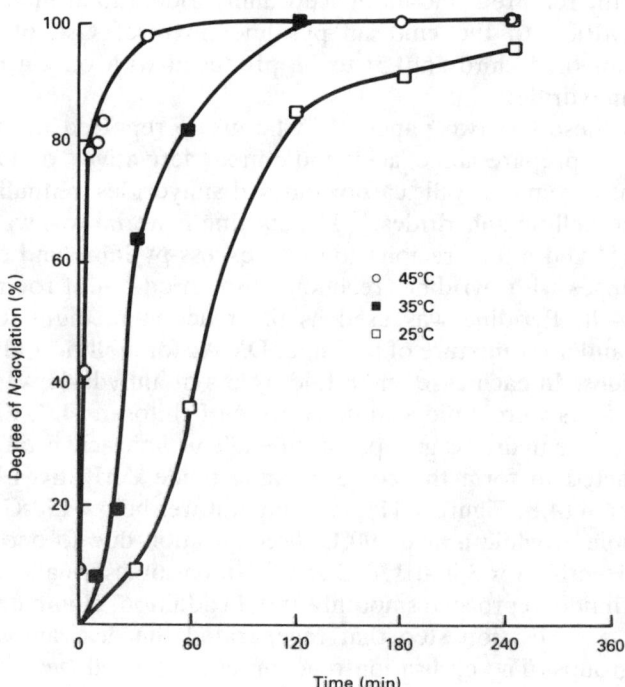

FIGURE 4.10 Rates of reaction for the N-butyrylation of chitosan film at different temperatures[74]

fresh portions of pyridine, then suspended in pyridine and acetic anhydride added (60 cm^3 of acetic anhydride for 0.3 g chitosan, a 700-fold excess based on the amine group concentration) and the mixture stirred for 3 minutes at room temperature. The solid, without intermediate drying, was steeped in a saturated NaHCO$_3$ solution to hydrolyse any O-acetyl groups and give chitin[1.0] as the final product.

In a subsequent paper[73] a swollen precipitate of chitosan[0.12] was prepared by addition of a solution in aqueous acetic acid–methanol to pyridine. The mixture was not solvent-exchanged and much smaller quantities of acetic anhydride were used so it is not surprising that complete N-acetylation was not achieved. The highest mole ratio used was 31:1 which gave chitosan[0.62] after a reaction time of 5 h, but despite incomplete N-acetylation some O-acetylation also occurred.

In a third paper[76] water-soluble chitosan was prepared by N-acetylation of a swollen precipitate of chitosan[0.1]. The required chitosan[0.50] was obtained by treatment of the precipitate with a 20 to 25-fold excess of acetic anhydride and a reaction time of 5 h. Better efficiency was attained if the precipitate was produced by pouring the chitosan solution into pyridine containing the required amount of acetic anhydride, rather than adding the acetic anhydride to the chitosan–pyridine mixture. Use of this latter technique enabled chitosan[0.5] to be produced with only a 6 to 7-fold excess of anhydride.

Prior to these last two papers[73, 76] the group reported the use of this technique to prepare amic acid and imide derivatives of chitosan by reaction with aromatic cyclic carboxylic acid anhydrides: phthalic, trimellitic and pyromellitic anhydrides.[77] The starting material was water-soluble chitosan[0.5] and after precipitation into excess pyridine and several solvent exchanges with pyridine, reaction was carried out at room temperature for 24 h. Pyridine was used as the reaction medium for phthalic anhydride and a 1:1 mixture of pyridine: DMAc for mellitic and pyromellitic anhydrides. In each case an 18-fold excess of anhydride was used.

The products were amic acid derivatives of chitosan (**4.7**, Figure 4.11) with some of the hydroxyl groups esterified. The amic acid groups could be further reacted to form the corresponding imide derivatives by heating under vacuum (**4.8**, Figure 4.11). At temperatures below 150°C cyclisation was incomplete, while above 190°C discolouration due to decomposition occurred. Heating for 3 h at 170°C and 0.1 mbar pressure allowed cyclisation to the imide to proceed smoothly and, in addition, eliminated the ester groups by a cyclisation step that regenerated the acid anhydrides and hydroxyl groups. The cyclisation reaction could be followed by IR spectroscopy.

Both the amic acid (**4.7a**) and imide (**4.8a**) derivatives from phthalic anhydride and the amic acid derivative (**4.7b**) from trimellitic anhydride were soluble in DMSO and these three, together with the imide from

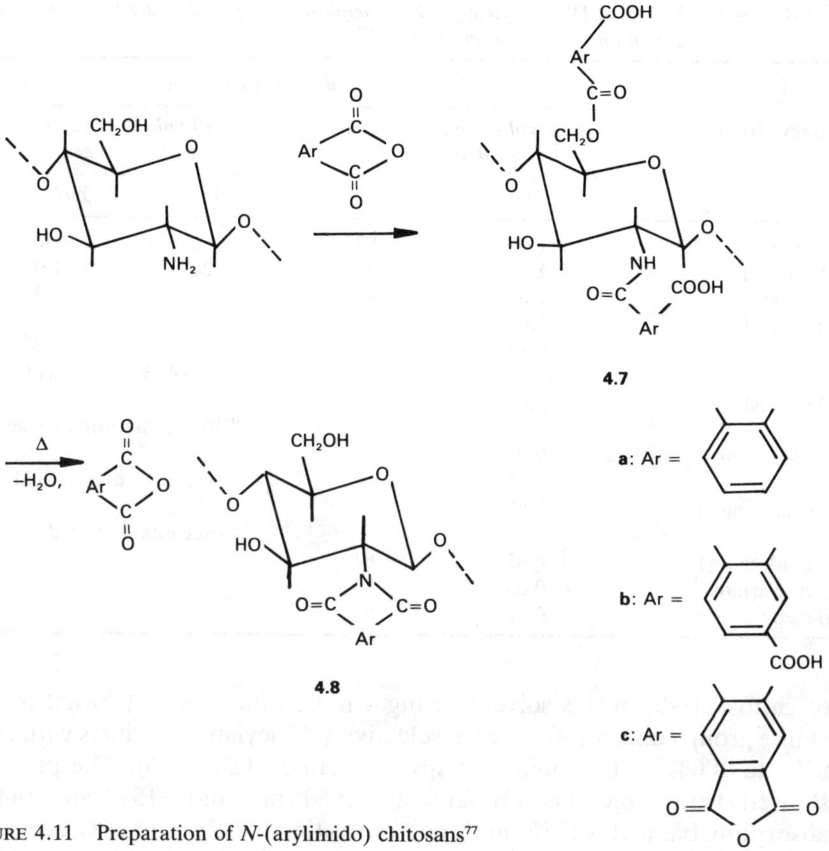

FIGURE 4.11   Preparation of N-(arylimido) chitosans[77]

trimellitic anhydride (**4.8b**) were soluble in 0.5M NaOH and saturated NaHCO$_3$ solutions. Both the amic acid and imide derivative from pyro-mellitic anhydride (**4.7c** and **4.8c**) were insoluble in all the solvents tested, indicating the probable formation of interchain crosslinks by the dianhy-dride. Derivatives **4.8b** and **4.8c** could be further reacted with amines such as p-chloroaniline.

The technique of N-acylation under homogeneous conditions was pion-eered by Hirano and much of the subsequent work has been carried out by his group. However a considerable number of their papers deal with the production of N-acylchitosan gels which are considered in Chapter 6.

Combining the selective N-acetylation of chitosan oligomers in aqueous methanol,[78] which is a non-solvent for chitosan, and the facile N- and O-acetylation of chitosan in aqueous acetic acid[79] led Hirano to investigate the acylation of chitosan in aqueous acetic acid-methanol mixtures.[80] Two solvent systems were examined, the first being a 1:9:40 mixture of acetic acid:water:methanol. Addition of a carboxylic acid anhydride to a solution

TABLE 4.6   *Values of DS and yield for N-acylchitosans prepared under homogeneous conditions in two solvent systems*[80]

N-acyl group	Solvent system			
	10 vol.-% aqueous acetic acid–methanol		10 vol.-% aqueous carboxylic acid	
	DS	Yield (%)	DS	Yield (%)
Acetyl	1.0	92	1.39[a]	93[a]
Propionyl	1.0	92	1.30[b]	104[a]
Butyryl	1.0	92	1.10[c]	103[c]
Hexanoyl	1.0	96		
Octanoyl	0.83	92		
			[a] 10 vol.-% acetic acid	
Decanoyl	1.0	93		
			[b] 10 vol.-% propionic acid	
Dodecanoyl	0.92	88		
			[c] 10 vol.-% butyric acid	
Tetradecanoyl[d]	0.87	77		
			[d] precipitate formed	
Hexadecanoyl[d]	0.82	87		
Octadecanoyl[d]	0.83	80		
Benzoyl	0.82	90		

of chitosan[~0] in this solvent, using a mole ratio of 2–3:1 based on the amine group concentration, gave selectively *N*-acylated products with from 82% to 100% of the amine groups acetylated (Table 4.6). The products showed strong amide I and II bands at ~ 1650 cm^{-1} and ~ 1540 cm^{-1} but no absorption band at ~ 1750 cm^{-1}, demonstrating the absence of *O*-acyl ester groups. The second solvent system was a 10 vol.-% solution of the appropriate carboxylic acid and in this case much higher mole ratios of acyl anhydride were required, 20–40:1 based on the amine group concentration. The products showed absorption bands at ~ 1750 cm^{-1} (C=O) and ~ 1240 cm^{-1} (C–O), indicating the presence of ester groups, in addition to the amide I and II bands (Table 4.6). The ester groups could be removed by treatment overnight at room temperature in 0.5M ethanolic KOH, the products after this treatment being identical with the appropriate derivatives produced in the first solvent. All of the dried *N*-acylchitosans were gelatinous and hygroscopic, insoluble in cold or boiling water, 50 vol.-% formic acid, 10 vol.-% acetic acid, acetic acid, formamide, DMSO, 50 vol.-% aqueous resorcinol and inorganic acids and alkalis. However the *N*-acetyl-, *N*-propionyl- and *N*-butyrylchitosans were soluble in formic acid.

In the second paper Hirano et al.[81] prepared mixed *N*-acyl derivatives of chitosan in a two-step process in aqueous acetic acid–methanol. In the first step 0.25–0.75 molar equivalents of anhydride were added for each mole of amine and the reaction left at room temperature overnight, after which the partially N-acylated chitosan was either isolated and purified or else 1.5–1.75 molar equivalents of a second acyl anhydride were added and the reaction continued for another period of time before isolating the products, many of which gelled during the second acylation step. A wide range of mixed N-acyl derivatives was prepared, for all of which complete N-acylation and no O-acylation was claimed. None of these second stage products were soluble in 10 vol.-% acetic acid which was the only solvent examined, although the first stage products, with DS <0.35, were all soluble.

Hirano's group have used this technique, or modifications of it, to produce a range of N-acylchitosans – N-formyl-, N-chloroacetyl-, N-glycyl-, N-(2-methylpropionyl)- and N-pentanoylchitosan – to study the effect of variation in the N-acyl group on the susceptibility to biodegradation by chitinase,[82] to prepare a slow-release aspirin carrier by N-acylating chitosan with 2-acetoxybenzoic anhydride,[83] and to prepare N-acyl derivatives of chitosan O-nitrate.[7] They have also studied the distribution of N-acetyl groups in partially N-acetylated chitosans, their results indicating that the groups are distributed uniformly along the chain.[84]

Another area that this group has looked at is the use of cyclic anhydrides for homogeneous N-acylation of chitosan[0.0].[85] A range of derivatives were prepared (**4.9**, Table 4.7) with the extent of N-acylation ranging from 43% obtained with diphenic anhydride (**4.9i**) to 80% with cis-tetrahydrophthalic andydride (**4.9g**). In all cases the remaining amine groups could be subsequently N-acetylated. The products were isolated as the methyl or ethyl ester, free acid or sodium salt. No crosslink or imide formation was observed (see Kurita et al.).[77] In another paper[69] the reaction between succinic anhydride and chitosan or glycol chitosan (hydroxyethyl chitosan) was studied in some detail. In all cases N-succinylation was incomplete, even with a molar ratio of 14.6:1 only 79% of the amine groups were acylated and there was no evidence of any O-acylation. The products from the reaction of chitosan, together with their solubilities, are given in Table 4.8. The N-succinyl derivatives of the glycol chitosans were soluble in all three solvents.

In the work of Hirano's group the N-acylation reactions have been allowed to go to completion and no attempt has been made by them to follow the course of partial N-acylations. A brief study of this aspect has been made by Aiba[64] who followed the degree of N-acetylation at 23°C in a 1:50:20 mixture of acetic acid:water:methanol containing 0.56 molar equivalents of acetic anhydride. This demonstrated that the reaction is quite fast, reaching completion in under 3 h. Other workers have used the technique of partial N-acetylation of chitosan in aqueous acetic acid–methanol to prepare water-soluble chitosan (section 2.5.7).

A water-soluble N-acylchitosan has been prepared by reaction of D-glucoheptonic acid-γ-lactone with chitosan in aqueous acetic acid–methanol.[86] A reaction time of 144 h was required and the product had a

TABLE 4.7 *N-acylchitosans produced by reaction with cyclic anhydrides under homogeneous conditions*[85]

4.9

	R		R
a	$-CH=CH-COOH$	f	
b	$-CH_2-CH(SAc)-COOH$	g	
c	$-\underset{\parallel}{\overset{}{C}}-CH_2COOH$ $\quad CH_2$	h	
d	$-CH_2CH_2CH_2COOH$		
e		i	

DS of ~0.75. Other workers[87] have reacted aldonic acid lactones of D-glucose, D-maltose and D-cellobiose with chitosan hydrochloride at 70°C using ethylene glycol containing triethylamine as the reaction medium. Surprisingly, in view of Yalpani and Hall's later work,[86] these products were stated to be insoluble in water.

A novel form of *N*-acylation, the formation of chitosan–polypeptide graft copolymers through reaction of chitosan with the *N*-carboxyanhydride (NCA) derivative of an amino acid, has been reported.[88, 89] Aqueous solutions of water-soluble chitosan[0.05] were treated at 0°C for 2 h with varying amounts of γ-methyl-L-glutamate *N*-carboxyanhydride in ethyl acetate and the resultant graft copolymers coagulated and washed with

TABLE 4.8   DS *values and solubilities of* N-succinylchitosans *and hydroxyethyl* N-succinylchitosans[69]

Substrate	Mole ratio	DS[a]	Solubility[b]		
	anhydride: amine group		Water	0.5M CH₃COOH	0.5M NaOH
Chitosan	0.24	0.12	−	+	−
	0.49	0.17	S	+	−
	0.74	0.29	±	+	S
	0.98	0.35	+	+	+
	3.67	0.63	+	±	+
	14.6	0.8	+	−	+
Hydroxyethyl chitosan	1.2	0.1	All soluble in all three solvents		
	2.9	0.22			
	5.9	0.53			
	8.8	0.65			
	14.7	0.77			

[a] Average result from saponification, free amine group determination and elemental analysis.
[b] −, insoluble; +, soluble; S, swollen.

aqueous DMF and acetone, and then dried (**4.10a**, Figure 4.12). Treatment of the copolymers with 1M Na₂CO₃ solution at 70°C for 1 h caused hydrolysis of either the *N*-acetyl groups on the chitosan backbone nor the peptide bonds. The resultant copolymers (**4.10b**, Figure 4.12) were purified and isolated by dialysis and freeze drying.

The average *DP* values of the polypeptide side chains were calculated from weight gain and IR spectroscopic results on the assumption that all the amine groups had undergone reaction. Very high grafting conversions were obtained (Table 4.9) and since no homopolypeptide was detected in any of the copolymerisation runs the grafting efficiency was thought to be almost 100%. The IR spectra of the unhydrolysed graft copolymers agree with the proposed structures and at *DP* values ≥ 10 a band at 610 cm⁻¹, due to the presence of the α-helix structure, becomes apparent. Evidence of such a structure is also obtained from X-ray diffraction studies where a peak at $2\theta = 8.3°$ is observed in the diffractogram of the *DP* = 10.4 sample and is more intense in that for a sample having *DP* = 20.

The initial products (**4.10a**) were found to be readily soluble in hexafluoro-2-propanol giving clear, viscous solutions from which tough, flexible, transparent films could be cast. They were also soluble in dichloroacetic acid but insoluble in DMF, DMAc, DMAc–LiCl, HMPA and DMSO. Some of these graft copolymers swelled in DMSO, the extent of swelling depending on the *DP* of the side chains and reaching a maximum at an average *DP* of 4. The decrease in swelling with further increase in side chain *DP* was ascribed to the formation of ordered structures by the polypeptide chains, including the adoption of an α-helix conformation.

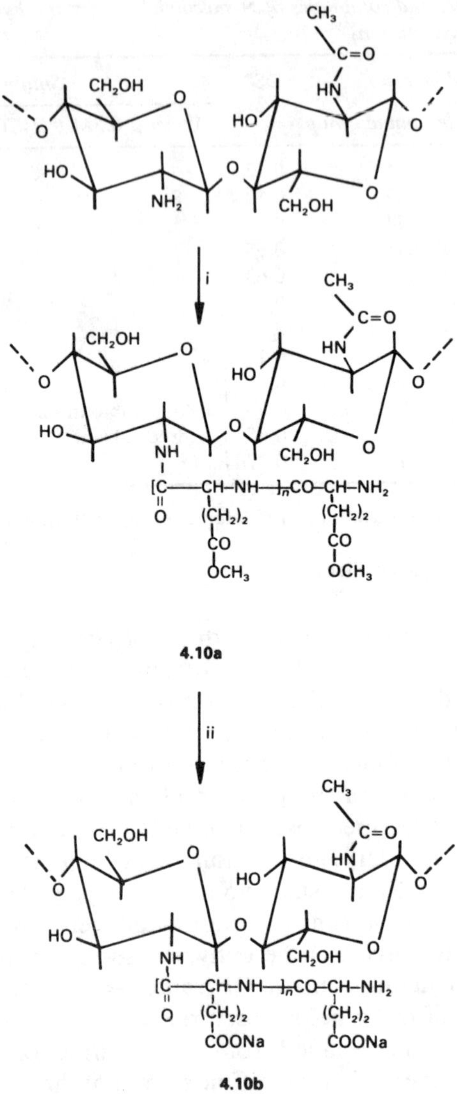

4.10a

4.10b

Reagents: i = γ-methyl L-glutamate NCA; ii = 1M Na$_2$CO$_3$, 70°C

FIGURE 4.12 Synthesis of chitosan[0.5]–polypeptide graft copolymer[88, 89]

After hydrolysis the copolymers (**4.10b**) were soluble in water, dichloro-acetic acid and *m*-cresol, but insoluble in dilute acetic acid.

Finally, both the heterogeneous swollen precipitate method and the homogeneous acetic acid–methanol method have been used[90] to produce *N*-nicotinylchitosans from water-soluble chitosan[0.5] and, in the case of the latter method, also from chitosan[0.1]. Following de-*O*-acylation with

TABLE 4.9   DP *values of polypeptide side chains in chitosan [0.5]–polypeptide graft copolymers*[89]

Mole ratio	DP of side chains		Conversion[a] (%)	
NCA:amine group	Weight gain	IR	Weight gain	IR
1.9	1.6	1.5	84	79
2.3	2.1	1.8	91	78
4.9	4.0	3.2	82	65
8.4	6.2	6.0	73	71
9.9	8.8	7.2	89	73
13.6	10.4	10.9	76	80
16.1	13.1	12.3	81	76
26.6	22.0	20.8	83	78

[a] Amount NCA polymerised/Amount of NCA added.

aqueous Na_2CO_3 the *N*-nicotinylchitosans (**4.11**) were quaternised with benzyl choloride and the resultant benzylpyridinium groups (**4.12**) reduced with alkaline sodium hydrosulphite to give products containing a number of 1,4-dihydronicotinamide groups (**4.13**) along the chain (Figure 4.13). The products were examined for their efficacy in the asymmetric reduction of ethyl benzoylformate to ethyl mandelate

$$C_6H_5-\overset{\overset{O}{\|}}{C}-\overset{\overset{O}{\|}}{C}-O-C_2H_5 \quad \xrightarrow{\textbf{4.13}/Mg(ClO_4)_2} \quad C_6H_5-\overset{*}{C}HOH-\overset{\overset{O}{\|}}{C}-O-C_2H_5$$

the reaction being carried out in acetonitrile in the presence of $Mg(ClO_4)_2$. The chitosan derivatives could be recovered easily by filtration at the end of the reaction, and regenerated by treatment with sodium hydrosulphite without any apparent loss in effectiveness.

The yield of ethyl mandelate was 6.5–8% when the ratio of moles of 1,4-dihydronicotinamide to moles of ethyl benzoylformate was 1.5, but increased to ~14% when this ratio was increased. The low yields may be due to the limited accessibility of the 1,4-dihydronicotinamide groups, acetonitrile having very limited swelling power for the polymers.

The optical purity of the ethyl mandelate produced by either of the derivatives from water-soluble chitosan[0.5] was much higher than of that produced by the 1,4-dihydronicotinimide derivative from chitosan[0.1], and also much higher than the values attained previously for NADH models immobilised on other polymers, the highest reported value[91] being 7.3%. The high asymmetric selectivity of the water-soluble derivatives led the authors to conclude that the *N*-acetyl groups on pyranose rings adjacent to the 1,4-dihydronicotinamide groups are involved in forming a chiral environment to accommodate the substrate. Details of the products and their efficiencies are given in Table 4.10.

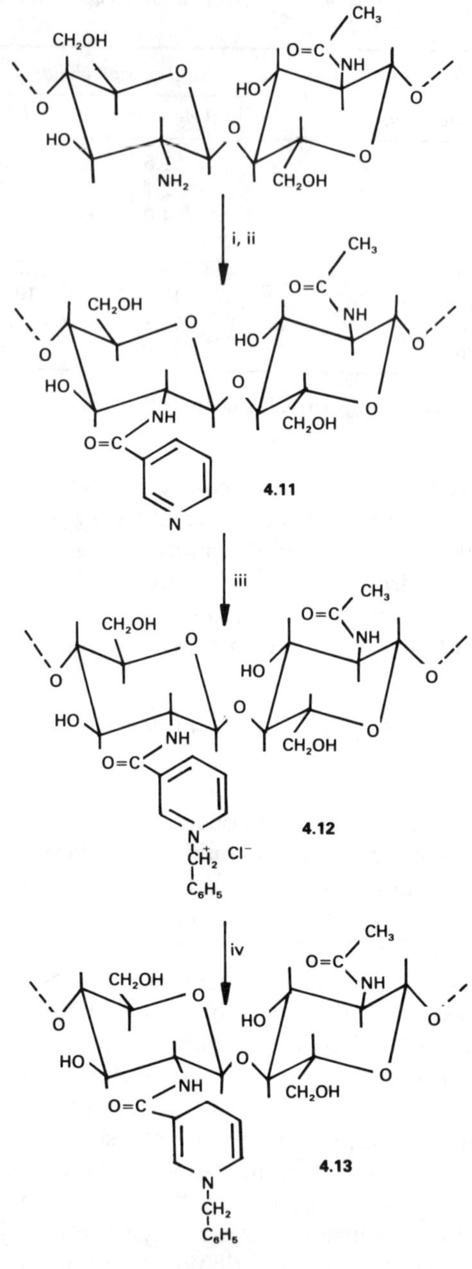

Reagents.

i = Nicotinic anhydride–
    pyridine
ii = 5 wt-% Na_2CO_3
iii = $C_6H_5CH_2Cl$–DMF
iv = $Na_2S_2O_6$–K_2CO_3

FIGURE 4.13   Preparation of N-(1,4-dihydronicotinyl) chitosan[90]

TABLE 4.10 *Asymmetric reduction of ethyl benzoylformate with N-(1, 4-dihydronicotinyl) chitosans*[90]

Chitosan derivative[a]	DS[b]	Mole ratio Reducing groups/Formate	Yield of ethyl mandalate (%)	Optical purity (%)
4.13a	0.17	1.5	6.5	33.3
4.13a	0.17	3.3	14.0	—
4.13b	0.14	1.3	8.0	21.6
4.13c	0.27	1.5	7.0	3.6
4.13c	0.27	6	14.2	—

[a] Derivative **4.13a** prepared from chitosan[0.5]; N-nicotinyl chitosan prepared in pyridine using swollen precipitate method.
Derivative **4.13b** prepared from chitosan[0.5]; N-nicotinyl chitosan prepared in solution in aqueous acetic acid–methanol.
Derivative **4.13c** prepared from chitosan[0.1]; N-nicotinyl chitosan prepared in solution in aqueous acetic acid–methanol.
[b] DS of 1,4-dihydronicotinamide groups as determined by oxidation–reduction titration under heterogeneous conditions.

4.2.2 O-Acyl derivatives of chitin and its N-acyl analogues

Although many of the techniques used for N-acylation of chitosan (section 4.2.1) also produce partial O-acylation, this is incidental to the main purpose of N-acylation and indeed is considered to be a disadvantage. Therefore only those reactions whose main objective is the O-acylation of an already N-acylated material will be considered in this section and the degree of substitution (DS) will refer to the extent of reaction of the hydroxyl groups alone. Hence a fully O-acetylated derivative will have a DS of 2.0.

O-*Acylation using carboxylic acids*

Esterification by direct reaction between a carboxylic acid and chitin has only been reported in a few instances. Kaifu et al.[92] treated chitin with methanesulphonic acid-formic acid at 0°C for 2 h, then overnight at −20°C, and isolated the product by precipitation on ice, washing, neutralisation, washing and drying. The DS increased with increase in the concentration of HCOOH in the reaction mixture (Table 4.11). Replacement of the methanesulphonic acid by concentrated H_2SO_4, and using a mole ratio of 34 for HCOOH:N-acetyl-D-glucosamine residues, gave a product having $DS = 1.1$, compared with 1.4 obtained when using the methanesulphonic acid. The O-formylchitin was soluble in acidic solvents such as formic acid and dichloroacetic acid. The analogous reactions with propionic acid or butyric acid and methanesulphonic acid gave O-propionylchitin ($DS = 1.0$)

TABLE 4.11 *Preparation of O-formylchitins in methanesulphonic acid–formic acid*[92]

Mole ratio[a]	0.5	1.0	5.0	5.3	18.0
DS	0.4	0.6	1.0	1.3	1.4

[a] Mole ratio of formic acid:N-acetyl-D-glucosamine residues.

and *O*-butyrylchitin ($DS = 0.3$). Gagnaire *et al.*[93] found that the dissolution of chitin in HCOOH occurs by reaction between chitin and the acid, forming *O*-formylchitin. The reaction is slow and the product had $DS = 1.45$, which could be increased to 1.8 by addition of $HClO_4$ as catalyst.

Heterogeneous O-acylation using acyl anhydrides

Acyl anhydrides are again the most common acylating agents used but in *O*-acylation reactions the addition of a catalyst is required. The catalyst may be either a strong acid or a tertiary amine and the reaction may be carried out heterogeneously or homogeneously. Schorigin and Hait[3] were the first to examine heterogeneous *O*-acetylation of chitin, the method used being to bubble $HCl_{(g)}$ into a suspension of chitin in acetic anhydride. A completely *O*-acetylated product was obtained after 120 h at 23°C and products having $DS > 1.2$ were found to be readily soluble in formic acid. A plot of the tabulated results is given in Figure 4.14, from which it appears that increasing the temperature of reaction has little effect on the rate of *O*-acetylation. Tokura and co-workers[56, 63] repeated this work using acetic anhydride saturated with HCl at 0°C. After 240 h a product having $DS = 1.6$, and soluble in both 99 vol.-% and 85 vol.-% formic acid, was obtained.

The use of other mineral acids as catalysts has been examined. Schorigin and Hait[3] reported that the acetic anhydride-acetic acid–H_2SO_4 system, used for the production of cellulose triacetate, was unsuccessful with chitin, giving a product having a $DS < 0.75$. Subsequent workers have investigated the use of $HClO_4$ which is an excellent catalyst for the preparation of cellulose triacetate. Nishi and co-workers[56, 63] obtained di-*O*-acetylchitin ($DS \sim 2.0$) by treating chitin for 3 h at 0°C with acetic anhydride containing glacial acetic acid and $HClO_4$. An anhydrous catalyst system was prepared by adding the concentrated $HClO_4$ to acetic anhydride and leaving the mixture overnight at 0°C. They also prepared partially *O*-acetylated chitins by varying the amount of acetic anhydride used and found[56] that a DS of 2.0 was obtained using a mole ratio of 3.7 for acetic anhydride to N-acetyl-D-glucosamine residues (Table 4.12). Despite these results their reported method for preparing di-*O*-acetylchitin uses a mole ratio of either 20[63] or 25,[56] considerably in excess of what would appear to be required. The same group prepared *O*-acetylchitin fibres, $DS = 1.0$, by treatment of chitin fibres in an acetic anhydride–$HClO_4$

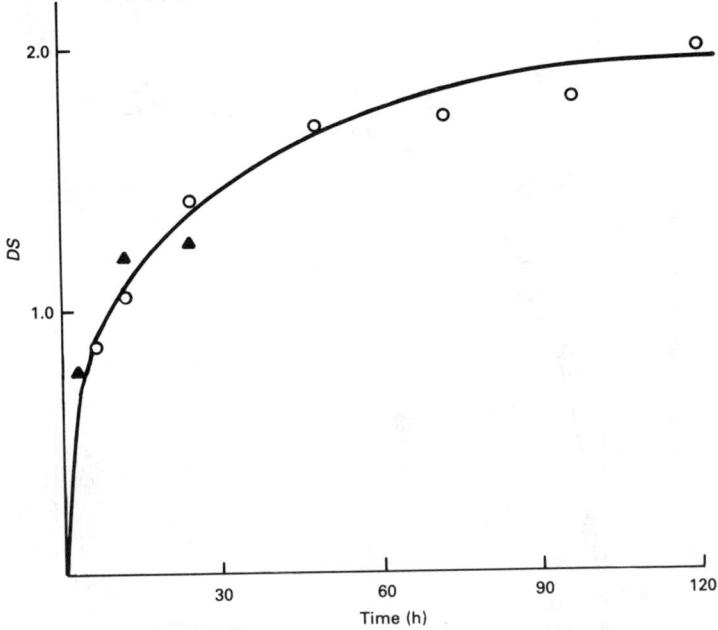

FIGURE 4.14   Rate of O-acetylation of chitin in acetic anhydride–HCl$_{(g)}^3$: ○, 23°C;
▲, 40°C

TABLE 4.12   *Preparation of O-acetylchitins in acetic anhydride–acetic acid–HClO$_4$[56]*

Mole ratio[a]	0.8	1.6	2.6	3.7
DS	0.4	0.6	1.2	2.0

[a] Mole ratio of acetic anhydride:N-acetyl-D-glucosamine residues.

mixture,[94] while Gagnaire et al.[93] obtained O-acetylchitin (DS = 1.8) by
reacting chitin with acetic anhydride–acetic acid–HClO$_4$.

Moore and Roberts[95] prepared several di-O-acetyl-N-acylchitosan products by reaction of N-butyryl-, N-hexanoyl- and N-decanoylchitosan with
acetic anhydride–glacial acetic acid–HClO$_4$, the acetylation being carried
out at room temperature for 72 h. The products were soluble in DMSO and
85 vol.-% HCOOH and swelled in DMF. Their IR spectra showed the
expected bands at 1735 cm^{-1} and 1230 cm^{-1} (C=O and C–O bands of the
ester groups respectively) as well as the amide I and II bands at 1660 and
1550 cm^{-1}. They also showed either a very weak band at 3450 cm^{-1} or a
shoulder on the high-frequency side of the amide N–H band at 3250 cm^{-1},
suggesting the presence of only a very small number of unreacted hydroxyl

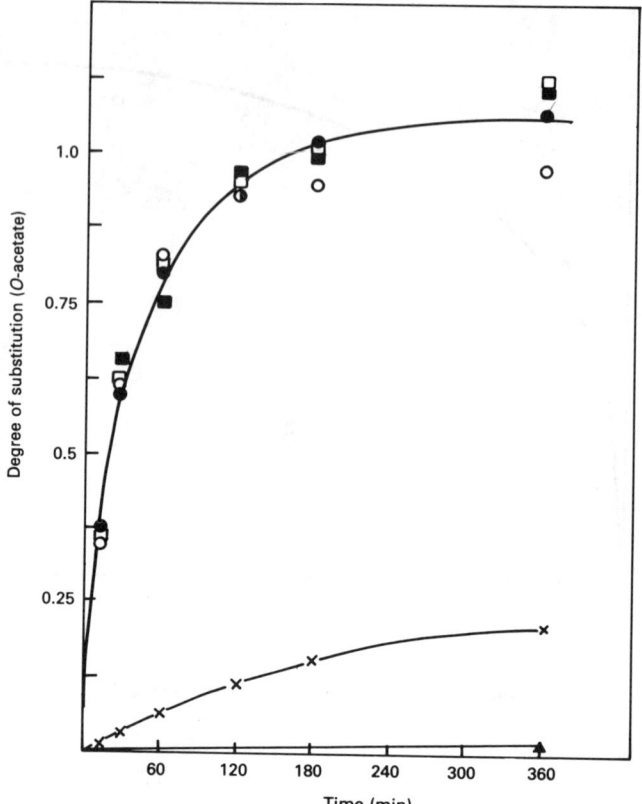

FIGURE 4.15   The influence of chain length on the rate of O-acetylation of
N-acylchitosans[75]: ▲, N-acetylchitosan; X, N-propionylchitosan;
○, N-butyrylchitosan; ●, N-hexanoylchitosan; □, N-octanoylchitosan;
■, N-decanoylchitosan

groups. Extending the reaction time brought little improvement in the
extent of O-acetylation. A di-O-propionyl-N-hexanoylchitosan, which had
similar solubility properties and IR spectrum, was prepared by the same
method.

The influence of the nature of the acyl group in N-acylchitosans on the
ease of O-acetylation in acetic anhydride–pyridine has been studied in
some detail.[74, 75] With linear aliphatic N-acyl derivatives there is a great
difference in reactivity between N-acetyl- and N-propionylchitosan on the
one hand and the other linear aliphatic N-acylchitosans (Figure 4.15) but
the rates for these latter derivatives rapidly decrease and have almost
levelled off to zero after 6 h, although there is a slight increase in the extent
of esterification when measured after 30 h. A similar abrupt increase in the

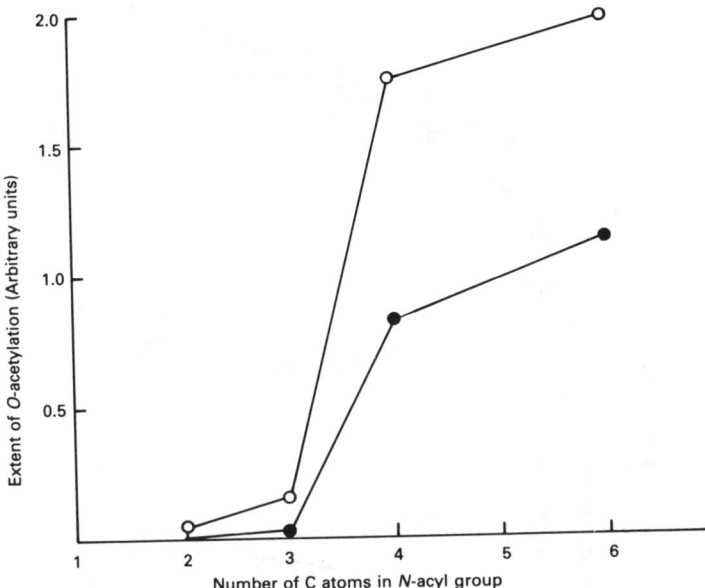

FIGURE 4.16   Extent of O-acetylation of N-acylchitosans as a function of the number of C
atoms in the N-acyl group, O-acetylation being carried out for 1 h at 25°C
in: ●, acetic anhydride–pyridine; ○, acetic anhydride–glacial acetic acid–
perchloric acid[74]

rate of O-acetylation occurs with N-butyrylchitosan and above in acetic
anhydride–acetic acid–HClO$_4$ system[74] (Figure 4.16). The rate curves for a
series of aromatic N-acylchitosans (N-benzoyl-, N-o-toluoyl-N-m-toluoyl-
and N-p-toluoylchitosan) were all very similar to the curve for the higher
linear aliphatic N-acylchitosans.

The effect of chain branching in aliphatic N-acylchitosans was studied
with two series: N-propionyl-, N-(2-methylpropionyl)-, N-(2,2-dimethyl-
propionyl)chitosan and N-butyryl-, N-(3-methylbutyryl)-, N-(3,3-dimethyl-
butyryl)chitosan. The results are plotted in Figure 4.17 and it can be seen
that in the first series the introduction of one CH$_3$– side chain increases the
rate of O-acetylation dramatically, bringing it close to that for the linear
N-acyl group having the same number of C atoms. However introduction
of a second CH$_3$– side chain reduced the rate of O-acetylation to a lower
value than that for the linear N-propionylchitosan. In the second series the
introduction of one CH$_3$– side chain has no effect on the rate, both
N-butyryl- and N-(3-methylbutyryl)chitosan reacting at the same rate,
but again introduction of a second CH$_3$– side chain reduces the rate of
O-acetylation to that shown by N-acetylchitosan. In all three types of
N-acyl groups – linear aliphatic, aromatic, branched aliphatic – the

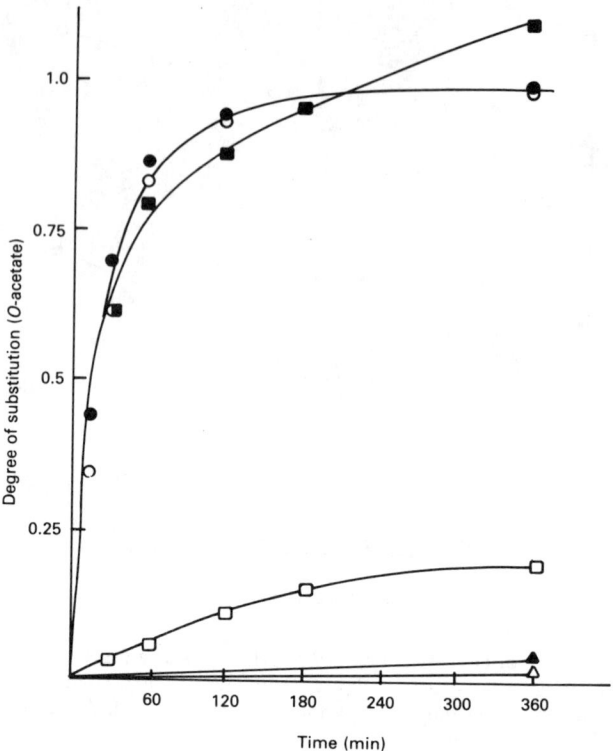

FIGURE 4.17   The influence of chain branching on the rate of O-acetylation of
N-acylchitosans:[75] ○, N-butyrylchitosan; ●, N-(3-methylbutyryl) chitosan;
△, N-(3, 3-dimethylbutyryl) chitosan; ■, N-(2-methylpropionyl) chitosan;
▲, N-(2, 2-dimethylpropionyl) chitosan, □, N-propionylchitosan

maximum extent of O-acetylation that could be obtained in reaction times
of up to 72 h was ~50% (DS ~1.0) and extending the reaction time to 1500
h with N-(2-methylpropionyl)chitosan only increased the DS by less than
0.1. However the extent of reaction could be increased to DS >1.8 by a
second, more vigorous O-acetylation treatment with acetic anhydride–
glacial acetic acid–$HClO_4$.

The lack of reactivity of α-chitin can at least in part be attributed to the
close packing of the polymer chains, which limits the accessibility of the
hydroxyl groups. Furthermore there is a highly developed network of both
intra- and interchain hydrogen bonds. Increasing the length of the N-acyl
group will disrupt this close packing and although it must be assumed that
N-propionyl groups can be accommodated without such disruption, introduction of N-butyryl- or larger N-acyl groups must increase the chain
separation sufficiently to prevent formation of the interchain C(6′)OH . . .

O(6) hydrogen bonds (Figure 1.11, section 1.4.2). In this respect it is of interest that the X-ray diffraction data of as-cast chitosan salt films, prepared from solutions of chitosan in various dilute aqueous organic acids, show an abrupt increase in the unit cell dimensions along the c axis on going from chitosan acetate and propionate to chitosan butyrate.[96] However, even with the increased accessibility due to the change in interchain spacing, only about 50% of the hydroxyl groups are available for acetylation under the reaction conditions used, the most likely explanation being the presence of the C(3')OH . . . O(5) bond which is only disrupted under more vigorous acetylation conditions. The almost total loss of reactivity on introduction of a second CH_3- side group in both the N-propionyl- and N-butyryl-series (Figure 4.17) may be attributed to intrachain steric hindrance of the C(3)OH and C(6')OH groups by the $-C(CH_3)_3$ group of the N-acyl group.

Homogeneous O-acylation reactions in methanesulphonic acid

A technique for preparing O-acylchitin derivatives using methanesulphonic acid as both solvent and acid catalyst has been developed by Tokura and co-workers. In their first reports[56, 63] chitin was O-acetylated to different extents by reaction for 20–24 h at 0°C with methanesulphonic acid–glacial acetic acid containing varying amounts of acetic anhydride. The relationship between the amount of acetic anhydride and the *DS* of the product is given in Table 4.13. Di-O-acetylchitin was prepared similarly using a methanesulphonic acid–acetic anhydride mixture (the glacial acetic acid being omitted) with a mole ratio of acetic anhydride:N-acetyl-D-glucosamine residue of 12. The preparation of di-O-acetylchitin was mainly a homogeneous reaction, a clear solution being obtained in the first 1–2 h provided at least 4 cm³ methanesulphonic acid were used per 1 g of chitin. In the preparation of the partially O-acetylated chitin series a clear, homogeneous reaction mixture was not obtained.

Elemental analysis of the starting chitin was in agreement with a formula containing 0.5 H_2O per N-acetyl-D-glucosamine residue, and thermal analysis showed that this bound water is released at temperatures >115°C. Di-O-acetylchitin was found to contain no bound H_2O and the elemental analysis figures of the partially O-acetylated chitins were in agreement with formulae based on the assumption that N-acetyl-D-

TABLE 4.13   *Preparation of O-acetylchitins in acetic anhydride–acetic acid–methanesulphonic acid*[56]

Mole ratio[a]	1.0	1.6	2.6	3.0	3.5
DS	1.0	1.1	1.4	1.6	1.7

[a] Mole ratio of acetic anhydride:N-acetyl-D-glucosamine residues.

TABLE 4.14   *Properties of O-acetylchitin fibres*[63]

		DS *of fibre-forming polymer*			
		1.1[a]	*1.6*[a]	*2.0*[a]	*2.0*[b]
Tenacity (g den^{-1})	(dry)	1.89	1.52	1.13	1.48
	(wet)	0.75	0.58	0.50	0.56
Elongation (%)	(dry)	11.7	5.3	7.5	7.0
	(wet)	22.7	12.9	14.1	15.4
Knot strength (g den^{-1})		0.64	0.39	0.14	0.31
Young's modulus (g den^{-1})		55.67	75.17	46.69	63.96
Moisture regain (%)		11.3	9.3	8.5	9.5

[a] *O*-acetylated in acetic anhydride–methanesulphonic acid.
[b] *O*-acetylated in acetic anhydride–acetic acid–HClO$_4$.

glucosamine and mono-*O*-acetyl-*N*-acetyl-D-glucosamine residues each bind 0.5 H$_2$O but di-*O*-acetyl-*N*-acetyl-D-glucosamine residues do not bind water.

The *O*-acetylchitins were soluble in methanesulphonic acid, dichloro-acetic acid and formic acid, the solubility in formic acid increasing with increase in *DS* as previously observed by Schorigin and Hait.[3] Fibres spun from the *O*-acetylchitins showed[63] that the properties tended to decrease with increase in the *DS*, while the values for fibres from di-*O*-acetylchitin prepared using acetic anhydride–acetic acid–HClO$_4$ were, in general, better than those for fibres from di-*O*-acetylchitin prepared with acetic anhydride–methanesulphonic acid (Table 4.14).

The use of the methanesulphonic acid–acyl anhydride system was ex-tended to the preparation of *O*-propionyl- and *O*-butyrylchitin.[92] The procedure was modified insofar as after reaction for 2 h at 0°C the resultant gel was held at −20°C overnight, and in most of the reaction mixtures a mixture of the anhydride and the corresponding acid was used with methanesulphonic acid, the carboxylic acid being added to keep the total volume of anhydride plus acid in the reaction mixture constant. The relationships between *DS* and the composition of the reaction media are given in Tables 4.15(a) and 4.15(b). The *O*-propionyl- and *O*-butyryl-chitins were soluble in dichloroacetic acid and formic acid, while *O*-butyryl-chitin having *DS* = 1.8 is soluble in dioxane, THF, acetone, methanol, acetonitrile, DMF and acetic acid.

The viscosity of a solution of chitin in methanesulphonic acid was monitored as a function of time at different temperatures. Although the viscosity number decreased by approximately 45% over 6 h at 40°C and 30% at 25°C, the corresponding decrease at 0°C, which is the highest temperature used in the process, was less than 10%. This indicates that the chitin chains undergo only limited hydrolysis under the reaction conditions.

TABLE 4.15 *Preparation of (a) O-propionylchitins and (b) O-butyrylchitins in acyl anhydride–carboxylic acid–methanesulphonic acid[92]*

(a) Propionic anhydride conc.[c] (vol.-%)	5	10	20	40	60	100
Mole ratio[d]	0.5	1.0	2.0	4.0	6.0	10
DS	1.0	1.0	1.0	1.5	1.6	1.9
(b) Butyric anhydride conc.[c] (vol.-%)	12.8	25.6	38.3	51.1	100	
Mole ratio[d]	1.0	2.0	3.0	4.0	8	
DS	0.4	1.0	1.5	1.6	1.8	

[c] Concentration (vol.-%) of acyl anhydride in carboxylic acid.
[d] Mole ratio of acyl anhydride:N-acetyl-D-glucosamine residues.

TABLE 4.16 *Preparation of O-benzoylchitins in benzoyl chloride–methanesulphonic acid[97]*

Mole ratio[a]	2	4	8	10	20
DS	0.3	0.6	1.2	1.7	1.6

[a] Mole ratio of benzoyl chloride:N-acetyl-D-glucosamine residues.

The preparation of aromatic O-acylchitin derivatives using methanesulphonic acid has also been investigated.[97] The O-benzoylation of chitin in this medium could not be achieved using benzoic anhydride since it decomposed in the solution, but could be prepared using benzoyl chloride (Table 4.16). In all examples given in Table 4.16 the reaction mixture was stirred at 0°C for 3 h then kept overnight at −20°C. Extending the initial time of stirring from 3 h to 4 h, and using a mole ratio of 10, increased the DS of the product slightly from 1.7 to 1.8, while reducing the stirring time to 1 h reduced the DS to 0.5.

The solubility of O-benzoylchitin of $DS = 1.8$ is good, being very soluble in formic acid, dichloroacetic acid, benzyl alcohol and DMSO, and fairly soluble in DMF, DMAc, HMPA and some mixed solvents such as tetrachloroethane or 1,2-dichloroethane with ethanol or 1-butanol. The mono-O-benzoylchitin contained bound water, 0.5 H_2O per anhydrosugar residue, but the product having $DS = 1.8$ did not contain any bound water. This is similar to the O-acetylchitins prepared by the same group.[56]

The process has been extended[98] to the preparation of substituted benzoyl esters of chitin; O-(p-chlorobenzoyl)-, O-(methylbenzoyl)-, O-(p-t-butylbenzoyl)- and O-(p-methoxybenzoyl)chitin. The maximum substitution levels obtained were $DS = 1.6$ for O-(p-methoxybenzoyl)chitin, $DS = 1.8$ for O-(p-chlorobenzoyl)chitin and O-(p-methylbenzoyl)chitin, and $DS = 2.0$ for O-(p-t-butylbenzoyl)chitin. These maximum values were obtained using a mole ratio of 6 for reagent:N-acetyl-D-glucosamine residue compared with a mole ratio of 10 for preparation of the most substituted *O*-benzoylchitin.[97]

The *O*-(*p*-chlorobenzoyl)chitin was less soluble than *O*-benzoylchitin, being soluble only in formic acid and dichloroacetic acid, while the *O*-(*p*-methylbenzoyl)chitin was soluble in methanol, ethanol, 1-butanol, benzyl alcohol, methylene chloride, chloroform, DMF, DMAc and THF, in addition to formic acid and dichloroacetic acids. The general order of increasing solubility for the series is

$$\text{chitin} < p\text{-Cl} < \text{benzoyl} < p\text{-CH}_3\text{O} < p\text{-C(CH}_3)_3 < p\text{-CH}_3$$

The use of methanesulphonic acid acyl chloride mixtures has also been used to prepare longer chain aliphatic *O*-acylchitins such as *O*-hexanoyl-, *O*-decanoyl- and *O*-dodecanoylchitin.[99] High levels of substitution were obtained, the products having *DS* values of 2.0, 1.8 and 1.9 respectively.

O-*Acylation reactions in other media*

Another acidic reaction system is the trichloroacetic acid–1,2-dichloro-ethane-acyl anhydride system used by Ando and Kataoko[100] to prepare *O*-propionyl-, *O*-butyryl-, *O*-hexanoyl- and *O*-decanoylchitin. However, although this is perhaps a simpler system than the methanesulphonic acid–acyl anhydride one, fully *O*-acylated derivatives could not be prepared.

Chitin can also be *O*-acylated under strongly alkaline conditions although the *DS* levels obtained are low. Alkali chitin, prepared using 40 wt-% NaOH and steeping temperatures of 0–5°C, was frozen at −20°C overnight in an evacuated container then acetic anhydride was introduced, a mole ratio of 4 based on the original amount of chitin in the alkali chitin being used. The mixture was shaken while cooled in ice then kept at room temperature for 24 h, with occasional shaking,[56, 63] to give a product having a *DS* = 0.3. A sample of *O*-propionylchitin, *DS* = 0.4, has also been prepared by the same general procedure.[92]

Chitin viscose, prepared by xanthation of alkali chitin, has been *O*-acety-lated by treatment with acetic anhydride at room temperature using a mole ratio of 5 for acetic anhydride:*N*-acetyl-D-glucosamine residue.[56] The product, which had a *DS* = 0.3, could be further *O*-acetylated by heating with anhydrous sodium acetate and acetic anhydride (mole ratio of 8.5) at 96°C for 24 h under reduced pressure, to give a product having a *DS* = 1.1. Although higher *DS* values may be obtained by using longer times or higher temperatures, decomposition and discolouration of the *O*-acetyl-chitin occurs.

The preparation of *O*-acylchitin derivatives by the reaction of acyl chlorides and anhydrides with chitin dissolved in DMAc–LiCl has been reported.[42] The reaction conditions and *DS* values are given in Table 4.17, pyridine being added to the reaction mixtures to act as an acid acceptor.

TABLE 4.17   *Preparation of O-acylchitins in DMAc–LiCl solutions[42]*

Acyl group	Reaction conditions	Mole ratio[a]	DS
CH$_3$CO-	2 h at 110°C	27	~2
C$_6$H$_5$CO-	3 h at 80°C	24	~1
ClCH$_2$-(CH$_2$)$_2$-CO-	3 h at 80°C	25	~2

[a] Mole ratio of reagent:N-acetyl-D-glucosamine residues.

Properties of O-acylchitins

The biocompatibility of O-acylchitins, and hence their potential for use as materials for blood contacting surfaces, has been investigated by measuring, *inter alia*, their critical surface tensions, clotting times, and plasma protein absorption.[101] The results (Table 4.18) indicate that of those O-acylchitins examined the di-O-acetylchitin shows most promise.

TABLE 4.18   *Properties of relevance to biomedical applications for films of chitin and O-acylchitins[101]*

O-acyl group	DS	γ_c (dyn cm^{-2})	Clotting time ratio		Albumin adsorption (μg cm^{-2})
			A^a	B^b	
(Chitin)–	0	29	1.3	0.7	0.02
Formyl	1.4	27	1.6	1.0	0.01
Acetyl	2.0	27	3.8	1.9	0.02
Propionyl	1.9	24	2.4	0.9	—
Butyryl	1.8	23	2.0	1.2	0.06
Hexanoyl	2.0	22	2.5	1.1	0.05
Decanoyl	1.8	21	2.5	1.1	0.06
Dodecanoyl	1.9	21	2.5	1.1	0.11
Benzoyl	1.9	27	2.6	1.0	0.04

[a] Sample: glass.
[b] Sample: siliconised glass.

The accessibilities of chitin, mono-O-acetylchitin, and di-O-acetylchitin to lysozyme, as determined by the weight loss as a function of time, have been found[15] to increase in the order:

chitin < mono-O-acetylchitin < di-O-acetylchitin

Finally, the molecular motion and dielectric relaxation behaviour of chitin and four O-acylchitins (O-acetyl-, O-butyryl-, O-hexanoyl- and O-decanoylchitin) have been studied.[102] Chitin and O-acetylchitin show only one peak in the plot of the temperature dependence of the loss permittivity, whereas those derivatives having longer O-acyl groups show

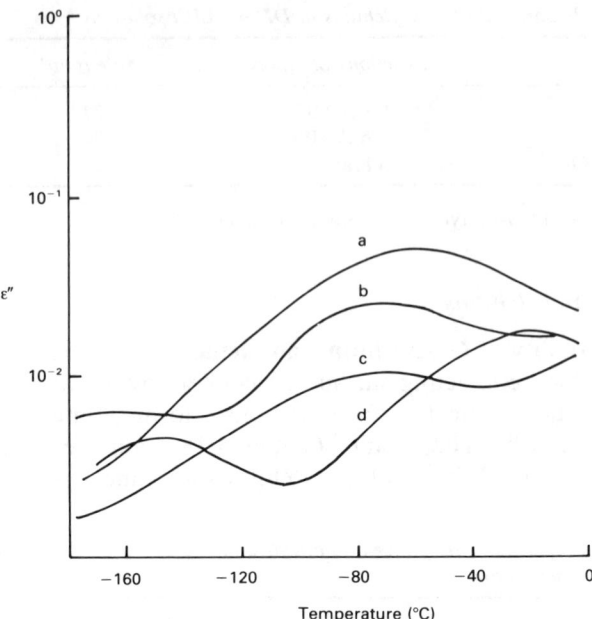

FIGURE 4.18   Loss permittivity, at 1 kHz, as a function of temperature for chitin and
O-acylchitins:[102] a, chitin; b, O-butyrylchitin; c, O-acetylchitin;
d, O-decanoylchitin. (The curve for O-hexanoylchitin has been omitted
for clarity)

two peaks (Figure 4.18). The peak corresponding to the higher temperature relaxation process, β, moves to higher temperatures with increase in
the size of the O-acyl group, excluding chitin itself, as does that for the
lower temperature γ process in the three derivatives which exhibit this
second relaxation process. The presence of two relaxation processes in
these three samples, but only one in chitin and O-acetylchitin, was confirmed by measurements of the temperature dependence of the second
moment of the NMR spectrum. The authors suggest that a comparison
with the side chain relaxations in poly(n-alkyl methacrylate)s[103, 104] indicates that the β relaxation is due to the motion of the O-acyl groups as a
whole, and hence is observed in all the samples, while the γ relaxation is
due to local motion of the O-acyl group and hence only arises with the
longer members of the series.

4.2.3   O-Acylation of chitosan and chitosan derivatives

As amine groups are normally more reactive than hydroxyl groups it is
necessary to protect the amine group in some way during the acylation step

if O-acylchitosans are to be produced. One method of protecting the amine group is Schiff's base formation, followed by O-acetylation using acetic anhydride–pyridine to prevent acid hydrolysis of the Schiff's base.[95, 105] Using a chitosan[0.24] sample it was found[105] that with Schiff's bases formed with aliphatic aldehydes O-acetylation proceeded rapidly, provided that the aldehyde contained 3 or more C atoms, but that complete O-acetylation was not achieved owing to hydrolysis of the Schiff's base and concomitant N-acetylation. However Schiff's base derivatives formed with aromatic aldehydes show little tendency to hydrolyse under the conditions used in the O-acetylation step and di-O-acetyl-N-arylidenechitosan is readily obtained. Salicylaldehyde is particularly efficacious in this reaction. The Schiff's base may then be hydrolysed under mild conditions to give di-O-acetylchitosan which is readily N-acetylated to produce di-O-acetylchitin.[95]

Recently the direct preparation of O-acetylchitosan has been attempted by carrying out the acetylation on chitosan dissolved in 90 wt-% HCOOH containing acetic anhydride and $HClO_4$, the assumption being that protonation of the amine group would prevent N-acetylation taking place.[106] ^{13}C NMR spectroscopy indicated that some N-formyl groups had been introduced as well as N-acetyl and O-acetyl groups. Fibres and films were produced from the acetylated material, which could be converted back to chitosan by treatment with sodium methoxide–methanol. Analysis of the product after this de-O-acetylation step showed that very little modification of the amine group had taken place, the initial chitosan[0.47] having been converted to chitosan[0.54].

Acetic anhydride–pyridine has been used[107] to esterify hydroxypropyl chitosan to give acetoxypropyl chitosan.

4.2.4   Simultaneous N- and O-acylation of chitosan

This can be achieved by use of acyl chlorides. Fujii et al.[108] treated chitosan under reflux with a large excess of dodecanoyl chloride in a mixture of dry pyridine and chloroform, additional acid chloride being added after 5 h. The product obtained after a reaction time of 9 h was soluble in chloroform, benzene, diethyl ether and pyridine. NMR spectroscopy showed it to have a DS of 4.0, suggesting the formation of –N(COR)$_2$ groups. This was confirmed by the IR spectrum which showed the absence of an amide II band at $\sim$1540 cm^{-1} and of an amide I band at $\sim$1660 cm^{-1}, but the presence of a new absorption band at 1710 cm^{-1}, which was presumed to be due to the –N(COR)$_2$, in addition to the ester band at 1745 cm^{-1}. However if the reaction was stopped after 1 h the product obtained was insoluble in chloroform, while the IR spectrum and elemental analysis showed it to be the expected tri-substituted di-O-dodecanoyl-N-dodecanoylchitosan. Analogous tetrasubstituted products, soluble in chloroform, were obtained on refluxing chitosan for 9 h in pyridine–chloroform containing hexanoyl chloride or tetradecanoyl chloride.

The reaction with dodecanoyl chloride was subsequently studied by Grant et al.[109] using substrates covering the range chitosan[0.42]–chitosan[0.14]. All products showed an IR absorption band at 1710 cm^{-1}, but elemental analysis indicated a progressive increase in DS with increase in the extent of deacetylation of the starting chitosan. Indeed the IR spectrum of the product from chitosan[0.42] showed significant peaks at 1670 cm^{-1} (amide I) and at 3450 cm^{-1}. The authors suggested that this trend of increasing reactivity is due to increasing disruption of the crystal structure, with accompanying increasing accessibility of the substrate, as the extent of deacetylation goes from 58% to 86%. They also noted that the procedure was improved if after 5 h the polymer is isolated, washed, dried and the reaction restarted with fresh solvent and acyl chloride, rather than merely adding fresh acyl chloride to the existing reaction mixture.

The same authors have butyrylated chitosan[0.14] dissolved in methane-sulphonic acid, using the method of Kaifu et al.[92] The products were not analysed, but those prepared using mole ratios of anhydride:anhydro-D-glucosamine residue of 7–12 were soluble in methanol and water.[110] The aqueous solubility is surprising but from the experimental details it would appear that no post-precipitation neutralisation treatment was given to the products, so that any unacylated amine group would be present as the methanesulphonic acid salt, and this could explain the aqueous solubility.

Berkovich et al. examined the reaction between maleic anhydride and chitosan.[111] From the range of solvents examined – dilute acid, DMAc, DMF, HMPA and formamide – only the latter worked. The reaction, which was carried out at 55°C for 3 h, was initially heterogeneous but as the reaction proceeded it became homogeneous. The product was found to be the fully substituted di-O-acyl-N-acylchitosan. After dissolving in water containing sufficient KOH to exactly neutralise the carboxylic acid groups, acrylamide and $(NH_4)_2S_2O_8$ were added and the mixture heated to induce polymerisation. A transparent, elastic gel formed which was shown to consist of a crosslinked graft copolymer with the polyacrylamide chains incorporating the C=C groups of the unsaturated N-acyl groups. The authors concluded that the maleic acid residues do not interact with each other since prolonged heating of the chitosan derivative with $(NH_4)_2S_2O_8$ in the absence of acrylamide failed to produce a gel.

4.2.5 Arylcarbamate and arylureido derivatives

Both of these types of derivative were first prepared by Moore[112] but prior to publication[95] McCormick et al.[113] reported on the preparation of arylcar-bamate derivatives of chitin by reaction of a range of isocyanates – phenyl isocyanate, p-tolyl isocyanate, and p-chlorophenyl isocyanate – with chitin dissolved in DMAc–LiCl (50 g dm^{-3}). Despite the homogeneous conditions complete substitution was not achieved, *DS* values of 1.74 and 1.5 being attained with phenyl isocyanate and *p*-tolyl isocyanate respectively.

A more detailed study was carried out by Moore and Roberts[95] who prepared di-*O*-arylcarbamate-*N*-acetylchitosans by reaction of *N*-benzylidene- or *N*-salicylidenechitosan (from chitosan[0.23]) with phenyl isocyanate or 1-naphthyl isocyanate in pyridine at 70°C to give the di-*O*-arylcarbamate-*N*-arylylidenechitosan. These, on dissolving in 85 wt-% HCOOH containing acetic anhydride, were rapidly converted to the di-*O*-arylcarbamate-*N*-acetylchitosans.

The IR spectra of the products showed that almost complete substitution had taken place, there being only a very minor shoulder on the high-frequency side of the NH band at ~3300 cm^{-1} in each case. Both derivatives were soluble in DMF, DMAc, DMSO and pyridine, while the di-*O*-(1-naphthylcarbamate)-*N*-acetylchitosan was also soluble in THF.

Direct reaction between chitosan[0.23] and either isocyanate in pyridine proceeded at a much slower rate than in the case of the Schiff's bases, requiring 120 h compared with 6 h for the latter. This was attributed to the greater accessibility of the Schiff's bases. That complete reaction had taken place was shown by the absence of any IR peak or shoulder above 3300 cm^{-1}. Both derivatives were soluble in DMF, DMSO and pyridine. In each of the di-*O*-arylcarbamate- and di-*O*-arylcarbamate-*N*-arylureidochitosan pairs the phenyl derivative had a considerably higher viscosity (η_{inh}) than had the 1-naphthyl analogue.

The aim of McCormick *et al.* in their earlier study was to evaluate the usefulness of chitin derivatives for controlled release of herbicides, and the preliminary studies[113] showed a satisfactory rate of release of aniline and substituted anilines from the chitin. They subsequently[114] prepared chitin containing carbamate-linked 'Metribuzin' [4-amino-6-(1,1-dimethylethyl)-3-methylthio-1,2,4-triazin-5(4H)one] a pre-emergence herbicide, through reaction of its chloroformamide (**4.14**) with chitin dissolved in DMAc–LiCl. Products having *DS* values of 0.01–0.39 were obtained but they showed no significant release of herbicide over the range pH 4–10. This was attributed to increased stability of the carbamate link in the Metribuzin derivative, relative to that in the phenylcarbamate derivative,[113] because of reduced electrophilic character of the former.

4.14

The authors commented on the dramatic decrease in viscosity caused by the introduction of even such low levels of the herbicide as substituents, the LVN being reduced from 870 cm^3 g^{-1} for the starting chitin to 151 cm^3 g^{-1} for a sample having $DS = 0.23$. This effect of arylcarbamate substituents on viscosity in DMAc–LiCl is also shown by less complex substituents. Thus the phenylcarbamate and p-tolylcarbamate derivatives reported previously were found[113] by osmometry to have $\bar{M}_n$ values of 4.5×10^5 and 3.9×10^5 respectively, while their $[\eta]$ values were 234 and 191 cm^3 g^{-1} which, by applying the viscosity equation constants determined[115] for chitin in DMAc–LiCl, give $\bar{M}_v$ values of 2.1×10^4 and 1.6×10^4 only.

Terbojevich et $al.$[42] have also prepared phenylcarbamate derivatives of chitin by reaction in solution in DMAc–LiCl containing pyridine. Products having DS values of 1.7–2.0, determined by elemental analysis, were obtained using mole ratios of phenyl isocyanates:N-acetyl-D-glucosamine of 7–12. Attempts to prepare samples having $DS \sim 1$ by selective substitution using lower mole ratios were unsuccessful and at a mole ratio of 1.25 the DS was only 0.2.

4.3 ETHERS AND N-ALKYL AND N-ARYL DERIVATIVES

4.3.1 Chitin ethers

Preparation of alkali chitin

Alkaline conditions are required for chitin etherification to generate alcoholate anions which react with suitable reagents including alkyl halides, sulphate diesters, epoxides and activated vinyl compounds (Figure 4.19). This requires that the initial step is preparation of alkali chitin and while the general principles discussed in section 4.1.4 apply, there are three distinct methods used to carry out this initial step.

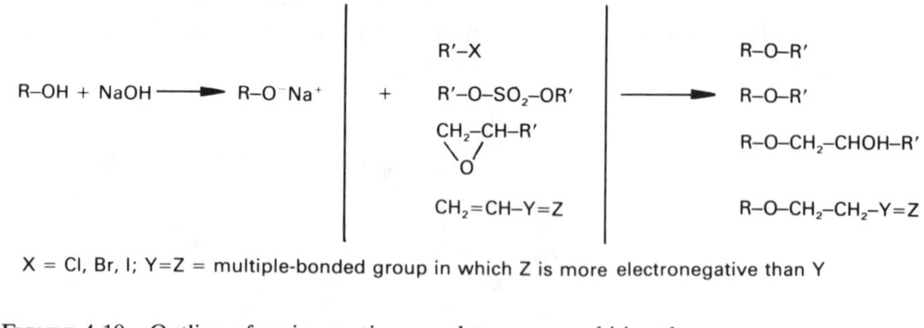

R–OH + NaOH ⟶ R–O⁻Na⁺   +   R'–X   R–O–R'

R'–O–SO$_2$–OR'   ⟶   R–O–R'

CH$_2$–CH–R'   R–O–CH$_2$–CHOH–R'
\ /
O

CH$_2$=CH–Y=Z   R–O–CH$_2$–CH$_2$–Y=Z

X = Cl, Br, I; Y=Z = multiple-bonded group in which Z is more electronegative than Y

FIGURE 4.19   Outline of main reactions used to prepare chitin ethers

In the first the chitin is treated with the required concentration of NaOH, the steeping being carried out either at ambient or at −20°C. Although it has been argued that the freezing process breaks up the micelle structure of chitin so that a more uniform reaction ensues,[56] a rapid-freezing process using liquid nitrogen or dry ice was found to be considerably less effective in this respect.[116] The steeping treatment may also be carried out under reduced pressure; this has been claimed both to deaerate the mixture[117] and to aid penetration of NaOH into the micelles,[58] while the addition of a surface-active agent such as sodium dodecyl sulphate has been stated to aid the latter process.[60]

The second method is similar to the normal process for production of carboxymethyl cellulose or hydroxyethyl cellulose, and involves slurrying the chitin in a water-miscible liquid such as 2-propanol (IPA) followed by addition of a concentrated solution (∼40 wt-%) of NaOH. The IPA or other diluent aids uniform distribution of NaOH throughout the chitin mass.

The third method involves steeping chitin in concentrated NaOH, at least 40 wt-%, with or without a freezing step, then adding ice to disperse the alkali chitin and give a homogeneous, viscous solution on dilution of the alkali concentration down to approximately 12–14 wt-%. The etherification reaction is then carried out under homogeneous conditions.

As NaOH solutions of 40 wt-% or higher are routinely used in the production of chitosan, it is likely that some deacetylation will occur during alkali chitin formation and also during the subsequent etherification reaction. It is partly for this reason that generally lower reaction temperatures are used for etherifying chitin than is the case with cellulose. However this is not an absolute rule and the presence of free amine groups, either initially present in the chitin or introduced during the steeping and/or etherification reaction, raises the possibility of reaction between the reagent and the free amine groups as well as with the R–O⁻ ions (section 4.3.4).

Alkyl ethers of chitin

Schorigin and Makarova-Semljanskaya[118] attempted the methylation of chitin using the Haworth methylation technique, dimethyl sulphate–NaOH, but only achieved an O-methyl chitin having a DS ∼1 after 45 repeated treatments. Wolfrom et al.[119] tried to overcome the inherent lower reactivity of chitin compared with cellulose by using reprecipitated chitosan and treating the swollen, hydrated, undried chitosan with dimethyl sulphate–NaOH, obtaining a product having a DS = 0.58 and which the authors accepted might include a limited number of N-methyl groups. This intermediate derivative was then acetylated, the product dissolved in DMF, and methylated with BaO–Ba(OH)$_2$–CH$_3$I, thus increasing the DS to 1.32. Repeating the acetylation/methylation cycle two more times, followed by a final re-acetylation, gave 3,6-di-*O*-methyl chitin having a $DS = 1.84$.

The next member of the sequence, *O*-ethyl chitin, was first prepared by Danilov and Plisko[54] by reaction of alkali chitin with ethyl chloride with a reaction time of 10 h and a temperature of up to 130°C. The product had good solubility in a range of organic solvents, as did the ethyl chitin prepared by Capozza[120] by reaction of alkali chitin with ethyl chloride at 130°C for 7 h, being soluble in benzene, toluene, xylene, methyl ethyl ketone, and some mixed solvents. Capozza claimed his product to be 6-*O*-ethyl chitin but it is more likely, in view of the reaction conditions and the organic solubility, to have had a $DS > 1$, in which case some 3-*O*-ethyl groups must also have been introduced.

Somorin *et al.*[97] attempted to prepare *O*-benzyl chitin by treatment of a suspension of alkali chitin in DMSO with benzyl chloride, but the product had only a $DS \leq 0.4$. However alkali chitin prepared by treating a suspension of di-*O*-acetylchitin in DMSO with 40 wt-% KOH at 4–12°C gave, on subsequent reaction at a mole ratio for benzyl chloride:*N*-acetyl-D-glucosamine residue of 10:1, an *O*-benzyl chitin having $DS = 0.75$, substitution only occurring at the C(6)OH group. The most extensive reports on the preparation and properties of *O*-alkyl chitins are those of Tokura *et al.*[59, 121] Alkali chitin was frozen for 16 h then suspended directly in the appropriate alkyl halide, using a mole ratio of 10:1 for alkyl halide:*N*-acetyl-D-glucosamine residue, and the reaction continued for 24 h at 12–14°C with intermittent stirring. The alkyl halides used were CH_3I, CH_3CH_2Br, $CH_3(CH_2)_3Br$, CH_3–CH_2–$CH(CH_3)Br$, $(CH_3)_3CBr$, $CH_3(CH_2)_4Br$, $(CH_3)_2CHCH_2CH_2Br$, $(CH_3)_3C$–CH_2Br and, rather surprisingly, all the derivatives had a DS of 0.4 as calculated from C, H, N analysis. Both ^{13}C and ^{1}H NMR spectroscopy indicated that the primary site for alkylation in each case was the C(6)OH group and deacetylation was considered not to have occurred. The products were spun into fibres from solution in formic acid–dichloroacetic acid, using ethyl acetate as the coagulating medium, and the tensile properties measured after being given a drawing process. X-ray analyses and moisture uptake measurements were also carried out. The unit cell dimensions for the two extremes are given in Table 4.19. The *O*-methyl and *O*-ethyl chitin fibres were very similar in respect of their unit

TABLE 4.19   *Cell dimensions (Å) of chitin and O-alkyl chitin fibres*[59,121]

Fibre	a	b	c
Chitin	4.7	10.3	10.5
O-butyl chitin	4.8	10.7	10.6
O-(*iso*-butyl) chitin	4.9	10.7	10.9
O-(*tert*-butyl) chitin	5.0	10.9	11.0

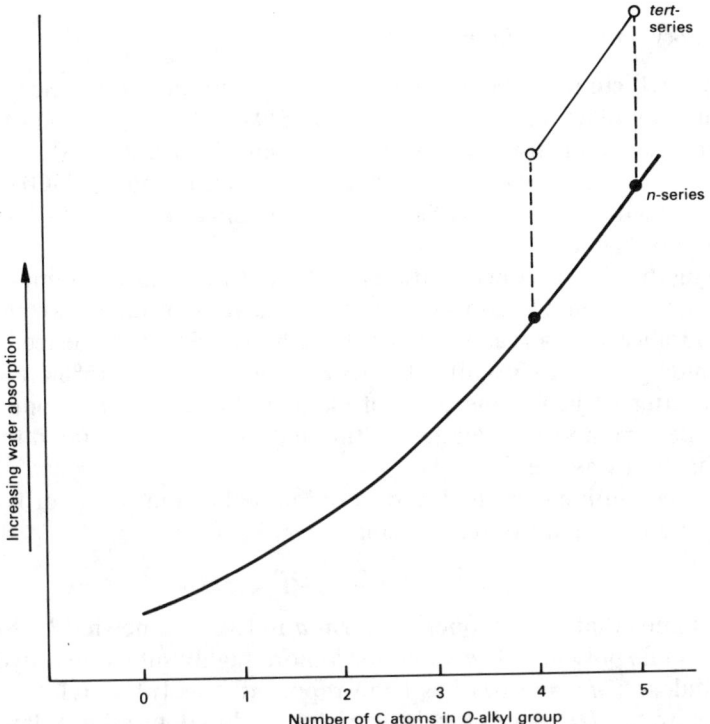

FIGURE 4.20   Schematic representation of increase in water absorption capacity with
              increase in (a) chain length and (b) branching of the alkyl group in O-alkyl
              chitins[59, 121]

cells to the chitin fibres, while the O-pentyl chitin fibres resembled the
O-butyl specimens. The volume of the unit cell increases by approximately
20% on going from chitin to O-tert-butyl chitin and no change was observed on annealing. Surprisingly the increase in the fibre repeat distance
along the b axis was similar to the increases along the other axes.
   The fibres swelled quickly on contact with water, the degree of swelling
increasing both with increase in the number of C atoms of the O-alkyl
substituent and with increase in the extent of branching (Figure 4.20). On
absorption of water the fibres no longer show birefringence when observed
under polarised light. The increase in swelling with increase in chain
branching is also reflected in the accessibilities of the fibres to attack by
lysozyme[15] where the order of increasing rate of degradation was found
to be

   chitin < n-butyl chitin < iso-butyl chitin < tert-butyl chitin.

Hydroxyalkyl ethers of chitin

Hydroxyethyl chitin (glycol chitin) was first prepared by Senju and Okimasu[122] by reaction of ethylene oxide (EO) with a dispersion of alkali chitin. It was found, using reaction conditions of 15 h at 33°C, that if a water-soluble chitin was to be obtained the concentration of NaOH in the solution needed to be ≥ 12 wt-%, and that the mole ratio of EO:N-acetyl-D-glucosamine needed to be $\geq 2.3{:}1$.

Keeping the NaOH concentration constant at 13 wt-%, the temperature at 33°C, and the mole ratio of EO at 6.9, gave a product having good water-solubility after a reaction time of 1–2 h. The degree of deacetylation of the amide group was 3% after 1 h, rising to 9% after 8 h, 15% after 16 h, and 38% after 64 h. On the basis of elemental analysis the product was found to be a mono-hydroxyethyl chitin, and the authors claimed that the site of reaction was the C(6)OH.

The same authors studied the solution behaviour of hydroxyethyl chitin[123] and obtained the relationship*

$$[\eta] = 1.2 \times DP$$

which assumes that the exponential term a is 1.0, a somewhat high value for a non-ionic polymer. The value for a more highly substituted hydroxyethyl cellulose ($MS = 1.67$) has been reported[124] as $[\eta] = 1.1 \times DP^{0.87}$. Furthermore the DP values for the samples used to determine K lay in the range 10–27, as determined by hypoiodite oxidation, so that extrapolation to DP values > 800 is liable to introduce a considerable error. Despite these two questions regarding the validity of the equation, the value obtained for the hydroxyethyl chitin, $DP = 851$, is equivalent to $\bar{M}_v = 2.1 \times 10^5$ which is a reasonable value since in neither the alkali chitin preparation, nor in the hydroxyethylation step, was any attempt made to exclude air and so reduce chain degradation.

Danilov and Plisko[54] produced hydroxyethyl chitin by the reaction of EO and an alkali chitin prepared by the freezing technique, obtaining products having MS values** of 1.4–2.9. The level of substitution was determined by nitration, followed by determination of the N content. Optimum conditions were stated to be a mole ratio of EO:N-acetyl-D-glucosamine of 16:1 and reaction conditions of 30 h at 70°C.

---

* The equation given was actually $[\eta] = 12 \times 10^{-4} \, DP$ with the LVN being expressed in $\mathrm{dm^3 \, g^{-1}}$. The constant K has been corrected for application with LVN values expressed in the more common units of $\mathrm{cm^3 \, g^{-1}}$.

** Unlike most other substitution reactions of chitin the maximum substitution per N-acetyl-D-glucosamine residue is not limited to 2.0 since the product, at any level of substitution, will have two –OH groups available for further reaction. Hence it is better to use the concept of moles of substituent (MS) rather than DS. This applies also to hydroxypropylation.

Yamada and Imoto[58] claimed that hydroxyethyl chitin is difficult to prepare using EO and used 2-chloroethanol (ethylene chlorohydrin) and a highly viscous alkali chitin solution containing 14 wt-% NaOH. After a reaction time of 0.5 h the temperature was allowed to rise to ambient, the solution let stand overnight then recooled to ~0°C and any free amine groups acetylated by dropwise additions of acetic anhydride. The product was used as a substrate for lysozyme and it was claimed that, within reasonably broad limits, the *MS* had no effect on the measured enzyme activity.

The use of hydroxyethyl chitin as a water-soluble substrate for studies on enzyme degradation of chitin, both chain cleavage and de-N-acetylation,[125] is quite common and Hirano has published two 'standard' procedures for preparing hydroxyethyl chitin suitable for use as a substrate in such studies.[126] Both methods involve the use of 2-chloroethanol and an N-acetylation step after hydroxyethylation, as first proposed by Yamada and Imoto,[58] but differ in the conditions used in the hydroxyethylation step. In one the alkali chitin is solubilised by addition of ice and the reaction takes place at room temperature overnight, and in the second the alkali chitin, after being given a freezing treatment, is suspended in IPA and hydroxyethylation is carried out at room temperature for 1 h.

Hydroxypropyl chitin has been obtained by reaction of alkali chitin with propylene oxide,[127, 128] both being claimed to be 6-O-hydroxypropyl chitin. However analysis of the hydrolysis products showed[129] the structure to contain 6-O-hydroxypropyl-, 3,6-di-O-hydroxypropyl-, 3-O-hydroxypropyl-, and unsubstituted N-acetyl-D-glucosamine residues in the mole ratio of 1.0:0.32:0.51:1.45, showing the substitution pattern to be quite irregular.

Dihydroxypropyl chitin has been prepared by the reaction of 3-chloropropan-1,2-diol (glycerol α-monochlorohydrin) with alkali chitin in an IPA slurry. The reaction was carried out at room temperature with a mole ratio of 3-chloropropan-1,2-diol:N-acetyl-D-glucosamine of 10:1, the reagent being added in portions until the reaction mixture was neutralised. The *MS* value was 0.9 and the ^{13}C NMR spectrum indicated that reaction had taken place at both C(6)OH and C(3)OH with the former predominating.[60, 116] The solubility of dihydroxypropyl chitin in aqueous solution decreased on standing at room temperature for 2–3 weeks. This gradual insolubilisation was attributed to the formation of interchain hydrogen bonds between the primary and secondary hydroxyl groups, introduced on the dihydroxypropyl substituent groups, and the –NHCOCH₃ groups. Support for this suggestion comes from the fact that it is possible to redissolve the insolubilised material by pretreatment with 6M urea solution for 12 h at room temperature.

In both the synthesis of hydroxyethyl chitin through the use of 2-chloroethanol, and of dihydroxypropyl chitin, the most probable reaction mechanism involves the formation of an epoxide intermediate so that the actual reaction with chitin is a nucleophilic addition reaction rather than a nucleophilic substitution one.

Ionic ethers of chitin

Ionic ether derivatives of chitin may also be prepared and chief among these is sodium carboxymethyl chitin, prepared by reaction between alkali chitin and monochloroacetic acid (MCA). Danilov and Plisko[130] investigated both the effects of NaOH concentration (40–50 wt-%) and steeping time (3–24 h) during the alkali chitin preparation step, and the duration (12–20 h) and temperature (20–60°C) of the carboxymethylation step. They concluded that the optimum conditions were a mole ratio of NaOH:N-acetyl-D-glucosamine: MCA of 6:1:3 and reaction conditions of 20 h at room temperature. They also prepared the free acid form, carboxymethyl chitin (H+), by treatment of the Na+ salt form with dilute HCl, and found that it swelled strongly in water, being partially soluble. No evidence of any deacetylation having occurred was found, the test for primary amine groups being negative. From a consideration of the products prepared by them Danilov and Plisko concluded that sodium carboxymethyl chitins having a Na+ content > 6.13%, which is equivalent to a DS of 0.7, have good water solubility. This differs from carboxymethyl celluloses where products having DS values of 0.4 are sufficiently soluble in water to be used as film-forming materials for textile sizes. This need for higher DS levels to achieve water solubility may reflect a less uniform substitution pattern in sodium carboxymethyl chitin due to the lower initial swelling of the chitin in NaOH.

Muzzarelli reports[129] that, prior to the publication of Danilov and Plisko, Okimasu[131] had determined the activation energy for carboxymethylation of chitin to be ~96–5 kJ mol^{-1}, which is very similar to that for the hydrolysis of monochloroacetic acid by NaOH, so that any change in temperature will affect both the carboxymethylation and hydrolysis reactions equally. However, increase in the NaOH concentration has a greater effect on the rate of carboxymethylation than on the side reaction, hence it is more efficient to use high NaOH concentrations.

Trujillo[132] recommended steeping the chitin in DMSO for 24 h, followed by washing with ethanol and drying, as a preliminary step before treating with concentrated NaOH to form alkali chitin. The carboxymethylation reaction was carried out by addition of the pressed alkali chitin to IPA containing MCA (mole ratio of MCA: N-acetyl-D-glucosamine residue = 3.2:1) and stirring for 1 h at room temperature. The product, which had DS = 0.99, was used as a substrate for a study of the degradation of chitin by lysozyme. Hackman and Goldberg[117] used this technique without the initial DMSO treatment, finding that it increased neither the yield nor the DS. The product obtained by Hackman and Goldberg had a

DS of 0.73 and a degree of deacetylation of 22%. Light scattering measurements in 0.5M and 2.5M NaCl solutions gave *DP* values of 7500 and 5295 respectively, indicating that some dissociation of polymer chains occurred at the higher NaCl concentration. Since the latter value was very similar to that obtained for the starting chitin in 5.55M LiSCN, *DP* = 5206, they concluded that no degradation had occurred during carboxymethylation, a result that seems unlikely in view of the fact that no attempt was made to exclude oxygen from either the alkali chitin preparation or the carboxymethylation reaction. Capozza[120] also used a pretreatment with DMSO, adding the DMSO-swollen chitin to IPA, followed by 30 wt-% NaOH and, after stirring for 1 h to allow the formation of alkali chitin, monochloroacetic acid. The reaction conditions used, 24 h at 55°C, seem somewhat excessive. The product had a *DS* of ~1.0 (by potentiometric titration) and again it was claimed to be exclusively substituted at the C(6)OH position.

Formation of sodium carboxymethyl chitin in an IPA slurry at room temperature has also been reported by Tokura and co-workers[60, 116] using an alkali chitin prepared by the freezing process, −20°C overnight, and with the addition of a surfactant. There are some contradictions between the two reports: in one[116] the *DS* of the water-soluble product was determined by ^{13}C NMR spectroscopy and potentiometric titration and given as ~0.4, while in the second[60] it was stated that a *DS* ≥ 0.6 is required for water-solubility, this latter conclusion being in reasonable agreement with that of Danilov and Plisko.[130] Another contradiction is in the molecular weights reported for a sample having [η] = 520 cm^3 g^{-1}: in the first report[116] the molecular weight is given as 6.6 × 10^4 while in the second[60] it is given as 1.63 × 10^5, in both cases the equation proposed by Kaneko *et al.*[133]

$$[\eta] = 7.92 \times 10^{-3}.M^{1.0}$$

being used. This difference in molecular weights may be due to the lower value being an $\bar{M}_n$ value, Kaneko *et al.* having determined the viscosity equation constants from osmometry studies on sodium carboxymethyl chitins,[133] while the larger value is stated to be an $\bar{M}_v$ value which would be expected to be approximately twice the $\bar{M}_n$ value assuming a normal type of molecular weight distribution. In both reports, as also in that of Danilov and Plisko,[130] sodium carboxymethyl chitin is described as a characteristic polyelectrolyte in aqueous solution, rectilinear plots of Viscosity number *versus* Concentration being obtained in solutions having a NaCl concentration ≥ 0.1M. Potentiometric titration showed the carboxyl groups to have a pK_a of 3.4 in 0.1M NaCl, while ^{13}C NMR spectroscopy indicated that carboxymethylation had occurred specifically at the C(6)OH group.

Two 'standard' procedures for preparing sodium carboxymethyl chitin, using analogous techniques to those described above for hydroxyethyl chitin, have been described by Hirano.[126]

In a number of studies carboxymethyl chitin has been further derivatised, principally by sulphation (section 4.1.3) but two recent papers by Tokura's group have dealt with other second-stage derivatisation processes. In the first[134] a carboxymethyl chitin, $DS = 0.38$, was allylated under alkaline conditions by treatment with allyl bromide in the presence of a trace of iodine. The product, O-allyl $DS = 1.48$, was soluble in methanol provided it was kept wet after preparation. A film cast from methanol and dried in air was found to have become insoluble, the loss in solubility being attributed to formation of interchain hydrogen bonds and crosslinks, the latter arising from free radicals generated at the allyl groups. FTIR spectroscopy indicated that irradiation with a mercury lamp in air led to a decrease in the C=C group concentration and an increase in the C=O group concentration, while the concentration of both groups remained relatively constant on irradiation under nitrogen, leading the authors to suggest that the radical formed induced oxidative cleavage of the C=C double bond in air.

In the second paper[135] carboxymethyl chitin (H^+) was treated with D-N-(4-aminoethyl)- and D-N-(2-aminobutyl)methamphetamine in the presence of a water-soluble carbodiimide to give the carboxymethyl chitin derivatives (**4.15a, b**). The spacer link length was found to be one of the major factors regulating antibody production on injection of these carboxymethyl chitin derivatives into rabbits, with the effectiveness of the 2-aminoethyl-derivative (**4.15a**) being very much less than that of the 2-aminobutyl-derivative (**4.15b**) where the spacer is approximately twice the length.

4.15

Another anionic O-alkyl derivative reported[136] is that obtained by the ring-opening reaction between alkali chitin and 1,3-propane sultone to give water-soluble derivatives. A cationic O-alkyl derivative has been obtained[137] by the reaction between alkali chitin and glycidyltrimethyl-ammonium chloride, but as pointed out by Lang and Clausen[138] the reaction does not give good yields because both the reagent and the product decompose under alkaline conditions through a Hoffmann-degradation reaction.

Other chitin ethers

Nucelophilic addition of alkali chitin to acrylonitrile to form O-cyanoethyl chitin has been studied by Tokura et al.[139] The reaction was carried out for 18 h at 12–14°C both in the presence and the absence of IPA as an inert diluent. Products with DS values up to 1.4 were obtained using an NaOH:N-acetyl-D-glucosamine mole ratio of 7:1. The reaction was followed by IR spectroscopy and the amide I band at 1620 cm^{-1} was observed to decrease with increase in the extent of O-cyanoethylation, and to have completely disappeared at a DS of 1.0, while the ^{13}C NMR spectrum showed the C(6)OH group to be the initial site of reaction.* Cyanoethyl-ation enhanced the solubility in HCOOH, the solubility increasing with increase in DS.

Kurita et al.[140] have prepared diethylaminoethyl chitin by reaction of chitin with N,N-diethylaminoethyl chloride either in organic solvent, in the presence of NaOH, or as a solution of alkali chitin. The organic solvents used were dioxane, DMSO and DMAc, and the better results obtained in DMSO and DMAc were attributed to the greater extent of swelling in these two liquids (Table 4.20). The alkali chitin solution contained 1.4 wt-% chitin and 12 wt-% NaOH, and although the reactions were carried out at room temperature (Table 4.21) some deacetylation of the chitin occurred and hence a post-diethylaminoethylation treatment with acetic anhydride–pyridine was given to ensure a completely N-acetylated product.

The main changes in the IR spectra were increases in intensity for both the 2965 cm^{-1} and 1000–1150 cm^{-1} bands, in agreement with substitution having occurred, while the NMR spectra showed peaks due to both the methyl and methylene groups of the substituent.

---

* There is some confusion in the interpretation of the IR spectroscopic results since in this paper, and at least two others,[56, 59] the authors incorrectly assign the β-chitin structure to their starting material, which was from crab shell, on the basis of the presence of the two amide I bands at 1660 cm^{-1} and 1620 cm^{-1}, which is in fact a characteristic of the α-chitin structure with β-chitin having only one amide I band in this region. They also state[56] that 'natural-type chitin has the α-structure and is supposed to be converted to the β-structure during processes to isolate chitin such as the procedure for removing protein or calcium' whereas in fact α-chitin is the more stable of the two forms (section 1.4.5).

TABLE 4.20   *Dependence of DS on solvent, temperature and time in the heterogeneous[a]*
              O-diethylaminoethylation of chitin[140]

Solvent	Temperature (°C)	Time (h)	DS
Dioxane	60	4	0.30
Dioxane	95	4	0.34
DMAc	RT	24	0.31
DMSO	RT	24	0.37
DMSO	60	4	0.77

[a] Reaction system contains 1 g chitin:7.5 cm^3 20 wt-% NaOH:50 cm^3 solvent and a mole ratio of 5:1 for N, N-diethylaminoethyl chloride hydrochloride:chitin.

TABLE 4.21   *Dependence of DS on mole ratio of N, N-diethylaminoethyl chloride*
              hydrochloride:chitin on reaction time in the homogeneous, room
              temperature O-diethylaminoethylation of alkali chitin[140]

Mole ratio	Time (h)	DS
1	24	0.30
5	24	0.56
10	24	0.78
10	2	0.31

At *DS* levels > 0.3 the products prepared in organic solvents showed considerable swelling in water, dichloroacetic acid and DMSO, while the products prepared under homogeneous conditions showed high solubility in water and dichloroacetic acid, and a remarkable degree of swelling in benzene, alcohols and polar solvents. The differences in solubility were attributed to differences in the distribution pattern of substituents introduced under heterogeneous and homogeneous conditions.

4.3.2 Chitosan ethers

Preparation of *O*-alkyl derivatives of chitosan may be carried out using two procedures: *O*-alkylation of chitin followed by de-*N*-acetylation and *O*-alkylation of a chitosan derivative in which the amine group is protected during the alkylation reaction.

Okimasu[131] prepared carboxymethyl chitosan by the first procedure, obtaining the product either as the sodium salt with the amine groups present in the free base form, or as the hydrochloride salt of the amine with the carboxymethyl groups present in the free acid form. The sensitivity to added electrolyte increased with increase in the extent of carboxymethylation. *O*-Carboxymethyl chitosan has also been prepared by Somorin *et al.*[116] who found that, unlike the carboxymethyl chitin precursor, the O-carboxymethyl chitosan was not degraded by lysozyme. Treatment of alkali chitin with epichlorohydrin at 0–15°C, followed by deacetylation, yielded O-hydroxyalkyl chitosans.[141]

The O-alkylation of chitosan in which the amine group is protected by Schiff's base formation has been studied by Plisko and co-workers.[142, 143] In the first paper N-salicylidenechitosan was treated with monochloroacetic acid in the presence of alkali, and of a diluent such as IPA or xylene, to give O-carboxymethyl-N-salicylidenechitosan which was converted to O-carboxymethyl chitosan by hydrolysis of the Schiff's base under acid conditions. The use of chloroalkanethio acids was also claimed. In the second paper[143] O-carboxymethyl chitosans having DS values ≤ 1.1 were prepared by the same procedure and, in addition, O-sulphoethyl chitosans were prepared by using sodium 2-chloroethane sulphonate in place of chloroacetic acid and a reaction temperature of 80°C. The maximum DS obtained was 0.31, the extent of reaction being improved by using an excess of reagent. After hydrolysis of the Schiff's base in 50 vol.-% acetic acid, the product was obtained by precipitation in acetone.

The same group also examined the direct O-alkylation of chitosan by its reaction with acrylonitrile[144] in the presence of 10 wt-% NaOH with the reaction temperature kept at 20°C. Under these conditions only O-cyanoethylation occurs, as was proved by NMR studies of the products. Furthermore no hydrolysis of –C≡N groups to –$CONH_2$ or –COONa groups occurred. Products having DS~ 1.9 were obtained after a reaction time of 24 h using an acrylonitrile:D-glucosamine mole ratio of 30:1. If the reaction temperature was increased to 70°C some N-substitution took place, together with some hydrolysis of nitrile groups to carboxylic acid groups.

Park *et al.*[145] have described the preparation of O-carboxymethyl chitosan by the simultaneous O-carboxymethylation and de-N-acetylation of chitin on treating a suspension of chitin in 42 wt-% NaOH with monochloroacetic acid at 30°C for 5 h. At the end of the reaction, the system had become homogeneous and the solution was neutralised to pH 7 with HCl and dialysed before isolating the product. The extent of deacetylation achieved was approximately 33% while the carboxymethyl DS was ~0.3, based on a Na^+ content of 1.42 meq g^{-1} for the basic O-carboxymethyl chitosan.

The product was soluble in water over the pH range 3–11 and, as might be expected, the solution viscosity was pH sensitive. Below a pH of approximately 8, the polymer exists as a polyampholyte and the viscosity is reduced as a result of coil contraction arising from intrachain ionic bonding between –$\overset{+}{N}H_3$ and –COO⁻ groups. As the pH is raised the concentration of –$\overset{+}{N}H_3$ groups decreases until, in the region of pH 8, the polymer becomes an anionic polyelectrolyte with an accompanying increase in viscosity due to coil expansion.

4.3.3   N-*Alkyl and* N-*aryl derivatives of chitosan*

Reaction with alkyl and aryl halides

There are several methods for the *N*-alkylation of chitosan. The most obvious is the reaction between chitosan and alkyl halides, a method first investigated by Nud'ga *et al.*[146] who investigated the reactions of chitosan with methyl and ethyl iodides in the presence of several tertiary amines, pyridine, dimethylpyridine (lutidine), trimethylpyridine (collidine) and triethylamine. They established that the base used must have a pK_a higher than that of chitosan, which is approximately 6.4 (section 5.1), that triethylamine was the best of the bases examined, and that a gradual addition of the base, rather than the initial addition of all of it to the reaction mixture, was beneficial. They obtained derivatives in which 26–78% of the amine groups had been quaternised, the basicity increasing with increase in the degree of alkylation. The approximate pK_a values were 7.0 for the *N,N,N*-trimethylchitosan derivative and 7.4 for the *N,N,N*-triethyl analogue.

Domard *et al.*[147] subsequently studied the reaction using *N*-methyl-2-pyrrolidone (NMP) as the reaction medium and chitosan[0.05] as substrate. However unlike Nud'ga *et al.*[146] they failed to obtain more than a very limited extent of quaternisation in the presence of triethylamine, even on reacting for 72 h at 36°C. Instead the reaction mixture turned a yellow-orange immediately on addition of triethylamine and the yellow precipitate was shown by NMR spectroscopy to be triethylmethylammonium iodide, evidence that the triethylamine is quaternised too rapidly for its use with chitosan. Indeed the extent of quaternisation was greater in the absence of triethylamine than in its presence. Domard *et al.*[147] therefore used an inorganic base (1.4M NaOH) to neutralise the acid produced during quaternisation. Additionally, NaI was added to the reaction mixture to screen the electrostatic charge on the polymer as quaternisation proceeds. The maximum degree of quaternisation, 64%, was achieved by three successive treatments of 3 h duration each, with the polymer isolated by precipitation between successive treatments. Extending the reaction time by a further 10 h gave no increase in the extent of reaction, but a decrease in the LVN of the product. The final precipitation step was preceded by passage through an anion exchange resin to convert the polyelectrolyte to the chloride salt which was stated to be more stable than the iodide salt.

Only one *N*-aryl derivative has been prepared, *N*-(2, 4-dinitrophenyl)-chitosan, formed by the reaction between chitosan and 2,4-dinitrofluorobenzene.[64, 125, 148] However even under homogeneous conditions only approximately 50% of the amine groups of a chitosan[0.16] sample had reacted after 72 h at room temperature[148] using a reaction system similar to that for homogeneous N-acylation (section 4.2.1.). This was attributed to steric hindrance, a view supported by the fact that the percentage of amine groups reacting with 2,4-dinitrofluorobenzene decreases with increase in the extent of deacetylation of the substrate,[148] since when the amine group concentration is very small steric hindrance between reaction sites is much less likely. The rate of reaction has also been studied[64] and at 23°C the reaction had reached equilibrium within 3 h.

Reductive alkylation

One of the most versatile methods of preparing N-alkyl derivatives of chitosan is reductive alkylation, a technique discovered by Borch *et al.*[149] and subsequently applied to chitosan* by Hall and Yalpani.[48] Although Borch *et al.* described the process as reductive amination[149] the alternative name of reductive alkylation[86] is more appropriate as the amine, chitosan, is clearly the substrate. The process involves the reaction between a primary or secondary amine and a carbonyl compound, with simultaneous reduction of the imminium species produced as an intermediate, reduction being carried out using sodium cyanoborohydride or sodium borohydride. The general reaction is outlined in Figure 4.21. The method depends on the fact that while the reduction of aldehydes and ketones is negligible at pH 6–7, that of the imminium species $\text{\textbackslash}C=\overset{+}{N}HR$ or $\text{\textbackslash}C=\overset{+}{N}R_2$ is rapid at this pH which is also the optimum pH for imminium formation. In practice the reaction can be carried out at pH 4–10, the only requirement appearing to be the presence of sufficient protons to generate the cationic imminium ion.[149]

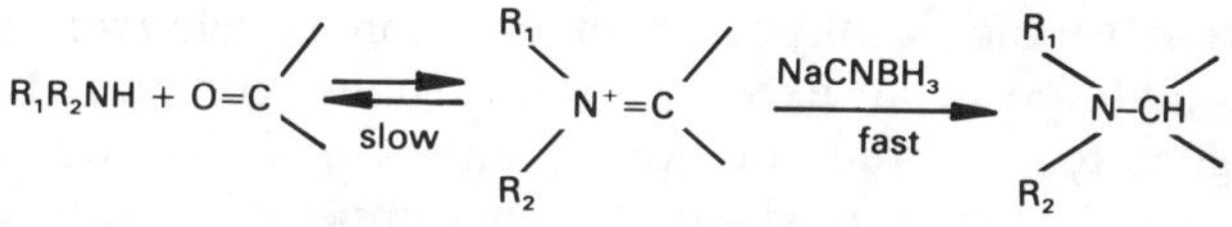

FIGURE 4.21   General reaction scheme for reductive alkylation[149]

---

* Although Hall and Yalpani were the first to exploit extensively the reductive alkylation of chitosan they were not the first to carry out borohydride reduction of Schiff's base derivatives of chitosan, Masri *et al.*[150] previously having prepared immobilised enzyme systems by addition of glyoxal, glutaraldehyde or dialdehyde starch to chitosan solutions containing enzymes and subsequently treating the gels with sodium borohydride to reduce the Schiff's base structures. Hall and Yalpani appear[48] to have been unaware of this previous application of the technique to Schiff's base derivatives of chitosan.

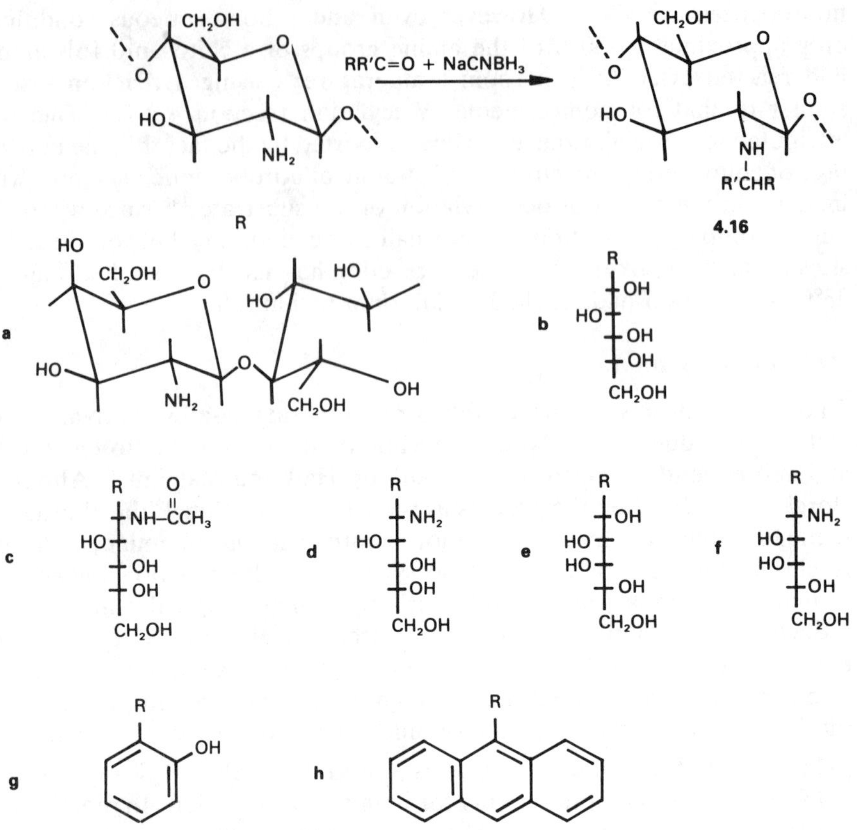

FIGURE 4.22   Representative products from the reductive alkylation of chitosan.[86]
In all structures shown R' = H

Application to the *N*-alkylation of chitosan is relatively simple, involving the addition of an excess of the carbonyl compound and of the
reducing agent to a stirred solution of chitosan in aqueous acetic acid
–methanol and allowing it to stand at room temperature, with stirring, for
up to 24 h. The versatility of the reaction was demonstrated by Yalpani and
Hall in a comprehensive paper[86] in which the nature of the groups R and
R' in (**4.16**) were varied widely (Figure 4.22). In many of their examples
RR'C=0 was a reducing sugar, either a mono-, di- or oligosaccharide, or a
ketose such as fructose, but the use of aliphatic or aromatic aldehydes and
ketones was also described, particularly their use as coreactants with
reducing sugars to tailor the hydrophobic character, solubility and other
properties of the products. The incorporation of other functional groups
through the use of carbonyl compounds containing amine or carboxylic
acid groups was also examined.

The reactions between reducing sugars and chitosan[0.15] were generally found to proceed smoothly, forming soft to very rigid gels at the end of the reactions. The extent of reaction of the amine groups, the DS, was normally high and ranged from 0.54 for maltotriose to 0.97 for D-galactose and N-acetyl-D-glucosamine.* In comparison the attempted preparation of some analogous Schiff's base derivatives using the same reaction system, except for the omission of the NaCNBH$_3$, yielded products having DS values ~0.1.

Modification of the free amine groups in a sample of chitin[0.90] by reductive alkylation with lactose was carried out on either a solution of the chitin[0.90] in hexafluoro-2-propanol or on a sample that had been activated by conversion to alkali chitin with subsequent treatment with a solution of lactose and NaCNBH$_3$ in 5 vol.-% acetic acid. DS values of ~0.1 were obtained by both procedures. Yalpani and Hall also linked dextran, streptomycin sulphate, specifically-oxidised cyclodextrins, and 4-oxy-2,2,6,6-tetramethylpiperidine-1-oxyl to chitosan, the latter giving a spin-labelled derivative.[86, 151]

All the branched chitosan derivatives were soluble in either neutral or slightly acidic (pH 5–6) aqueous media, solubility under neutral conditions being achieved at a DS of 0.14 when lactose was the carbonyl compound used (4.16a). A number of the products, those from D-glucose (4.16b), N-acetyl-D-glucosamine (4.16c), D-glucosamine (4.16d), D-galactose (4.16e), D-galactosamine (4.16f) and maltotriose, formed gels in aqueous solutions. The cellobiose and melibiose derivatives were soluble in 50 vol.-% aqueous ethanol, while the product from the coreaction of lactose and propionaldehyde, at a mole ratio of 1:1.3, was water-soluble and also compatible with acetone, ether, ethanol and chloroform. Solutions of the lactose derivative were compatible with a number of anionic species, including sulphate, chromate and phosphate, which precipitate chitosan itself from solution, and the cellobiose derivative forms rigid white gels when mixed with solutions of sodium alginate.

A large number of other carbonyl compounds have subsequently been used in this technique including: salicylaldehyde, which gives a derivative (4.16g) having increased metal ion binding capacity;[152] 9-anthraldehyde,[153] which gives a fluorescent product (4.16h); poly(ethylene glycol) containing

---

* The DS values assigned are those that best fit the elemental analysis results and the structures include, where necessary, bound water. However although starting with chitosan[0.15] the products are calculated to be derived from chitosan[0.02]–chitosan[0.05]. This explains how a product having a DS value of 0.97 can be derived from chitosan[0.15]. Similarly the products from the chitin[0.90] sample were calculated to be derived from chitin[0.80]. Since the reaction conditions used in the reductive alkylation reaction are unlikely to cause de-N-acetylation, the results illustrate the inherent difficulty of determining structures of chitin and its derivatives by elemental analysis.

a terminal aldehyde group[154] and with $\bar{M}_w = 8 \times 10^3$; and o-phthalaldehyde.[154] The preparation of branched chitosans having N-acetyl-D-glucosamine oligosaccharide (chitin oligosaccharide) side chains has been studied.[155] It was found that the reaction was much more efficient when the chitin oligosaccharides were prepared by partial depolymerisation of chitosan with nitrous acid (section 5.4.3) followed by N-acetylation, so that the terminal aldehyde-bearing unit is a 2,5-anhydro-D-mannose unit rather than the normal 4-O-substituted N-acetyl-D-glucosamine. The formation of membranes from N-alkylchitosans prepared by reductive alkylation has been reported. These membranes swelled more in water than did membranes of chitosan and this was attributed to the reduction in interchain hydrogen bonding caused by the N-alkyl substituents.[156]

As the reaction also works with secondary amines the initial products can undergo a second reaction to yield tertiary amines,[149] provided sufficient carbonyl compound is present, and Muzzarelli and Tanfani[157] have made use of this fact to prepare N,N-dimethylchitosan (N-permethylated chitosan) in addition to N-methylchitosan, by reductive alkylation with formaldehyde. A mole ratio for HCHO:–NH$_2$ of ~36:1 was used, compared with a mole ratio of 2:1 for synthesis of the mono-N-methyl derivative, and reduction was carried out with NaBH$_4$ rather than NaCNBH$_3$, the former having been rejected by Yalpani[158] on the grounds that the acid conditions required for removing excess reagent would cause degradation of the polymer chains. The N,N-dimethylchitosan was quaternised by reaction with CH$_3$I–acetonitrile, the reaction being carried out at 35°C for 30 h, and the product calculated to contain, from element analysis: CH$_3$CO–, 40%; $-\overset{+}{\mathrm{N}}$(CH$_3$)$_3$I$^-$, 35%; –NHCH$_3$ and –N(CH$_3$)$_2$, 25%. This represents conversion of 60% of the amine groups originally present to quaternary groups, but despite this the product was not soluble in water. This is most probably due to the long blocks of anhydro-N-acetyl-D-glucosamine residues present in the starting chitosan[0.42], compared with the results of Domard et al.[147] who obtained water-soluble products at quaternisation levels of $\geq$ 25% with chitosan[0.05] as starting material.

The use of reductive alkylation to prepare N-(carboxyalkyl)chitosans by reaction with aldehydo acids and keto acids has been extensively reported by Muzzarelli et al.,[159–169] primarily with the aim of studying their metal binding properties. A list of some of those prepared is given in Table 4.22. The general procedure used is to dissolve the chitosan in an aqueous solution of the aldehydo or keto acid and, after allowing time for Schiff's base formation to take place, carry out the reduction step by adding either NaCNBH$_3$ or NaBH$_4$. Although complete substitution of the amine groups can be obtained with glyoxylic acid, leading to N-(carboxymethyl)chitosan (4.17a), a maximum of ~30% substitution appears attainable in the other reactions and this has been attributed to the bulky nature of the reagent molecules.[165]

TABLE 4.22   *Representative N-(carboxylalkyl) chitosan derivatives*[164]

CH₂OH structure — 4.17 (structure diagram)

Structure represented as 4.17 showing chitosan ring with CH₂OH, HO, NH–CH(R)–COOH groups.

Structure	Reagent	R	Product
a	Glyoxylic acid	H–	Glycine glucan
b	Pyruvic acid	CH₃–	Alanine glucan
c	β-Hydroxypyruvic acid	HOCH₂–	Serine glucan
d	p-Hydroxyphenylpyruvic acid	HO—⟨benzene⟩—	Tyrosine glucan
e	β-Phenylpyruvic acid	⟨benzene⟩-CH₂-	Phenylalanine glucan
f	2-Ketoglutaric acid	HOOC–(CH₂)₂–	Glutamate glucan
g	2-Ketomethylbutyric acid	CH₃S.(CH₂)₂–	Methionine glucan
h	2-Keto-4-methylpentanoic acid	(CH₃)₂CH.CH₂–	Leucine glucan
i	Oxalactic acid	HOOC–CH₂–	Aspartate glucan

The two derivatives that have been studied most closely are N-(carboxymethyl)chitosan and N-(carboxybutyl)chitosan, the latter being prepared from levulinic acid.[168] Preparation of N-(carboxymethyl)chitosan was carried out[159, 160] by treating solutions of chitosan[0.42] in aqueous glyoxylic acid with $NaCNBH_3$, using mole ratios of glyoxylic acid:–NH_2 of ~2:1, 0.65:1, 0.32:1. One noticeable feature of the reaction was the considerable increase in solution viscosity a few seconds after the addition of the last of the reducing agent.

IR spectra of samples isolated at different pH values showed a progressive decrease in the intensity of the COOH band at 1730 cm^{-1} and an increase in the COO$^-$ band at 1580 cm^{-1}. The titration curve for fully substituted N-(carboxymethyl)chitosan, prepared using 2 mole equivalents of glyoxylic acid, shows three inflection points, at pH 2.0, 4.1 and 9.5, that at pH 4.1 representing the isoelectric point. The molecular weight of the product, when compared with that for the starting chitosan, confirmed that complete substitution of the amine groups had occurred,[166] in agreement with the titration results[160] and the ^{13}C NMR spectrum.[162]

N-(Carboxylmethyl)chitosan is biocompatible and it has been suggested[167] that its rheological and hydrophilic properties make it suitable as an ingredient in cosmetic creams and ointments. It has been shown to induce neovascularisation when inoculated into the cornea and to reduce inflammation on the cornea tissue.[169]

N-(Carboxybutyl)chitosan was prepared similarly[168] using a levulinic acid: $-NH_2$ mole ratio of 1.5:1, $NaBH_4$ as reducing agent, and five different commercial chitosans covering the range chitosan[0.13]–chitosan[0.42]. The extent of reaction, determined from the increase in molecular weight measured by laser light scattering, was found to be 26–28% for all samples, the division of the remaining 72–74% of the residues between $C(2)–NH_2$ and $C(2)–NHCOCH_3$ depending on the extent of deacetylation of the starting chitosan. The IR spectrum showed amide bands over the 1670–1600 cm^{-1} region, $–NH–$ bands at 1550 cm^{-1}, and the COOH band at 1700 cm^{-1} (levulinic acid itself exhibits its COOH band at 1700 cm^{-1}) while the ^{13}C NMR spectrum showed the CH_3 signal at 24 ppm, the CH_2 signals at 35 and 40 ppm, and the N–CH signal at 33 ppm. The authors state that the NMR and IR spectra do not rule out the possibility of lactam formation, in agreement with earlier work which indicated that 4-amino derivatives of levulinic acid readily form lactams.[170] The freeze-dried product was shown to be completely amorphous by X-ray spectrometry.

Concentrated solution viscosities are greater than those of the initial chitosans and are practically independent of pH over the range 2.9–5.0; above pH 5.5 they decrease significantly with increasing pH, finally precipitating out as voluminous solids at pH $\geq$ 8.5.

In solution N-(carboxybutyl)chitosan is compatible with gelatine and poly(acrylic acid), although the latter forms an insoluble complex with chitosan itself, but forms precipitates or gels with carboxymethyl guar, carageenan and sodium alginate. It is not precipitated by sulphate or phosphate ions nor by Mg^{2+}, Ca^{2+}, Cr^{3+}, or Co^{2+} ions, although Fe^{2+}, Ni^{2+}, Cu^{2+}, Zn^{2+}, Cd^{2+} and Ag^+ ions give insoluble products.

Further derivatisation of N-(carboxymethyl)chitosan has been reported, both sulphation[46, 47] (section 4.1.3) and etherification to give O-hydroxyethyl-N-(carboxymethyl)chitosan.[58, 166] The latter product was obtained by treating N-(carboxymethyl)chitosan with 42 wt-% NaOH, adding crushed ice to the alkali-swollen solid, then reacting with 2-chloroethanol under homogeneous conditions. The ^{13}C NMR spectrum indicates that hydroxyethylation occurs primarily at the C(6)OH and that there is complete removal of all the residual N-acetyl groups present in the N-(carboxymethyl)chitosan.[166] Molecular weight measurements showed a decrease in chain length compared with the starting N-(carboxymethyl)chitosan, as would be expected in view of the alkaline conditions used in the reaction.

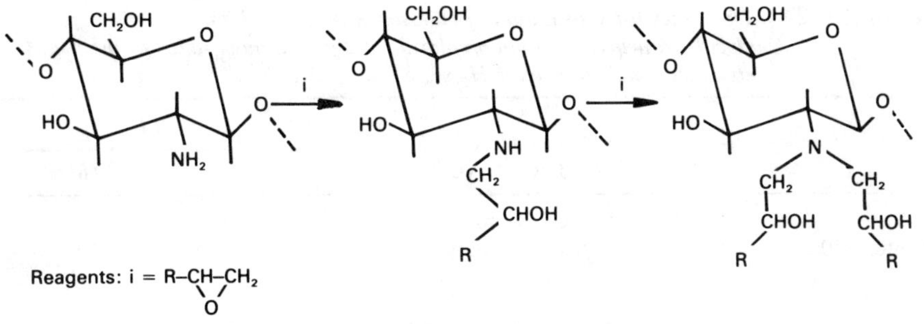

FIGURE 4.23   Scheme of reaction of epoxides with chitosan under neutral conditions

Reaction with epoxides

While O-hydroxyalkyl derivatives of chitin are formed by reaction of alkali chitin with epoxides or their precursors (section 4.3.1) they are not formed under neutral or acid conditions since the hydroxyl groups are not sufficiently nucleophilic in character to induce ring opening. However amine groups are sufficiently strong nucleophiles to react with epoxides and so the reaction offers a simple route for selective N-alkylation of chitosan. Unlike the case of reaction under alkaline conditions where poly(propylene oxide) chains can be formed, leading to MS values > 2.0 (section 4.3.1), reaction with amine groups under neutral or alkaline conditions does not give rise to chain formation so that the maximum extent of substitution is limited to two propylene oxide units per D-glucosamine residue (Figure 4.23). A number of such derivatives have been reported, mainly in patents (Table 4.23).

The reaction of chitosan with glycidol (2,3-epoxypropanol, GCD) and glycidyltrimethylammonium chloride (GTMAC), both separately and as coreactants, has been reported.[177] Two chitosan samples were derivatised, a low-molecular-weight chitosan[0.14] and a high-molecular-weight chitosan[0.24], and in all runs the mole ratio of total epoxide:–NH_2 was 3:1. The epoxides were added to a suspension of chitosan in water and the

TABLE 4.23   *Products from the reaction of chitosan with epoxides*

R (in Figure 4.23)	Reference
–H	171, 172
–CH$_3$	171, 173, 174
–CH$_2$CH$_3$	175
–CH$_2$CH$_2$OH	175–177
–CH$_2$.$\overset{+}{N}$(CH$_3$)$_3$Cl$^-$	172, 173, 176, 177
–(CH$_2$)$_{11}$CH$_3$	178

TABLE 4.24 DS *values for substitution of chitosan with glycidol and glycidyltrimethylammonium chloride at different mole ratios of the two epoxides (total epoxide:–NH$_2$ mole ratio is 3:1)*[177]

	Mole ratio of GCD:GTMAC	DS	
		GCD	GTMAC
Chitosan[0.14]	GCD only	1.9	—
	GTMAC only		1.0
	2.0	1.5	0.4
Chitosan[0.24]	4.0	1.76	0.15
	2.65	1.33	0.17
	2.0	1.31	0.24
	1.61	1.23	0.28
	1.31	1.30	0.28
	1.0	1.17	0.36
	0.8	1.04	0.40

reaction carried out for 6 h at 80°C, by which time a highly viscous solution had formed, following which the products were isolated by precipitation in acetone. The results are given in Table 4.24 and show that the efficiency of the reaction between chitosan and GCD is considerably greater than that for the reaction with GTMAC. All the products were soluble in water, and those having a DS < 0.25 for substitution with GTMAC were also soluble in solutions containing an anion surfactant (120 g dm^{-3}).

A second paper has examined the effects of pH and reaction temperature on the extent and position of reaction of propylene oxide with chitosan,[174] using dimethoxymethane as the reaction medium and keeping the mole ratio of propylene oxide:D-glucosamine residue constant at 12:1. In the first set of experiments (Table 4.25) the temperature and time were kept constant at 40°C and 24 h respectively, and the effect of carrying out the reaction under alkaline, neutral or acid conditions examined.

The results in Table 4.25 show that the extent of reaction is considerably greater under mildly alkaline conditions than it is under neutral, acid, or strongly alkaline conditions, but that under all the alkaline conditions investigated only about one-third of the propylene oxide molecules that react do so with the amine group. In comparison, under neutral or acid conditions almost 100% react with the amine groups.

In the second set of experiments the effect of temperature was examined under alkaline, neutral and acid conditions (Table 4.26). Increasing the reaction temperature causes an increase in the overall extent of substitution for both alkaline and acid conditions, but with a considerable decrease in the fraction of reacting propylene oxide molecules that react with the amine group. However increasing the temperature for the neutral reaction system causes a much smaller increase in the overall extent of

TABLE 4.25   *Effect of catalyst on the extent and position of substitution in the hydroxypropylation of chitosan at 40°C[174]*

Catalyst	Total MS[a]	Substitution distribution		% of substitution on −NH₂
		O	N	
1.25M NaOH[b]	1.40	0.99	0.41	29
2.5M NaOH[b]	1.40	0.95	0.46	33
5.0M NaOH[b]	0.50	0.35	0.17	33
5.0M NaOH[c]	0.46	0.30	0.16	35
None	0.25	0.01	0.24	96
1M HCl	0.62	0.02	0.60	96

[a] *MS* is used in preference to *DS* since poly(propylene oxide) side chains may be produced under alkaline conditions.
[b] The NaOH was added to a suspension of chitosan in dimethoxyethane.
[c] The 5.0M NaOH added to chitosan to form alkali chitosan which is suspended in dimethoxyethane.

TABLE 4.26   *Effect of temperature on the extent and position of substitution in the hydroxypropylation of chitosan under alkaline, neutral and acid conditions[174]*

Catalyst	Temperature (°C)	Total MS	Substitution distribution		% of substitution on −NH₂
			O	N	
5.0M NaOH[a]	40	0.46	0.30	0.16	35
5.0M NaOH[a]	75	3.69	3.26	0.43	12
5.0M NaOH[a]	90	4.18	3.12	0.36	9
None	40	0.25	0.01	0.24	96
None	75	0.33	0.33	0.00	100
None	90	0.40	0.40	0.00	100
1M HCl	40	0.62	0.02	0.60	96
1M HCl	75	1.65	0.30	1.35	82
1M HCl	90	1.66	0.46	1.20	72

[a] Alkali chitosan prepared first then added to the dimethoxyethane.

reaction but the reaction becomes specific for the amine group, no O-hydroxypropylation occurring.

The use of longer chain epoxides gives chitosan derivatives having detergent properties.[178] Reacting 1,2-epoxytetradecane with chitosan in 0.1M acetic acid–methanol gave a product that exhibited strong surface activity, reducing the surface tension of 0.1M acetic acid to < 36 dynes cm⁻¹, and good foam stability.[178] One interesting feature of its behaviour was that, unlike conventional surfactants, the presence of low-molecular-weight electrolyte enhanced the foam stability rather than decreasing it.

Reaction with sultones

Sultones undergo a ring-opening reaction with nucleophilic reagents such as amines

$$R\text{-}NH_2 + R'\underset{SO_2}{\overset{O}{\diagdown}} \longrightarrow R\text{-}NH\text{-}R'\text{-}SO_3H \longrightarrow R\text{-}N(R'\text{-}SO_3H)_2$$

and the reaction between chitosan[0.20] and two sultones – 1,3-propane sultone and 1,8-naphthol sultone – has been studied.[179] The conditions for the reactions are given in Table 4.27. Small samples of chitosan film were included in the reaction mixtures and their IR spectra recorded to monitor the reaction. The film treated with 1,3-propane sultone showed new bands at 1470, 880 and 740 cm^{-1} and an increase in the 2900 cm^{-1} band, all attributable to the introduction of $-(CH_2)_3-$ groups. In addition there was a shift in the –NH– band from 1595 cm^{-1} to 1540 cm^{-1}. The 1,8-naphthol sultone derivative showed bands typical of aromatic compounds – 1605 and 1510 cm^{-1} (C=C) and 3080, 830, 765 and 665 cm^{-1} (C–H).

TABLE 4.27   *Reaction of chitosan^a with sultones^{179}*

	Mole ratio^b	Time (h)	DS
1,3–Propane sultone	3.5:1	48	0.58
	8.75:1	168	1.2
1,8–Naphthol sultone	3:1	48	—
	7.5:1	168	—

^{a} 1 g chitosan[0.20] suspended in 100 cm^3 methanol; reaction temperature, 50°C.
^{b} Mole ratio of sultone:–NH$_2$.

The more highly substituted 1,3-propane sultone derivative was soluble in water, 0.1M acetic acid, 50 vol.-% formic acid, DMSO, DMF and CH$_2$Cl$_2$, and was swollen in pryidine, while the less highly substituted one was soluble in 0.1M acetic acid, 50 vol.-% formic acid and DMSO, and partially soluble in DMF. The 1,8-naphthol sultone derivatives were insoluble in all solvents examined.

4.3.4 Preparation of N-,O-carboxymethyl derivatives

Hayes[180] has described the preparation of *N-,O*-carboxymethyl chitosan by treatment of chitosan as a slurry in a suitable diluent such as IPA, butanol or methyl ethyl ketone, with concentrated NaOH and monochloroacetic acid. Typical mole ratios for chitosan:NaOH:MCA are 1:5:2.5 with the reaction being carried out at 70°C. The ^{13}C NMR spectrum of a sample prepared using labelled Cl–^{13}CH$_2$COONa confirmed the presence of both –NH–^{13}CH$_2$COONa and –O–^{13}CH$_2$COONa groups, with the latter predominating, and a total combined DS of ~1.0. A similar product was also produced from chitin, using 16M NaOH rather than 10M NaOH and a NaOH:N-acetyl-D-glucosamine mole ratio of 9:1 to bring about an initial deacetylation of the chitin, followed by treatment with monochloroacetic acid as described above.

Muzzarelli and Giacomelli[181] have proposed an alternative two-stage process from chitosan in which the chitosan is initially slurried in 42 wt-% NaOH and treated with monochloroacetic acid at 20°C for 24 h. The reaction mixture is then neutralised and a reductive alkylation reaction is carried out with glyoxylic acid (section 4.3.3) also at 20°C. The advantage claimed for this method is that the reactions are carried out in water and without heating.[181] However a disadvantage is that the reaction becomes a homogeneous one during the first stage so that the final product has to be precipitated out from solution before it can be isolated, whereas with the Hayes' method[180] it only requires to be filtered off and can be easily washed with aqueous methanol.

N-,O-Carboxymethyl chitosan is soluble at 2 > pH > 6. Between pH 2 and 6 it precipitates out as a gelatinous solid.[182] It complexes with heavy metal ions and large organic cations, forming precipitates.

The work of Hayes calls into question the conclusion of Park et al.[145] (section 4.3.2) that their process of simultaneous deacetylation and carboxymethylation leads to selectively O-carboxymethylated chitosan. The carboxymethylation reaction will continue after a reasonable degree of deacetylation has taken place and it is therefore reasonable to assume that N-carboxymethylation will also take place, particularly during the later stages of the reaction.

Relevant information regarding the relative strengths of –O$^-$ and –NH$_2$ groups as nucleophiles is very sparse. The Schotten–Baumann reaction in classical organic chemistry involves a nucleophilic substitution reaction in which aromatic amines function more efficiently as nucleophiles than do the OH$^-$ ions present, while the results of Hayes[180] clearly demonstrate that in another nucleophilic substitution reaction the amine groups of chitosan can function effectively, if not so effectively as the –O$^-$ ions of alkali chitosan, in the reaction with monochloroacetic acid, although at a higher temperature than that used by Park et al.[145] However the effect of increasing the reaction temperature in a nucleophilic addition reaction involving the –O$^-$ and –NH$_2$ groups of alkali chitosan has been shown by Maresch et al.[174] to favour reaction of the –O$^-$ at the expense of the –NH$_2$ group (Table 4.26). Hence if both N- and O-carboxymethylation occur at 70°C,[180] the relative extent of N-carboxymethylation should be greater at 30°C.

Muzzarelli and Giacomelli's two-stage process[181] duplicates the method of Park *et al.*[145] in the first stage, except that chitosan rather than chitin is used, and it is therefore very probable that the product at the end of the first stage is an N-,O-carboxymethyl chitosan rather than an O-carboxymethyl derivative as is implied by the second stage reaction used.

4.4 SCHIFF'S BASE DERIVATIVES OF CHITOSAN

Schiff's base derivatives of chitosan have not been studied to the same extent as have N-acylchitosans or ethers of chitin and chitosan, presumably because of the much lower stability of Schiff's bases. Indeed it is their ease of hydrolysis under acid conditions that has led to their use as amine-protecting groups during derivatisation of the hydroxyl groups, such as the use of N-salicylidenechitosan in the synthesis of O-carboxymethyl and O-sulphoethyl chitosan [142, 143] (section 4.3.2) and di-O-acetylchitosan[95] (section 4.2.3).

Moore and Roberts obtained Schiff's base derivatives from the reaction of chitosan[0.24] film with a range of aliphatic aldehydes, both linear – acetaldehyde to decanal – and branched, and aromatic aldehydes.[105] Schiff's base formation was carried out by treating films presteeped in methanol with methanolic solutions of the aldehydes at room temperature for 24 h. The IR spectra of Schiff's bases from aromatic aldehydes showed a very strong band at 1650–1630 cm^{-1} arising from the C=N stretch,[183] together with the expected aromatic ring bands at ~1600 and 1500 cm^{-1} and the C–H bands in the 900–650 cm^{-1} region. The C=N stretching band is very weak in aliphatic Schiff's bases[183] and, although the general shape of the amide I band was different with some of the products, no specific band assignable to the C=N stretch could be identified owing to the masking effect of the amide I band, and the most noticeable difference in the spectra was the increase of the C–H band at ~2900 cm^{-1}. Conversion of a film of N-benzylidenechitosan to N-salicylidenechitosan by treating the former with a methanolic solution of salicylaldehyde could be followed by monitoring the change in the IR absorption bands in the 900–650 cm^{-1} region or the change from colourless to yellow. Heterogeneous reaction of chitosan with a solution of salicylaldehyde in methanol and measurement of the reflectance value at 410 nm, which is λ_{max} for N-salicylidenechitosan, has been used as a method for determining the amount of chitosan adsorbed/deposited on paper handsheets.[66]

The Schiff's base derivatives, both aliphatic and aromatic, appear to be more reactive to further derivatisation through chemical reactions of the hydroxyl groups than are the corresponding N-acylchitosans containing the same number of C atoms. This was attributed to intramolecular steric hindrance of either the C(3)OH or C(6')OH group by the C(2)NHCOR

TABLE 4.28   *Reaction of chitosan with aldehydes[187]*

Chitosan	Aldehyde	Reaction system[a]	DS[b]
Chitosan[0.10]	Hexanal	1	0.98
	Octanal	1	0.86
	Decanal	1	1.00
	Dodecanal	1	0.89
Chitosan[0.50]	Hexanal	2	0.90
	Decanal	2	0.74
	Dodecanal	2	0.60

[a] Reaction system 1: 12.5 g dm^{-3} solution of chitosan in 50 vol.-% methanol–aqueous acetic acid.
   Reaction system 2: swollen precipitate in pyridine.
[b] Based on the extent of reaction of the available, non-acetylated amine groups.

group,[105] but a more feasible explanation is that it is due to the difference in accessibilities of individual chains. The *N*-acyl groups, when large enough, can increase chain separation in the *c* axis direction but have a much smaller effect in the *a* axis direction since interchain C(2$_1$)NH . . .O=C(7$_3$) hydrogen bonds can still form. However replacement of *N*-acyl groups by *N*-alkylidene or *N*-arylidene groups removes the possibility of these interchain hydrogen bonds, thereby increasing accessibility in the *a* axis direction as well as in the *c* axis direction. The X-ray diffraction patterns of both *N*-benzylidene- and *N*-salicylidenechitosan were diffuse, indicating that they were amorphous.[105]

Chitosan also reacts readily with aldehydes in solution,[152, 184–187] forming Schiff's base derivatives which frequently gel (section 6.3.2). Reaction without gelation may be achieved by using a low enough concentration either of chitosan[184] or of the aldehyde.[186] Kurita et al.[187] examined the reaction between chitosan and long chain aldehydes – hexanal, octanal, decanal and dodecanal – using both chitosan[0.1] and water-soluble chitosan[0.50]. The reactions with chitosan[0.10] were carried out under homogeneous conditions in aqueous acetic acid–methanol (50 vol.-%) while the chitosan[0.50] was used in the form of a highly swollen precipitate in pyridine, similar to that used previously in *N*-acylation studies[68] (section 4.2.1). The results are given in Table 4.28, the DS values being calculated from the elemental analysis results. The IR spectra showed the expected increase in the C–H stretching bands at ~2900 cm^{-1} and also, unlike the previous IR study,[105] the C=N stretching band at ~1650 cm^{-1}. The products were swollen considerably by DMAc, DMSO and *N*-methyl-2-pyrrolidone, and were at least partially soluble in hexamethylphosphoramide and DMAc–LiCl (50 g dm^{-3}). They were also all soluble in dichloroacetic acid and surprisingly appeared, from their IR spectra, to suffer no hydrolysis on standing in solution in this solvent for up to 24 h.[187] An X-ray diffraction study showed that they were all amorphous in character and DSC measurements indicated a glass transition temperature at 126°C for the Schiff's base from chitosan[0.10] and decanal.

The homogeneous reaction between chitosan[0.16] and salicylaldehyde in aqueous acetic acid–methanol has been followed spectrophotometrically[188] using λ_{max} at 410 nm. The reaction was found to be second order, being first order in respect to the concentration of both chitosan and salicylaldehyde, and shows an increase in the rate constant during the course of the reaction. This change in rate constant was not observed in the formation of N-salicylidene-D-glucosamine and was attributed to polymer chain expansion due to increasing compatibility with the solvent – 80 vol.-% methanol in aqueous acetic acid – as Schiff's base formation increases.

The extent of reaction at equilibrium was independent of salicylaldehyde concentration above a mole ratio for salicylaldehyde:–NH_2 of 3:1, indicating complete reaction of the chitosan under these conditions. This has enabled the reaction to be used to determine the concentration of chitosan in solution.[189]

Crosslinks between chitosan molecules can be introduced by the use of dialdehydes, particularly glutaraldehyde. Although the homogeneous reaction between chitosan and glutaraldehyde normally produces a gel (section 6.3.2) Muzzarelli et al.[190] found that crosslinking without the formation of a 3-dimensional network, and hence gelation, could be achieved by using a sufficiently low concentration of chitosan, no more than 3.0 g dm^{-3} in 4 vol.-% acetic acid, and a glutaraldehyde concentration of $\sim 16.5 \text{ g dm}^{-3}$. The product could be isolated by precipitation with acetone and redissolved in aqueous acetic acid provided the precipitate was not allowed to dry, as this led to insolubilisation through the formation of a fully crosslinked structure. The formation of a limited number of crosslinks in solution was demonstrated by light scattering, the molecular weight increasing from 1.2×10^5 for the original chitosan to 5.2×10^6 for the precipitated, redissolved and dialysed material after 48 h in solution, reaching 1.1×10^7 for the same solution after 3 months.

Much use has been made of the reaction between chitosan, or chitin having a proportion of free amine groups, and glutaraldehyde for immobilising enzymes on these polymers. The enzymes immobilised include β-galactosidase (EC 3.2.1.23),[150, 191–193] glucose isomerase,[194] papain (EC 3.4.22.2),[195] amyloglucosidase (EC 3.2.1.3),[192, 196, 197] glucose oxidase (EC 1.1.3.4),[198] catalase (EC 1.11.1.6),[198] urease (EC 3.5.1.5),[199] α-galactosidase (EC 3.2.1.22),[197, 200, 201] invertase (EC 3.2.1.26),[193, 196, 202] penicillinase (EC 3.5.2.6),[203] α-amylase (EC 3.2.1.1)[196] and β-amylase (EC 3.2.1.2).[196] The most common procedure for immobilisation involves the pretreatment of chitin with glutaraldehyde, then washing to remove the excess glutaraldehyde and treating with the enzyme.[191, 192, 194, 198–200, 202] However in a number of studies the enzyme was adsorbed directly onto the chitin then treated with glutaraldehyde,[193, 195, 203] but it has not been shown whether the enzymes are linked to the chitin support through glutaraldehyde or are insolubilised on the chitin surface by glutaraldehyde crosslinks between enzyme molecules, or whether both methods of immobilisation operate. The use of the chitosan–glutaraldehyde system has been used for gel entrapment of enzymes[150, 193] and for immobilising them on crosslinked chitosan particles or beads.[193, 197, 201, 203]

Glutaraldehyde–crosslinked chitosan has been utilised in other areas such as metal ion sorption where crosslinking was introduced to prevent the chitosan from dissolving in acidic waste liquors, to increase its stability, and to reduce its degradation on extended and repeated use.[204, 205] Although its metal ion binding capacity is reduced by such crosslinking, it was considered that the improvement in physical properties outweighed the loss in capacity. In both these studies the chitosan was crosslinked under heterogeneous conditions, thereby retaining the native physical structure and ultrastructure; the former is advantageous in any filtration step and the latter aids solvent penetration owing to the numerous internal spaces and surfaces.[205] Glutaraldehyde has also been used to crosslink chitosan[206] and chitosan–poly(vinyl alcohol) membranes[207, 208] and to insolubilise chitosan after deposition on the surface of wool fibres.[209] The aim in this last study was to improve the durability, to repeated washings, of the chitosan coating which imparted shrink resistance to the wool fabric by masking the surface scales of the wool fibres. Two procedures were used: in the first the fabric was impregnated with chitosan solution, dried, then crosslinked with glutaraldehyde; in the second glutaraldehyde in the form of its bisulphite adduct was added to the chitosan solution prior to its application to the fabric, crosslinking being initiated subsequently by heating.

4.5 MISCELLANEOUS DERIVATIVES

4.5.1 Reaction with halogenated heterocyclic ring systems

These compounds undergo nucleophilic substitution and hence will react with $-O^-$ and $-NH_2$ groups (Figure 4.24). Their ease of reaction depends on the number of N atoms in the heterocyclic ring and on the nature of the other groups attached to it.[210] Dichloro- and monochloro-s-triazine ring systems were the basis of the first commercially successful fibre-reactive dyes, the former being the more reactive of the two types, readily undergoing reaction at room temperature, while the monochloro-s-triazine ring requires temperatures ~85°C. Although the products are technically ethers and secondary amines, containing C–O–C and C–NH–C links respectively, they are much

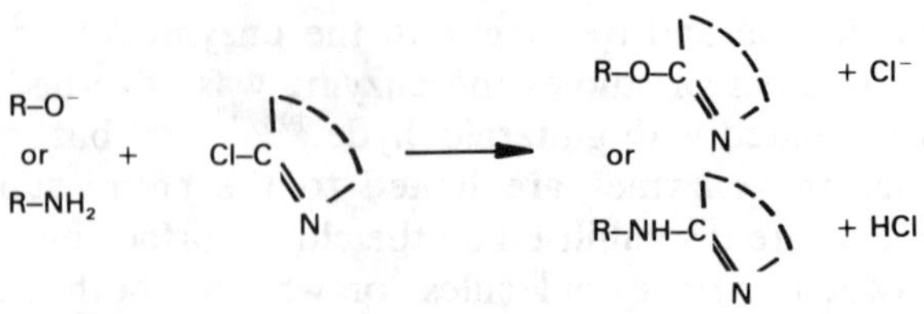

FIGURE 4.24   Nucleophilic substitution reactions of a halogenated heterocyclic substrate
with an alkoxide ion or amine group

more susceptible to hydrolysis under alkaline or acid conditions than are
ethers and secondary amines. In fact the stabilities of the bonds involved are
closer to those of esters and amides because of the adjacent –C=N– group,[210]
and it is for this reason, together with the mild conditions used in their
formation, that they are not included in section 4.3.

The reaction between a reactive dye and chitin or chitosan has been the
subject of a few studies, the first being that of Krichevshii and Sador[211] who
investigated the reaction of the dichloro-s-triazinyl dye Procion Yellow
M-R (C.I. Reactive Yellow 4) with chitosan. Reaction with the –NH$_2$
groups occurred under both alkaline and mildly acid conditions but the
–OH groups only reacted under alkaline conditions, the undissociated
hydroxyl group being too weak a nucleophile. A patent issued shortly after
their work gave information on the reaction of a range of dichloro-s-
triazinyl dyes with chitosan.[212] Hackman and Goldberg[213] dyed chitin by
reaction with Procion Red M-G (C.I. Reactive Red 8) and Remazol
Brilliant Violet 5R (C.I. Reactive Violet 5) under alkaline conditions and
used the product as a substrate for enzymatic degradation studies, following the degradation by monitoring the release of colour into solution.

More recently Allan and Hirabayashi have reacted chitosan with
chloro-s-triazine-based dyes with the objective of forming polymeric
dyes.[214] They found that at low levels of substitution they were soluble in
dilute acetic acid and could be applied to glass fibres to produce substantive
colours. However at higher levels of substitution the polymeric dyes were
insoluble and this was attributed to the formation of interchain ionic
crosslinks between protonated amine groups and the sulphonic acid solu-
bilising groups of the reactive dyes. The authors also examined the preparation of tinted contact lenses by dyeing the preformed lens, formed by
molding from a chitosan film, with a blue reactive dye.

The reaction of chitosan with the halogenated heterocyclic compounds
4-azido-2,3,5,6-tetrafluoropyridine (**4.18a**) and 4-azido-3,5-dichloro-2,
6-difluoropyridine (**4.18b**) produces photoactivatable polymers suitable for
binding proteins.[215] Incorporation of a spacer link was achieved by reaction
of the tetrafluoro compound with 4-aminobutanol followed by formation of
the tosylate derivative which subsequently underwent reaction with the
amine groups in chitosan (Figure 4.25). A *DS* of ~0.9 was achieved with

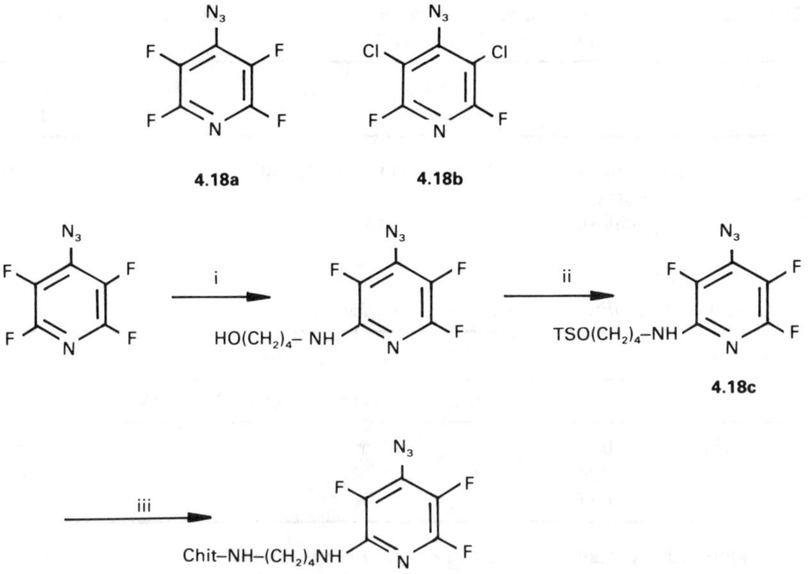

Reagents: i = $HO(CH_2)_4NH_2$ – TEA; ii = TsCl; iii = chitosan – TEA

FIGURE 4.25 Reagents used in the preparation of photoactivatable chitosan for immobilising proteins[215]

4.18a and **4.18b** on reacting at 80°C for 72 h but with **4.18c**, which is not stable, a *DS* of only ~0.13 was obtained. However this latter product, unlike those prepared from the other two reagents, was found to be soluble in dilute aqueous acid. Two factors may contribute to this: the lower *DS* value and the fact that on reaction with **4.18c** the C(2)–N retains its basic amine character. Since the photoactivation step can only occur at the surface, the amount of enzyme bound, β-galactosidase (EC 3.2.1.23), is very dependent on the particle size, as is the retained specific activity of the bound enzyme.

4.5.2 Tosyl and trityl derivatives

Kurita *et al.* have recently initiated a study of multi-step chemical transformations of chitin and chitosan with the aim of carrying out regioselective derivatisations. One of the key intermediates for such modifications of chitin is the 6-*O*-tosyl derivative. This was synthesised in a two-phase system consisting of an aqueous alkali chitin solution and a chloroform solution of toluene sulphonyl chloride (tosyl chloride), with a reaction time of 1 h at 0°C followed by 4 h at room temperature.[216, 217] The increase in *DS* with increase in mole ratio of tosyl chloride is given in Table 4.29;

TABLE 4.29   *Tosylation of chitin in a two-phase system[a] – effect of tosyl chloride concentration on extent of reaction[216]*

Mole ratio[b]	5	10	15	20	30
DS	0.15	0.58	0.87	1.10	1.15

[a] Solution of alkali chitin containing 1.4 wt-% chitin[0.82] and 12 wt-% NaOH; tosyl chloride dissolved in chloroform.
[b] Mole ratio of tosyl chloride:N-acetyl-D-glucosamine.

TABLE 4.30   *Solubility of O-tosyl chitins and iodo-chitin[217]*

Chitin derivative		Solubility[a]				
	DS	$CHCl_3$	DMF	DMSO	HMPA	NMP
O-Tosyl chitin	0.43	±	±	±	±	±
O-Tosyl chitin	0.55	±	+	+	±	+
Iodo-chitin	0.66[b]	−	+	+	−	+

[a] (+): soluble; (±): partially soluble or swollen; (−): insoluble.
[b] The iodo-chitin was prepared from an O-tosyl chitin sample having a DS of 0.66.

substitution was assumed to take place at the C(6)OH except for those derivatives having $DS > 1.0$. The DS levels obtained are in contrast to those obtained by Terbojevich et al.[42] for the reaction between tosyl chloride and chitin dissolved in DMAc–LiCl (50 g dm^{-3}) where a product having a DS of ~0.15 was obtained after 24 h reaction at 25°C. There is no value given for the mole ratio of tosyl chloride: N-acetyl-D-glucosamine, but for the analogous syntheses of acyl esters mole ratios of 24–27 were used.[42]

Replacement of the tosyl group by iodine to give 6-iodo-chitin was readily achieved by heating with excess NaI in DMSO – the low DS values obtained by Terbojevich et al.[42] were attributed to displacement of the tosylate group, once formed, by Cl$^-$ ion. Both 6-O-tosyl chitins and 6-iodo-chitins with DS values > 0.5 were readily soluble in a range of polar solvents[217] in addition to their expected solubilities in formic acid and dichloroacetic acid (Table 4.30).

Lewis acids such as $SnCl_4$ react with iodo-chitin, abstracting the iodide to form a carbonium ion capable of initiating cationic polymerisation of vinyl monomers such as styrene[216] (Figure 4.26). Details of the reaction conditions and the copolymers produced are given in Table 4.31. The IR spectrum of the graft copolymers shows bands characteristic of polystyrene as well as those characteristic of chitin.

The carbonium ion formed with $SnCl_4$ will also react with glucals, 1,2-unsaturated D-glucose derivatives,[217] but copolymers are not obtained since glucals do not undergo homopolymerisation (Figure 4.27). Branched

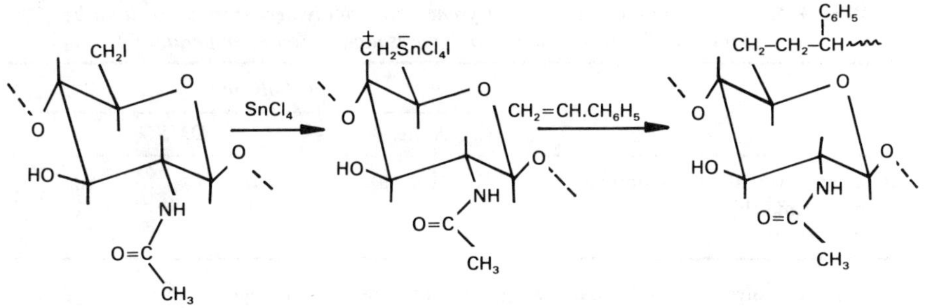

FIGURE 4.26 Formation of a chitin–polystyrene graft copolymer by cationic
polymerisation[216]

TABLE 4.31 *Formation of chitin–polystyrene graft copolymers from iodo-chitin by cationic polymerisation at 10°C[216]*

	Solvent				
	Nitromethane		Nitrobenzene		
Mole ratio SnCl$_4$:iodo-chitin	2.5	5.0	1.25	2.5	5.0
Reaction time (h)	1	1	5	5	5
Grafting (%)[a]	60	207	42	52	300

[a] Weight of polystyrene side chain/weight of chitin, as a percentage.

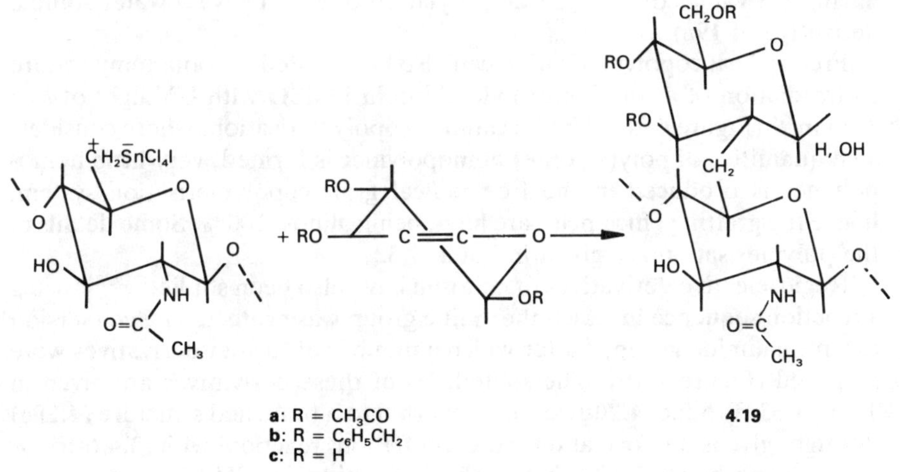

a: R = CH$_3$CO
b: R = C$_6$H$_5$CH$_2$
c: R = H

4.19

FIGURE 4.27 Formation of branched derivatives of chitin by reaction of glucals with
chitin carbonium ions[217]

TABLE 4.32   *Formation of chitin–polystyrene graft copolymers from iodo-chitin by photoinitiated free radical polymerisation at room temperature[216]*

	Solvent		
	None	DMSO	
Mole ratio styrene:iodo-chitin	150	50	150
Reaction time (h)	1	1	1
Grafting (%)[a]	0	52	63

[a] Weight of polystyrene side chains/weight of chitin, as a percentage.

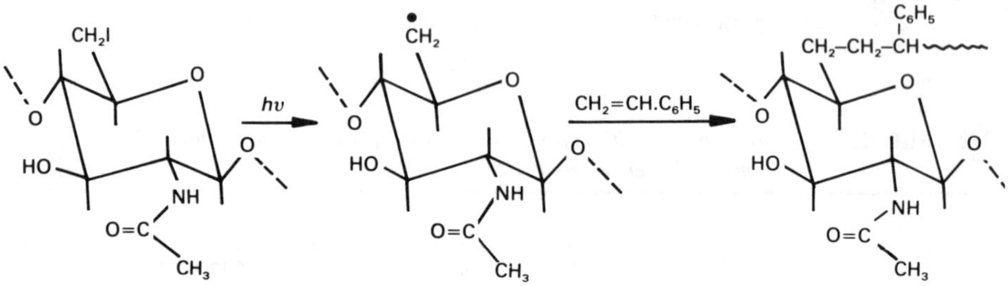

FIGURE 4.28   Formation of a chitin–polystyrene graft copolymer from iodo-chitin by free radical polymerisation[216]

chitins in which R = CH$_3$CO (**4.19a**) and R = C$_6$H$_5$CH$_2$ (**4.19b**) could be formed, with the former being produced more readily giving products having *DS* values of 0.44, de-*O*-acetylation of which gave a water-soluble derivative (**4.19c**).

Free radical copolymerisation can also be initiated at room temperature by irradiation of a solution of iodo-chitin in DMSO with UV light of λ = 308 nm[216] (Figure 4.28). Unlike cationic copolymerisation, where considerable quantities of poly(styrene) homopolymer is formed, very little homopolymer is produced in the free radical graft copolymerisation system, hence the grafting efficiencies are high, being almost 100%. Some details of the polymerisation are given in Table 4.32.

Regioselective derivatives of chitosan have also been studied[217, 218] using a reaction sequence in which the amine group was protected by conversion to a phthalimide group,[77] after which a number of further derivatives were prepared (Figure 4.29). The solubilities of these derivatives are given in Table 4.33. Product **4.20c** can be converted to a branched structure (**4.20e**) through glycoside formation at C(3)OH on reaction with 3,4,6-tri-*O*-acetyl-*N*-(phthalimido)-D-glucosaminyl bromide in CH$_2$Cl$_2$ in the presence of collidine and silver triflate.[217]

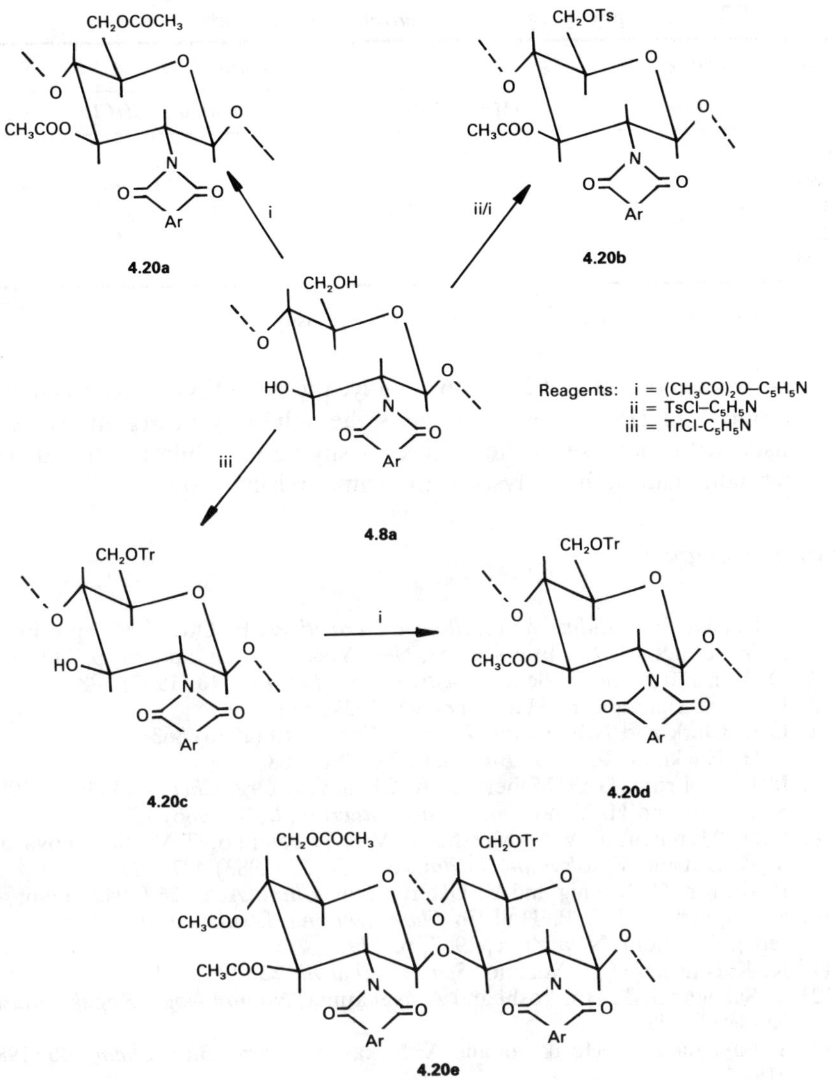

FIGURE 4.29 Regioselective chemical modifications of N-phthalimidochitosan[218]

4.5.3 Trimethylsilyl derivatives of chitin

These may be prepared by treating formamide-swollen chitin (5 g chitin suspended in 200 cm³ formamide at 80°C) with a large excess of hexamethyldisilazone for 2 h.[219] Although the cellulose and dextran I derivatives

TABLE 4.33 *Solubility of regioselective derivatives of chitosan[218]*

Chitosan derivative	Solubility[a]					
	DMF	DMAc	DMSO	Pyridine	$CHCl_3$	CH_2Cl_2
4.8	+	+	+	+	−	−
4.20a	+	+	+	+	+	+
4.20b	±	±	±	+	−	−
4.20c	+	+	+	+	±	+
4.20d	+	+	+	+	+	+

[a] (+): soluble; (±): partially soluble or swollen; (−): insoluble.

had *DS* values of 3.0 and 2.2 respectively, the trimethylsilyl derivative of chitin only had *DS* = 0.6. In all cases the solubility in organic solvents increased with increase in the degree of silylation, while treatment with water readily caused hydrolysis of the trimethylsilyl group.

REFERENCES

1. J. Barsha, in *Cellulose and Cellulose Derivatives*, E. Ott, H.M. Spurlin and M.W. Grafflin (eds), Interscience, New York, vol. 2, 2nd edn, p. 713.
2. O. Von Fürth and E. Scholl, *Beitr. Chem. Physiol.*, **10** (1907) 188.
3. P.P. Schorigin and E. Hait, *Ber.*, **67** (1934) 1712.
4. G.L. Clark and A.F. Smith, *J. Phys. Chem.*, **40** (1936) 863.
5. R.H. Hackman, *Aust. J. Biol. Sci.*, **7** (1954) 168.
6. M.L. Wolfrom, G.G. Maher and A. Chaney, *J. Org. Chem.*, **23** (1957) 1990.
7. S. Hirano and H. Yano, *Int. J. Biol. Macromol.*, **8** (1986) 153.
8. G.N. Marchenko, V.N. Marsheva, V.I. Kovalenko, T.M. Safronova and V.M. Datsun, *Vysokomol. Soedin., Ser. B*, **25** (1983) 427.
9. P. Karrer, H. Koenig and E. Usteri, *Helv. Chim. Acta*, **26** (1943) 1296.
10. S. Coppick and W.P. Hall, in *Flameproofing Textile Fabrics*, R.W. Little (ed.), Reinhold, New York, 1947, p. 179.
11. K. Katsuura and H. Mizuno, *Sen-i Gakkaishi*, **22** (1966) 510.
12. T. Sakaguchi, T. Horikoshi and A. Nakajima, *Nippon Nogei Kagaku Kaishi*, **53** (1979) 149.
13. T. Sakaguchi, T. Horikoshi and A. Nakajima, *Agric. Biol. Chem.*, **45** (1981) 2191.
14. B. Laszkiewicz, *J. Therm. Anal.*, **30** (1985) 889.
15. S. Tokura, N. Nishi, S. Nishimura and Y. Ikeuchi, in *Chitin, Chitosan, and Related Enzymes*, J.P. Zikakis (ed.), Academic Press, New York, 1984, p. 303.
16. N. Nishi, S. Nishimura, Y. Maekita, A. Ebina, A. Tsutsumi and S. Tokura, in *Chitin in Nature and Technology*, R.A.A. Muzzarelli, C. Jeuniaux and G.W. Gooday (eds), Plenum Press, New York, 1986, p. 297.
17. N. Nishi, A. Ebina, S. Nishimura, A. Tsutsumi, O. Hasegawa and S. Tokura, *Int. J. Biol. Macromol.*, **8** (1986) 311.
18. N. Nishi, Y. Maekita, S. Nishimura, O. Hasegawa and S. Tokura, *Int. J. Biol. Macromol.*, **9** (1987) 109.

19. S.E. Lasker and S.S. Stivala, *Arch. Biochem. Biophys.*, **115** (1966) 360.
20. G.A.F. Roberts and J.G. Domszy, *Int. J. Biol. Macromol.*, **4** (1982) 374.
21. M.L. Wolfrom and W.H. McNeely, *J. Amer. Chem. Soc.*, **67** (1945) 748.
22. M.L. Wolfrom, R. Montgomery, J.V. Karabinos and P. Rathgeb, *J. Amer. Chem. Soc.*, **72** (1950) 5796.
23. J. Doczi, A. Fischman and J.A. King, *J. Amer. Chem. Soc.*, **75** (1953) 1512.
24. M.L. Wolfrom, T.M. Shen and C.G. Summers, *J. Amer. Chem. Soc.*, **75** (1953) 1519.
25. I.B. Cushing, R.V. Davis, E.J. Kratovil and D.W. MacCorquodale, *J. Amer. Chem. Soc.*, **76** (1954) 4590.
26. I.B. Cushing, *US Patent 2,755,275* (1956).
27. R.V. Jones, *US Patent 2,689,244* (1954).
28. The Upjohn Company, *UK Patent 746,870* (1956).
29. F. Hoffmann–La Roche and Co., *UK Patent 777,204* (1957).
30. K. Vogler, *US Patent 2,831,851* (1958).
31. M.L. Wolfrom and T.M. Shen Han, *J. Amer. Chem. Soc.*, **81** (1959) 1764.
32. M.L. Wolfrom, *US Patent 2,832,766* (1958).
33. K. Nagasawa, Y. Tohira, Y. Inoue and N. Tanoura, *Carbohydr. Res.*, **18** (1971) 95.
34. K. Nagasawa and Y. Inoue, *Chem. Pharm. Bull.*, **19** (1971) 2617.
35. K. Nagasawa and N. Tanoura, *Chem. Pharm. Bull.*, **20** (1972) 157.
36. R.A. Gibbons and M.L. Wolfrom, *Arch. Biochem. Biophys.*, **98** (1962) 374.
37. A.M. Naggi, G. Torri, T. Compagnoni and B. Casu, in ref. 16, p. 371.
38. B. Focher, A. Massoli, G. Torri, A. Gervasini and F. Morazzoni, in ref. 16, p. 306.
39. B. Focher, A. Massoli, G. Torri, A. Gervasini and F. Morazzoni, *Makromol. Chem.*, **187** (1986) 2609.
40. S.S. Stivala, A. Patel and S. Patel, in ref. 16, p. 379.
41. S.S. Stivala, M. Herbst, O. Kratky and I. Pilz, *Arch. Biochem. Biophys.*, **127** (1968) 795.
42. M. Terbojevich, A. Cosani, C. Carraro and G. Torri, in *Chitin and Chitosan*, G. Skjåk-Braek, T. Anthonsen and P. Sandford (eds), Elsevier, London, 1989, p. 407.
43. D. Horton and E.K. Just, *Carbohydr. Res.*, **29** (1973) 173.
44. O. Okiei, S. Nishimura, O. Somorin, N. Nishi and S. Tokura, in ref. 16, p. 453.
45. S. Hirano, J. Kinugawa and A. Nishioka, in ref. 16, p. 461.
46. R.A.A. Muzzarelli, F. Tanfani, M. Emanuelli, D.P. Pace, E. Chiurazzi and M. Piani, *Carbohydr. Res.*, **126** (1984) 225.
47. R.A.A. Muzzarelli, F. Tanfani, M. Emanuelli, E. Chiurazzi and M. Piani, in ref. 16, p. 469.
48. L.D. Hall and M. Yalpani, *J. Chem. Soc., Chem. Commun.*, (1980) 1153.
49. N. Seno, I. Matsumoto, A. Doi and K. Anno, in *Chitin and Chitosan*, S. Hirano and S. Tokura (eds), The Japanese Society of Chitin and Chitosan, Tottori, 1982, p. 159.
50. C.J.B. Thor, *US Patent 2,168,374* (1939).
51. C.J.B. Thor, *US Patent 2,168,375* (1939).
52. C.J.B. Thor, *US Patent 2,217,823* (1940).
53. G. Saito, *Cellulose Chemie*, **18** (1940) 106.
54. S.N. Danilov and E.A. Plisko, *Zhur. Obsch. Khim.*, **28** (1958) 2217.
55. J. Noguchi, O. Wade, H. Seo, S. Tokura and N. Nishi, *Kobunshi Kagaku*, **30** (1973) 320.

56. N. Nishi, J. Noguchi, S. Tokura and H. Shiota, *Polym. J.*, **11** (1979) 27.
57. S. Tokura, N. Nishi and J. Noguchi, *Polym. J.*, **11** (1979) 781.
58. H. Yamada and T. Imoto, *Carbohydr. Res.*, **92** (1981) 160.
59. S. Tokura, J. Yoshida, N. Nishi and T. Hiraoki, *Polym. J.*, **14** (1982) 527.
60. S. Tokura, N. Nishi, A. Tsutsumi and O. Somorin, *Polym. J.*, **15** (1983) 485.
61. G. Sundara Rajulu and N. Gowri, in *Proceedings 1st International Conference on Chitin/Chitosan (1977)*, R.A.A. Muzzarelli and E.R. Pariser (eds), MIT Sea Grant Program Report MITSG 78–7, 1978, p. 430.
62. L.L. Balassa and J.F. Prudden, in ref. 61, p. 296.
63. J. Noguchi, S. Tokura and N. Nishi, in ref. 61, p. 315.
64. S. Aiba, *Int. J. Biol. Macromol.*, **8** (1986) 173.
65. G.W. Rigby, *US Patent 2,040,880* (1936).
66. J.G. Domszy, Ph.D. Thesis (CNAA), Trent Polytechnic, UK, 1983.
67. T. Yaku and I. Yamashita, *Japan Patent 7,319,213* (through *Chem Abs.*, **80** (1973) 72291v).
68. K. Kurita, T. Sannan and Y. Iwakura, *Makromol. Chem.*, **178** (1977) 2595.
69. R. Yamaguchi, Y. Arai, T. Ito and S. Hirano, *Carbohydr. Res.*, **88** (1981) 172.
70. G.A.F. Roberts and K.E. Taylor, *Makromol. Chem., Rapid Commun.*, **10** (1989) 339.
71. G.C. Windridge and E.C. Jorgensen, *J. Amer. Chem. Soc.*, **93** (1971) 6318.
72. P.V. Sundaram, *Biochem. Biophys. Res. Commun.*, **61** (1974) 717.
73. K. Kurita, S. Chikaoka, M. Kamiya and Y. Koyama, *Bull. Chem. Soc. Japan*, **61** (1988) 927.
74. G.K. Moore and G.A.F. Roberts, in ref. 61, p. 421.
75. G.K. Moore and G.A.F. Roberts, *Int. J. Biol. Macromol.*, **3** (1981) 292.
76. K. Kurita, Y. Koyama, S. Nishimura and M. Kamiya, *Chem. Letters*, (1989) 1597.
77. K. Kurita, H. Ichikawa, S. Ishizeki, H. Fujisaki and Y. Iwakura, *Makromol. Chem.*, **183** (1982) 1161.
78. S.A. Barker, A.B. Foster, M. Stacey and J.M. Webber, *J. Chem. Soc.*, (1958) 2218.
79. S. Hirano, S. Kondo and Y. Ohe, *Polymer*, **16** (1975) 622.
80. S. Hirano, Y. Ohe and H. Ono, *Carbohydr. Res.*, **47** (1976) 315.
81. S. Hirano, O. Miura and R. Yamaguchi, *Agric. Biol. Chem.*, **41** (1977) 1755.
82. S. Hirano and Y. Yagi, *Carbohydr. Res.*, **83** (1980) 103.
83. S. Hirano and Y. Ohe, *Carbohydr. Polymers*, **4** (1984) 15.
84. S. Hirano, S. Tsuneyasu and Y. Kondo, *Agric. Biol. Chem.*, **45** (1981) 1335.
85. S. Hirano and T. Moriyasu, *Carbohydr. Res.*, **92** (1981) 323.
86. M. Yalpani and L.D. Hall, *Macromolecules*, **17** (1984) 272.
87. W.N. Emmerling and B. Pfannemüller, *Makromol. Chem.*, **184** (1983) 1441.
88. K. Kurita, M. Kanari and Y. Koyama, *Polym. Bull.*, **14** (1985) 511.
89. K. Kurita, A. Yoshida and Y. Koyama, *Macromolecules*, **21** (1985) 1579.
90. K. Kurita, Y. Koyama, K. Murakami, S. Yoshida and N. Chau, *Polym. J.*, **18** (1986) 673.
91. S. Shinkai, H. Tsuji, T. Sone and O. Manabe, *J. Pol. Sci., Pol. Lett. Ed.*, **19** (1981) 17.
92. K. Kaifu, N. Nishi, T. Komai, S. Tokura and O. Somorin, *Polym. J.*, **13** (1981) 241.
93. D. Gagnaire, J. Saint-Germain and M. Vincendon, *Makromol. Chem.*, **183** (1982) 593.
94. S. Tokura, N. Nishi, O. Somorin and J. Noguchi, *Polym. J.*, **12** (1980) 695.

95. G.K. Moore and G.A.F. Roberts, *Int. J. Biol. Macromol.*, **4** (1982) 246.
96. K. Sakurai, T. Shibano and T. Takahashi, *Fukui Daigaku Kogakubu Kenkyui Hokoku*, **33** (1985) 71.
97. O. Somorin, N. Nishi, S. Tokura and J. Noguchi, *Polym. J.*, **11** (1979) 391.
98. N. Nishi, H. Ohnuma, S. Nishimura, O. Somorin and S. Tokura, *Polym. J.*, **14** (1982) 919.
99. K. Kaifu, N. Nishi and T. Komai, *J. Pol. Sci., Pol. Chem. Ed.*, **19** (1981) 2361.
100. T. Ando and S. Kataoko, *Kobunshi Ronbunshu*, **37** (1980) 1.
101. T. Komai, K. Kaifu, M. Matsushita, I. Koshino and T. Kon, in ref. 16, p. 497.
102. M. Kakizaki, T. Shoji, A. Tsutsumi and T. Hideshima, in ref. 16, p. 398.
103. Y. Kawamura, S. Nagai, J. Hirose and Y. Wada, *J. Pol. Sci.*, *A-2*, **7** (1969) 1559.
104. T. Tetsutani, M. Kakizaki and T. Hideshima, *Polym. J.*, **14** (1982) 305.
105. G.K. Moore and G.A.F. Roberts, *Int. J. Biol. Macromol.*, **3** (1981) 337.
106. S.M. Hudson and U.B. Kim, paper presented at *199th Amer. Chem. Soc. Meeting (Cellulose, Paper and Textile Division) Boston, 1990.*
107. K. Ogura, T. Kanamoto, T. Sannan, K. Tanaka and Y. Iwakura, in ref. 49, p. 39.
108. S. Fujii, H. Kumagi and N. Noda, *Carbohydr. Res.*, **83** (1980) 389.
109. S. Grant, H.S. Blair and G. McKay, *Makromol. Chem.*, **190** (1989) 2279.
110. S. Grant, H.S. Blair and G. McKay, *Polym. Commun.*, **29** (1988) 342.
111. L.A. Berkovich, M.G. Tsyurupa and V.A. Davankov, *J. Pol. Sci., Polym. Chem. Ed.*, **21** (1983) 1281.
112. G.K. Moore, Ph.D. Thesis (CNAA), Trent Polytechnic, UK, 1978.
113. C.L. McCormick, D.K. Lichatowich, J.A. Pelezo and K.W. Anderson, *Amer. Chem. Soc. Polymer Preprints*, **21** (1980) 109.
114. C.L. McCormick and K.W. Anderson, in ref. 15, p. 41.
115. M. Terbojevich, C. Carraro, A. Cosani and E. Marsano, *Carbohydr. Res.*, **180** (1988) 73.
116. O. Somorin, N. Nishi and S. Tokura, in ref. 49, p. 54.
117. R.H. Hackman and M. Goldberg, *Carbohydr. Res.*, **38** (1974) 35.
118. P.P. Schorigin and N.N. Makarova-Semljanskaya, *Ber.*, **68** (1935) 969.
119. M.L. Wolfrom, J.R. Vercellotti and D. Horton, *J. Org. Chem.*, **29** (1964) 547.
120. R.C. Capozza, *German Patent 2,505,305* (1975).
121. S. Tokura and N. Nishi, in ref. 49, p. 244.
122. R. Senju and S. Okimasu, *J. Agric. Chem. Soc. Japan*, **23** (1950) 432.
123. S. Okimasu and R. Senju, *J. Agric. Chem. Soc. Japan*, **23** (1950) 437.
124. W. Brown, D. Henley and J. Öhman, *Makromol. Chem.*, **64** (1963) 49.
125. Y. Araki and E. Ito, *Eur. J. Biochem.*, **55** (1975) 71.
126. S. Hirano, in *Methods in Enzymology*, W.A. Wood and S.T. Kellogg (eds), Academic Press, New York, 1988, vol. 161, p. 408.
127. T. Sannan and H. Sobue, *Repts Progr. Polym. Phys. Japan*, **18** (1974) 701.
128. K. Shimahara, N. Nagahata and Y. Takiguchi, *Seikei Daigaku Kogakubu Kogaku Okoku*, **18** (1974) 1371.
129. R.A.A. Muzzarelli, *Chitin*, Pergamon, Oxford, 1977, p. 120.
130. S.N. Danilov and E.A. Plisko, *Zhur. Obsch. Khim.*, **31** (1961) 469.
131. S. Okimasu, *Nippon Nogei Kagaku*, **32** (1958) 383.
132. R. Trujillo, *Carbohydr. Res.*, **7** (1968) 483.
133. M. Kaneko, Y. Inoue and S. Tokura, *Rep. Prog. Polym. Phys. Japan*, **25** (1982) 759.

134. K. Ohmiya, Y. Uraki, N. Nishi, A. Tsutsumi and S. Tokura, in ref. 42. p. 439.
135. S. Baba, Y. Uraki, Y. Miura and S. Tokura in ref. 42, p. 703.
136. G. Maresch, J. Titze and G. Lang, *German Patent 3,432,227* (1983).
137. T. Amiya and R. Tsushima, *Japan Patent 61,34,004* (through *Chem. Abs.*, **105** (1985) P135843j).
138. G. Lang and T. Clausen, in ref. 42, p. 139.
139. S. Tokura, N. Nishi, A. Tsutsumi and O. Somorin, *Polym. J.*, **15** (1983) 485.
140. K. Kurita, S. Inoue and Y. Koyama, *Polym. Bull.*, **21** (1989) 13.
141. J. Noguchi, K. Arato and T. Komai, *Kogyo Kagaku Zasshi*, **72** (1969) 796.
142. E.A. Plisko, L.A. Nud'ga and S.N. Danilov, *USSR Patent 325,234* (1972).
143. L.A. Nud'ga, E.A. Plisko and S.N. Danilov, *Zhur. Obsch. Khim.*, **43** (1973) 2752.
144. L.A. Nud'ga, E.A. Plisko and S.N. Danilov, *Zhur. Obsch. Khim.*, **45** (1975) 1145.
145. J.V. Park, D.M. Park and K.K. Park, *Polymer (Korea)*, **10** (1986) 641.
146. L.A. Nud'ga, E.A. Plisko and S.N. Danilov, *Zhur. Obsch. Khim.*, **43** (1973) 2756.
147. A. Domard, M. Rinaudo and C. Terrassin, *Int. J. Biol. Macromol.*, **8** (1986) 105.
148. J.G. Domszy and G.A.F. Roberts, *Int. J. Biol. Macromol.*, **7** (1985) 45.
149. R.F. Borch, M.D. Bernstein and H.D. Durst, *J. Amer. Chem. Soc.*, **93** (1971) 2897.
150. M.S. Masri, V.G. Randall and W.L. Stanley, *Amer. Chem. Soc., Polym. Preprints*, **16** (1975) 70; also in ref. 61, p. 364.
151. M. Yalpani and L.D. Hall, *Can. J. Chem.*, **62** (1984) 975.
152. L.D. Hall and M. Yalpani, *Carbohydr. Res.*, **83** (1980) C5.
153. M. Yalpani and L.D. Hall, *Can. J. Chem.*, **59** (1981) 2934.
154. R.A.A. Muzzarelli, F. Tanfani, S. Mariotti and M. Emanuelli, *Carbohydr. Polym.*, **2** (1982) 145.
155. B. Casu, M. Colombo, T. Compagnoni, A. Naggi, E. Pivari and G. Torri, in ref. 16, p. 309.
156. R.A.A. Muzzarelli, F. Tanfani, M. Emanuelli and S. Mariotti, *J. Membrane Sci.*, **16** (1983) 295.
157. R.A.A. Muzzarelli and F. Tanfani, *Carbohydr. Polym.*, **5** (1985) 297.
158. M. Yalpani, *Tetrahedron*, **41** (1985) 2957.
159. R.A.A. Muzzarelli, F. Tanfani, M. Emanuelli and S. Mariotti, *Carbohydr. Res.*, **107** (1982) 199.
160. R.A.A. Muzzarelli and F. Tanfani, in ref. 49, p. 45.
161. R.A.A. Muzzarelli, F. Tanfani, and M. Emanuelli, *Carbohydr. Polym.*, **4** (1984) 137.
162. R.A.A. Muzzarelli, F. Tanfani, M. Emanuelli and L. Bolognini, *Biotech. Bioeng.*, **27** (1985) 1115.
163. R.A.A. Muzzarelli, *Carbohydr. Polym.*, **5** (1985) 85.
164. R.A.A. Muzzarelli, in ref. 16, p. 321.
165. R.A.A. Muzzarelli and A. Zattoni, *Int. J. Biol. Macromol.*, **8** (1986) 137.
166. R.A.A. Muzzarelli, C. Lough and M. Emanuelli, *Carbohydr. Res.*, **164** (1987) 433.
167. R.A.A. Muzzarelli, *Carbohydr. Polym.*, **8** (1988) 1.
168. R.A.A. Muzzarelli, M. Weckx, O. Filippini and C. Lough, *Carbohydr. Polym.*, **11** (1989) 307.

169. G. Biagini, A. Pugnaloni, G. Frongia, G. Gazzanelli, C. Lough and R.A.A. Muzzarelli, in ref. 42, p. 671.
170. R.H. Leonard, *Ind. Eng. Chem.*, **48** (1956) 1330.
171. G. Lang, G. Maresch, H. Wendel, E. Konrad, H. Lenz and J. Titze, *German Patent 3,541,305* (1985).
172. G. Lang, H. Wendel, E. Konrad, *German Patent 3,502,833* (1985).
173. G. Lang, H. Wendel, E. Konrad, *German Patent 3,501,891* (1985).
174. G. Maresch, T. Clausen and G. Lang, in ref. 42, p. 389.
175. G. Lang, H. Wendel and E. Konrad, *German Patent 3,223,423* (1982).
176. G. Lang, H. Wendel and E. Konrad, *German Patent 3,245,784* (1982).
177. G. Lang, E. Konrad and H. Wendel, in ref. 16, p. 303.
178. G.A.F. Roberts and K.E. Taylor, paper presented at *5th International Conference on Chitin/Chitosan, Princeton, USA, 1991*.
179. S.L. Herlihy, B.Sc. (Applied Chem.) Dissertation, Nottingham Polytechnic, UK, 1987.
180. E.R. Hayes, *US Patent 4,619,995* (1986).
181. R.A.A. Muzzarelli and G. Giacomelli, *Carbohydr. Polym.*, **7** (1987) 87.
182. D.H. Davies, C.M. Elson and E.R. Hayes, in ref. 42, p. 467.
183. F.H. Suydam, *Anal. Chem.*, **35** (1963) 193.
184. P. Broussignac, *Chim. Ind. Genie Chem.*, **99** (1968) 1241.
185. S. Hirano, R. Yamaguchi, N. Matsuda, O. Miura and Y. Kondo, *Agric. Biol. Chem.*, **41** (1977) 1547.
186. S. Hirano, N. Matsuda, O. Miura and T. Tanaka, *Carbohydr. Res.*, **71** (1979) 344.
187. K. Kurita, M. Ishiguro and T. Kitajima, *Int. J. Biol. Macromol.*, **10** (1988) 124.
188. J.G. Domszy, G.A.F. Roberts and F.A. Wood, in ref. 16, p. 311.
189. G.A.F. Roberts and S. Ude, unpublished work.
190. R.A.A. Muzzarelli, G. Barontini and R. Rocchetti, *Biotech. Bioeng.*, **18** (1976) 1445.
191. W.L. Stanley, G.G. Watters, B. Chan and J.M. Mercer, *Biotech. Bioeng.*, **17** (1975) 315.
192. W.L. Stanley, G.G. Watters, S.H. Kelly and A.C. Olsen, *Biotech. Bioeng.*, **20** (1978) 135.
193. M.S. Masri, V.G. Randall and W.L. Stanley, in ref. 61, p. 364.
194. W.L. Stanley, G.G. Watters, S.H. Kelly, J.A. Garibaldi and J.C. Schade, *Biotech. Bioeng.*, **18** (1976) 439.
195. J.W. Finley, W.L. Stanley and G.G. Watters, *Biotech. Bioeng.*, **19** (1977) 1895.
196. J. Synowiecki, in ref. 16, p. 417.
197. A. Ohtakara, G. Mukerjee and M. Mitsutomi, in ref. 42, p. 643.
198. W.H. Liu, in ref. 49, p. 144.
199. L. Iyengar and A.V.S. Prabhakara Rao, in ref. 49, p. 149.
200. A. Ohtakara, M. Mitsutomi and Y. Uchida, in ref. 16, p. 409.
201. A. Ohtakara and M. Mitsutomi, *J. Ferment. Technol.*, **65** (1987) 493.
202. A. Illanes, R. Chamy and M.E. Zuniga, in ref, 16, p. 411.
203. J. Braun, P. LeChanu and F. LeGoffic, in ref. 16, p. 416.
204. M.S. Masri, V.G. Randall and A.G. Pittman, *Amer. Chem. Soc., Polym. Preprints*, **19(1)** (1978) 483.
205. M.S. Masri, and V.G. Randall, in ref. 61, p. 277.
206. T. Uragami, in ref. 42, p. 783.

207. T. Uragami, Y. Yoshida and M. Sugihara, in ref. 49, p. 221.
208. T. Uragami, Y. Yoshida and M. Sugihara, *Makromol. Chem., Rapid Commun.*, **4** (1983) 99.
209. M.S. Masri, V.G. Randall and A.G. Pittman, in ref. 61, p. 306.
210. E. Siegel, in *The Chemistry of Synthetic Dyes*, K. Venkataraman (ed.), Academic Press, New York, 1972, vol. 6, p. 1.
211. G.E. Krichevshii and F.I. Sadov, *Izv. Vyss. Uchib. Zavadenii Tekknol. Tekstil. Prom.*, **3** (1961) 102.
212. Ciba S.A., *Belgium Patent 609,054* (1962).
213. R.H. Hackman and M. Goldberg, *Anal. Biochem.*, **8** (1964) 397.
214. G.G. Allan and Y. Hirabayashi, *Cellul. Chem. Technol.*, **18** (1984) 83.
215. S. Sicsic, J. Léonil, J. Braun and F. LeGoffic, in ref. 16, p. 420.
216. K. Kurita and S. Inoue, in ref. 42, p. 365.
217. K. Kurita, S. Nishimura, Y. Koyama, S. Inoue and O. Kohogo, paper presented at *199th Amer. Chem. Soc. Meeting, Boston, USA, 1990*.
218. S. Nishimura, O. Kohogo, K. Kurita, C. Vittavatvong and H. Kuzuhara, *Chem. Letters*, (1990) 243.
219. R.E. Harmon, K.K. De and S.K. Gupta, *Carbohydr. Res.*, **31** (1973) 407.

5

Chemical Behaviour of Chitin and Chitosan

5.1 ACID–BASE PROPERTIES OF CHITIN AND CHITOSAN

The dissociation constant K_b, of an amine group is obtained from the equilibrium

$$-NH_2 + H_2O \rightleftharpoons -\overset{+}{N}H_3 + OH^-$$

$$\therefore K_b = \frac{[-\overset{+}{N}H_3][OH^-]}{[-NH_2]} \quad \text{and} \quad pK_b = -\log K_b$$

while the dissociation constant of the conjugate acid is obtained from the equilibrium

$$-\overset{+}{N}H_3 + H_2O \rightleftharpoons -NH_2 + H_3\overset{+}{O}$$

$$\therefore K_a = \frac{[-NH_2][H_3\overset{+}{O}]}{[-\overset{+}{N}H_3]} \quad \text{and} \quad pK_a = -\log K_a$$

In the case of polyelectrolytes the dissociation constants are not in fact constant but depend on the degree of dissociation at which they are determined. With polymeric carboxylic acids the polyanion will attract H^+ ions, making dissociation more difficult, thereby decreasing K_a and increasing pK_a, while with poly(amines) the ease of dissociation of conjugate acid groups, $-\overset{+}{N}H_3$, will be increased by the presence of adjacent $-\overset{+}{N}H_3$ groups, thereby increasing K_a and decreasing pK_a. Thus the pK_a of chitosan will depend on the charge density of the polymer and hence will depend on the extent of neutralisation of the charged groups and on the degree of N-acetylation for samples having the same fraction of the $-\overset{+}{N}H_3$ groups neutralised. Early evidence for this can be found in Terayama's classic paper on colloid titration,[1] for although he did not calculate pK_a values for the chitosan samples used it is possible to calculate from his results that at a degree of neutralisation of 0.5 the pK_a for a chitosan[0.29] sample is ~6.4 while it is ~6.0 for a chitosan[0.05].

Several workers have reported pK_a values for chitosan. The first was Doczi[2] who gave pK_a ~6.2. (Doczi actually gave a value of ~7.8 for the pK_b of chitosan.) Noguchi et al.[3] gave a pK_a value of 6.3, as did Nud'ga et al.,[4] while Muzzarelli et al.[5] reported the pK_a to be 6.8. None of these workers appears to have taken into consideration the effect of chain charge density.

The first to do so were Park et al.[6] using a chitosan[0.36] sample. They evaluated their titration results by the Katchalsky–Spitinik equation[7]

$$pH = pK_a + n \log [\alpha/(1-\alpha)]$$

where α is the degree of neutralisation and n is an empirical parameter related to the free energy change during titration. They concluded that the pK_a was 6.1 when $\alpha < 0.72$ and 6.7 when $\alpha > 0.72$, the value being increased on addition of a neutral salt such as NaCl. They also observed a break in the plot of pH versus Log $[\alpha/(1 - \alpha)]$ at $\alpha = 0.72$ and concluded that this indicates a change in the chain conformation at this degree of neutralisation. Since this α value coincides with the pH value of 6.6, which is close to the pH at which the polymer starts to precipitate, they suggested that the conformational changes allow the $-NH_2$ groups present to form intermolecular hydrogen bonds, leading to formation of a precipitate, when $\alpha > 0.72$.

The most detailed study has been carried out by Domard[8] using potentiometric and CD measurements. The variation of the pK_a value during titration was calculated using Katchalsky's equation[9]

$$pK_a = pH + \log \left(\frac{1-\alpha}{\alpha} \right) = pK_0 - \frac{\varepsilon \Delta \Psi(\alpha)}{K_T}$$

where $\Delta\Psi$ is the difference in electrostatic potential between the surface of the polyion and the reference. Extrapolation of the pK_a values to $\alpha = 1$, where the polymer becomes uncharged and hence the electrostatic potential becomes zero, enables the value of the intrinsic dissociation constant of the ionisable groups, pK_0, to be estimated. The value obtained, ~6.5, is independent of the degree of N-acetylation whereas the pK_a value is very dependent on this factor (Figure 5.1).

The pK_a value for the monomer, D-glucosamine, has been reported to be 7.5[10] and 7.8,[6] and the much lower values obtained for chitosan have

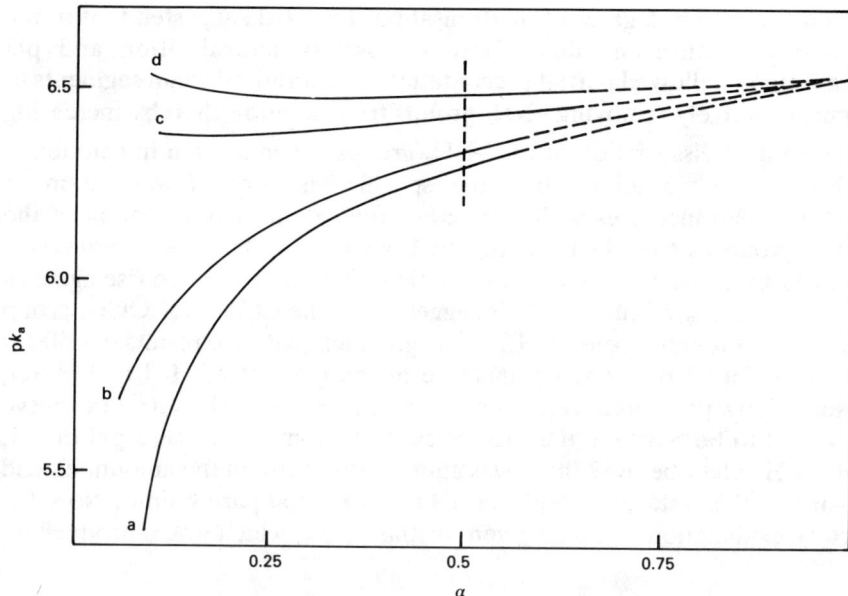

FIGURE 5.1   Variation of pK_a as a function of α for chitosans having different levels of
N-acetylation: a, chitosan[0.0]; b, chitosan[0.06]; c, chitosan[0.19];
d, chitosan[0.25]. All solutions had an amine group concentration of
5×10^{-4} mol dm^{-3} in 0.1M HClO$_4$[8]

been attributed[6] to the strong electrostatic interaction between adjacent
$-\overset{+}{N}H_3$ groups. It is therefore of interest that the extrapolated pK_a value
for $\alpha = 1$, that is pK_0, where there would be no adjacent $-\overset{+}{N}H_3$ groups,[8] is
still considerably lower than that for D-glucosamine. However the pK_a for
the methyl glycopyranoside is lower than that for the parent D-glucosamine
while substitution around the ring lowers it even further,[10] and this may
account for the low value for the pK_0 for chitosan compared with the pK_a
of D-glucosamine.

Acid–base titrations of chitosan solutions may be carried out using
either an excess[11] or stoichiometric amounts[8] of acid. Alternatively precipitated, purified salts may be used[12, 13] (section 3.2.3) or the chitosan may be
progressively titrated with HCl, using conductivity measurements to determine the stoichiometry.[14] A detailed study of the titration of acid solutions
of chitosan[0.0] has shown[8] that in the pH range 6–7 the actual pH values
obtained depend on the conditions used in the titration, the values obtained by a continuous titration procedure being considerably higher than
those obtained by allowing the system to stand, with stirring, for 36 h
before measuring the pH. Differences greater than 0.5 of a pH unit were

obtained at some degrees of neutralisation. Domard suggested that in the second procedure the delay between partial neutralisation and pH measurement allowed partial precipitation of neutralised chain segments to occur, effectively removing $-NH_2$ groups from solution thereby increasing the extent of dissociation of the $-\overset{+}{N}H_3$ groups which remain in solution.

Park et al.[6] observed that the specific viscosity of a solution of chitosan[0.36] increases with increase in the extent of protonation of the amine group as the pH is decreased from 6 to 4. The specific viscosity remains approximately constant until pH < 3.5, then starts to rise again as the pH is decreased further. This suggests that the $C(2)-NHCOCH_3$ group becomes protonated below pH 3.5, in agreement with the results of Giles et al.[15] who found that a sharp increase in the amount of HCl and H_2SO_4 absorbed by powdered chitin occurs at pH < ~4. The titration curve appeared to be reaching the theoretical maximum uptake at a pH of 2.5, but at pH values below 2 there was another sharp rise in the amount of acid adsorbed, this latter increase proceeding without apparent limit. No satisfactory explanation could be given for this result, which was reproducible.

5.2 COMPLEX FORMATION WITH METAL IONS

5.2.1 Introduction

The ability of chitosan to form complexes with metal ions, particularly transition metal and post-transition metal ions, is well documented and the reader is referred to the two comprehensive reviews by Muzzarelli[16, 17] for compilations of a considerable amount of data. The studies that have been carried out may be roughly divided into two groups: those whose primary aim is to determine whether or not chitosan will complex with a given ion, and those whose aim is to gain some understanding of the processes involved. Unfortunately the former exceed the latter in number.

5.2.2 Influence of chemical and physical structure

Comparison of results from different research workers is frequently difficult owing to the different techniques and/or conditions used in the experiments, and some of these differences may be crucial. For example, differences in the mode of agitation can have a considerable effect, not only on the rate of adsorption but also on the total amount taken up.[18] Also the physical state of the chitosan – flake, ground powder, reprecipitated powder or precipitated film – appears to have an effect on the total capacity[18, 19] in addition to any rate effects which might be anticipated because of different surface-to-volume ratios.

Another major problem in attempting comparisons is that there is frequently insufficient characterisation of the chitosan sample used. The evidence currently available supports the concept that chitosan–metal ion complex formation occurs primarily through the amine groups functioning as ligands, so that the extent of deacetylation of the chitosan is a critical factor in controlling the level of metal ion uptake, but all too frequently this information is not given. Lack of characterisation has also led to some confusion regarding the ability of chitin to complex with metal ions. Since chitin as isolated in the laboratory normally contains free amine groups (section 1.2) it would also be expected to complex with metal ions, but to a much smaller extent than occurs with chitosan, and this indeed is what is found. However the term chitin may refer to structures covering a considerable range of amine group contents (section 1.2) and indeed in one study[20] the adsorption of metal ions by chitin was investigated using a sample of 'chitin' characterised as having 7.68% N, a value which indicates (section 3.2.4) that it was chitin[0.50] and close to the borderline between chitin and chitosan.

Whether 'pure' chitin[1.0] would also complex with those metal ions is not yet conclusively established but there are some indications that it would not. Yaku and Koshijima[21] found that chitin film prepared by N-acetylation of chitosan film did not become coloured on immersion in solutions containing Cu(II), Co(II) and Ni(II) ions, although films prepared from both chitosan and partially N-acetylated chitosans did so. Supporting this is the work of Micera et al.[22] who studied the interactions in solution of D-glucosamine and N-acetyl-D-glucosamine with Cu(II) ions, using potentiometric and spectroscopic techniques, and concluded that N-acetyl-D-glucosamine did not complex with Cu(II).

While the adsorption capacity for metal ions depends on the amine group concentration of the chitosan, it is not the total concentration but the concentration of accessible amine groups that is important, since these are the ones available for complex formation. This fact has been clearly demonstrated by Kurita et at.[23] in a comparison of the adsorption behaviour of two series of chitin/chitosan substrates, one prepared by heterogeneous deacetylation and one by homogeneous deacetylation (section 2.5.7). The samples ranged from chitin[0.85] to chitosan[0.05] for the series produced heterogeneously and from chitin[0.68] to chitosan[0.12] for the one produced homogeneously. For the heterogeneous series the concentration of both Hg(II) and Cu(II) ions adsorbed under standard conditions increased with increasing extent of deacetylation, at first quite rapidly but considerably more slowly above an extent of deacetylation of 40–60%. The break in the curve approximates to the stage at which the disruption to the crystal structure due to deacetylation begins to be counteracted by the formation of crystalline regions of chitosan, thus restricting the accessibility of the amine groups to the metal ions.

In contrast to this the homogeneous series showed an initial rapid rise in

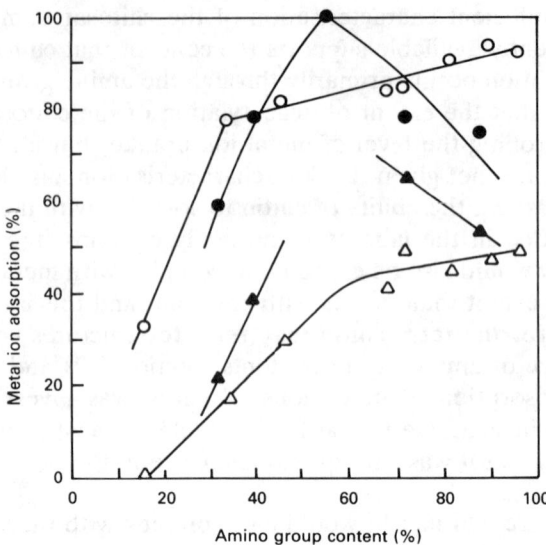

FIGURE 5.2 Sorption of Hg(II) and Cu(II) on chitosans prepared by heterogeneous and homogeneous deacetylation. ○, Hg(II); △, Cu(II) – both adsorbed on heterogeneously deacetylated material. ●, Hg(II); ▲, Cu(II) – both adsorbed on homogeneously deacetylated material[23]

metal ion adsorption with increase in extent of deacetylation but this peaked at about chitosan[0.45], after which further deacetylation gave products having decreased adsorption capacities (Figure 5.2). This behaviour was attributed to the solubility characteristics of homogeneously deacetylated materials having 45–55% deacetylation,[24] leading to a very high degree of swelling when in contact with the metal ion solutions and a much greater accessibility. Indeed the product from the interaction between the homogeneously deacetylated chitosan[0.45] and Cu(II) ions could not be isolated from the aqueous solution. Crosslinking of water-soluble chitosan[0.5] in solution with glutaraldehyde was subsequently studied by Kurita et al.[25] in order to overcome the problem of complex solubility. Surprisingly the adsorption of Cu(II) ions initially increased with increase in the amount of glutaraldehyde used in the crosslinking stage, up to a mole ratio of –CHO:–NH$_2$ of 1:1, then decreased rapidly (Figure 5.3). The initial increase in metal ion adsorption was attributed to the low levels of crosslinking in the precipitates preventing formation of closely packed chain arrangements without any great reduction in the swelling capacity. At higher levels of crosslinking the precipitates had lower swelling capacities, and hence lower accessibilities, because of the more extensive three-dimensional network and also of the more hydrophobic character with increase in the glutaraldehyde content. In a subsequent paper[26] the

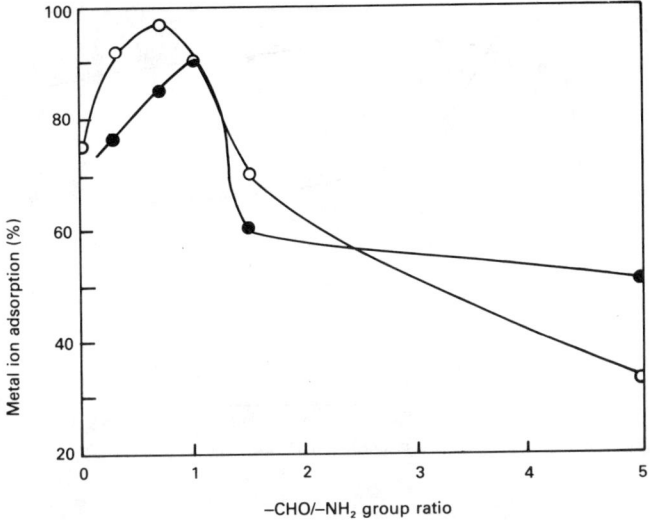

FIGURE 5.3   Adsorption of Cu(II) by water-soluble chitosan[0.5] (●) and chitosan[0.05]
(○) as a function of the extent of crosslinking with glutaraldehyde[25, 26]

adsorption capacity for Cu(II) ions of chitosan[0.05] samples that had been
crosslinked in solution with glutaraldehyde was examined and found to
follow closely the pattern established previously with crosslinked water-
soluble chitosan,[25] except that the maximum adsorption capacity was
achieved at a –CHO:–NH$_2$ mole ratio of 0.7:1 rather than 1:1 (Figure 5.3).
X-ray diffraction showed that the product formed by crosslinking with a
–CHO:–NH$_2$ mole ratio of 0.7:1 was almost totally amorphous, whereas
uncrosslinked reprecipitated chitosan showed a sharp peak at $2\theta = 20°$
indicating a considerable degree of crystallinity.

Increased adsorption capacity due to increased accessibility on *N*-
acylation of chitosan has also been demonstrated by Kurita *et al.*[27–29] The
introduction of a small number of *N*-nonoyl groups into chitosan[0.12]
initially increased the amount of Cu(II) ion adsorbed under standard
conditions, reaching a peak at a degree of *N*-nonoylation of ~5% and
decreasing with further increase in the extent of *N*-nonoylation (Figure
5.4). Similarly *N*-acetylation of chitosan[0.12] using the same preparation
process – treatment of a highly swollen precipitate of chitosan in pyridine
with an acyl chloride (section 4.2.1) – also gave products having a greater
adsorption capacity for Cu(II) ions, although in the case of *N*-acetylation
the maximum adsorption occurs at a higher degree of *N*-acylation, at
chitosan[0.28], and the subsequent decrease occurs more slowly (Figure
5.4); in fact the capacity of chitosan[0.71] produced by this method is still
considerably in excess of that of the starting chitosan[0.12] despite the

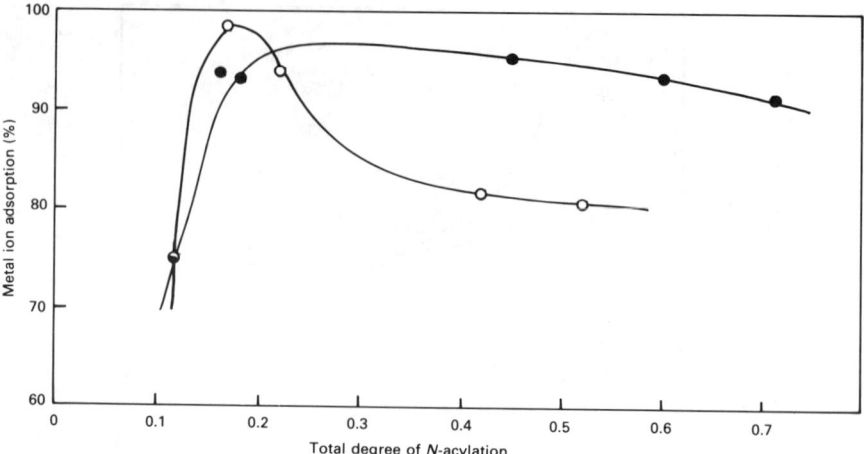

FIGURE 5.4   Adsorption of Cu(II) ions as a function of the degree of N-acylation: O, adsorption on N-nonanoylchitosan[0.12] derivatives; ●, adsorption on N-acetylchitosan[0.12][27-29]

much greater amine group content of the latter. The more rapid reduction in capacity with increase in the level of substitution with the N-nonoylchitosans was considered to be due to the onset of crystallinity in the more highly substituted samples, and this was confirmed by X-ray diffraction studies. However the reduction in crystallinity achieved by the initial N-acetylation to chitosan[0.16] is similar to that for the more highly substituted chitosan[0.71] hence there is only a very small decrease between the levels of adsorption for the two samples, possibly because in both cases there is a considerable excess of –NH₂ groups over the Cu(II) ions in the adsorption tests, and adsorption of approximately 90% of the Cu(II) ions only involves about 5% of the amine groups in the chitosan[0.16] sample and 15% of those of the chitosan[0.71] sample.

Another instance of greater adsorption capacities being shown by chitosans of lower free amine group content has been reported by Nishi *et al.*[30] who found that a sample of chitosan[0.55] adsorbed higher concentrations of transition metal ions than did a chitosan[0.03] sample (Table 5.1). This was interpreted in terms of more effective binding sites in the former involving N-acetyl groups on adjacent anhydrosugar rings rather than on greater accessibility because of greater swelling. The samples in this work[30] were deacetylated under heterogeneous conditions, but using IPA as the reaction medium, and it is possible that the presence of this diluent affects the pattern of deacetylation, tending to give a more random pattern as occurs with homogeneous deacetylation.[24]

TABLE 5.1   *Adsorption of transition metal ions at pH 7.4 by chitosans having different levels of deacetylation*[30]

Substrate	Metal ion adsorption $\times 10^3$ (mol g^{-1})			
	Ni(II)	Cu(II)	Zn(II)	Cd(II)
Chitosan[0.55]	3.5	5.3	5.5	6.5
Chitosan[0.03]	2.3	4.8	3.2	4.9

However the results reported for the chitosan[0.55] sample are peculiar in that the concentrations of adsorbed metal ions, 3.5–6.5 × 10^{-3} mol g^{-1}, are in all cases greater than the total concentration of amine groups in the sample, 2.4 × 10^{-3} mol g^{-1}. This suggests that some other mechanism of adsorption must be operating in addition to complex formation through the amine group ligands and in view of the pH used in the adsorption experiments, 7.4, one possible mechanism would be that of complex formation followed by hydrolysis, similar to that proposed by Tsezos and Volesky[31] for adsorption of U(VI) by *Rhizopus arrhizus* (section 5.2.4).

5.2.3   Studies on chitosan–metal ion complexation

As a general rule it may be said that chitosan forms complexes with transition metal and post-transition metal ions, but does not form complexes with the ions of the alkali and alkaline earth metals. Many studies have been carried out on the different capacities of chitosan for different metal ions, and a major problem in attempting comparisons between different sets of results is that of knowing whether or not the reported measurements were obtained at equilibrium. Muzzarelli and Tubertini[32] studied the rate of uptake of a number of transition metal ions by an excess of chitosan and their results indicate that equilibrium had been reached in all cases after 45 h contact time, or after 30 h with a number of the metal ions. Koshijima *et al.*[33] found that equilibrium was reached in the chitosan–Cu(II) system after 50 h at 30°C while McKay *et al.*[34] used contact times of 168 h to ensure equilibrium conditions in the same system.

All the above used ground chitosan but in a study using chitosan film, which will have a much smaller surface-to-volume ratio, Averbach measured concentration profiles of Cu(II) ions across the film thickness (~100 μm) and found that even after contact times of up to 200 h no Cu(II) ions had reached the membrane centre.[35] This slow diffusion into chitosan film agrees with the earlier finding of Muzzarelli *et al.*[5] who, in an attempt to measure the permeability of an 8 μm thick chitosan film to Cu(II) ions from a CuSO$_4$ solution, found no evidence of diffusion through the film even after a contact time of 50 h. It also explains the observation of Blair and Ho[36] that the adsorption of Cu(II) ions by a series of chitosan films treated overnight in $CuSO_4$ solution, when expressed as g kg^{-1} chitosan, decreased with increase in film thickness. The authors referred to these adsorption values as equilibrium adsorption values but in view of the findings of Muzzarelli et al.[5] and Averbach[35] this seems unlikely. Furthermore a true equilibrium adsorption value should be independent of film thickness. Unfortunately in many of the studies of metal ion adsorption the contact time is not given, making it impossible to determine the likelihood of equilibrium having been established. In view of the different rates of uptake for different metal ions,[32] measurements under non-equilibrium conditions could account for some of the conflicting results in the literature.

From their study Muzzarelli and Tubertini[32] concluded that the sequence of the uptake of the 1st transition metal series follows that of the Irving–Williams series,[37] though this is not strictly correct since for the later part of the series Muzzarelli and Tubertini found the order Zn(II) < Ni(II) < Cu(II) while the Irving–Williams series gives Ni(II) < Cu(II) < Zn(II).

However, the order found for the adsorption of metal ions by chitosan appears to vary between different studies. Two such studies are those of Koshijima et al.[33] and of Masri et al.,[38] in both of which an excess of metal ions was used, the mole ratios for M^{x+}:NH$_2$ being approximately 1.6:1 and 4:1 respectively. The order found for the metal ions common to both series is:

Koshijima et al.[33]

$$Cr(III) < Co(II) < Pb(II) < Mn(II) < Cd(II) < Ag(I) < Ni(II) <$$
$$Fe(II) < Cu(II) < Hg(II)$$

and Masri et al.[38]

$$Cr(III) < Fe(II) < Mn(II) < Co(II) < Cd(II) < Cu(II) < Ni(II) <$$
$$Ag(I) < Pb(II) < Hg(II).$$

Only Cr(III), Cd(II), Ni(II) and Hg(II) occupy the same position in both series, while both Pb(II) and Fe(II) are the most highly displaced in respect of their relative positions in the two series, the remaining ions being displaced to lesser degrees. In neither case is the Irving–Williams series followed.

In addition to the metal ions given above, chitosan has been shown to complex with inter alia Pd(II), Au(III) and Pt(IV),[38] Ti(III) and Fe(III),[39] and UO_2^{2+}.[40] Of these, complex formation occurs most readily with Pd(II), Hg(II), Pt(IV) and UO_2^{2+}, and least readily with Co(II), Mn(II) and Cr(III). In the 'hard' and 'soft' acid and base concept proposed by Pearson[41] chitosan would be expected to be classified as a hard base since the most likely groups to be involved in complex formation, R–NH$_2$,

R–OH and R–O⁻, are so classified. Furthermore Pearson proposed the empirical rule that the most stable complexes are those formed between either hard bases and hard acids or between soft bases and soft acids, so that chitosan would be expected to complex most readily with hard acids. However Pd(II), Hg(II) and Pt(IV) are classified as soft acids while Cr(III) and Mn(II) are classified as hard acids and Co(II) is 'borderline',[42] as are most of the intervening metal ions. Of the ions that complex most readily with chitosan only UO_2^{2+} is classified as hard. This suggests that complex formation between metal ions and chitosan is not as straightforward as with simple ligands and does not follow Pearson's rule.

Very few systematic studies of the physical characteristics of chitosan as a complex-forming polymer have been carried out. Hauer[43] reported a detailed study of a chitosan[0.19] sample having a total exchange capacity of 4.8 meq g⁻¹ and a volume capacity of 3.3 meq cm⁻³. Although the particles had a flat angular shape rather than the preferred spherical one, the fraction column void volume was found to be very similar to that for a column of spherical particles, 0.39 compared with 0.4 for the latter. Chitosan was found to exhibit very little volume change on immersion in acid (0.5M H_2SO_4) or alkali (1M NaOH) so that desorption of metal ions and resin regeneration do not lead to either plugging or breaking of the column owing to excessive swelling or contraction, respectively, with change in environment. This is unlike many commercial ion exchange resins which show marked volume changes.

Porosity measurements showed that chitosan has a single-phase homogeneous gel matrix structure rather than a two-phase heterogeneous (macroporous) one, and hence should have large pores giving comparatively rapid ion adsorption; a conclusion that was supported by Hauer's study of Ag(I) ion adsorption kinetics where the behaviour of chitosan was found to be closer to that of a commercial resin having a homogeneous gel structure than to that of one having a macroporous structure.[43] The adsorption of Cu(II) ions was shown to be very pH sensitive, showing a 10-fold increase in Cu(II) ion uptake on raising the pH from 3.4 to 4.8, no further increase being observed on further raising the pH to 10. Adsorption of $Cr_2O_7^{2-}$ ion was relatively independent of pH, being 4.85 mmol g⁻¹ at pH 2.4–3.4 and 4.45 mmol g⁻¹ at pH 5.0–5.7, suggesting that at the lower pH values a considerable fraction of dichromate is held by electrostatic attraction rather than by complex formation.

A detailed study of the adsorption of Cu(II) ions on chitosan has been carried out by McKay et al.[34] Equilibrium isotherms were determined and the curves analysed in terms of both the Langmuir and Freundlich equations, the former equation best describing the isotherms in the concentration range studied. The adsorption capacity was found to be independent of particle size but to decrease with increase in temperature.

The kinetics of the process were studied assuming three consecutive steps: mass transfer of solute from the bulk solution to the surface of the particle, intraparticle diffusion, and adsorption on a site in the interior of the particle – this latter step being assumed to be very rapid so that either of the first two steps will be rate-determining depending on the reaction conditions. The external mass transfer coefficient (β_L) increased with increase in agitation up to 200 rpm, after which it remained constant with increasing agitation. This indicates that the mass transfer of Cu(II) ions from the solution bulk to particle surface controls the uptake at low levels of agitation but that above a given level of agitation the rate of uptake is controlled by diffusion within the particle.

Plots of Cu(II) adsorbed *versus* $t^{0.5}$ gave linear plots over the 10–60% adsorption range, the slopes of which are proportional to the square root of the diffusion coefficient within the particles. The slopes, designated as the intraparticle diffusion coefficients K, were found to increase with increase in solution concentration of Cu(II) ion according to the equation

$$K = 1.7 \, C_0^{0.5}$$

where C_0 = initial Cu(II) ion concentration, up to a concentration of 0.2 g dm^{-3}. K also increased with increase in temperature and to some extent with agitation, but was independent of particle size. The information obtained should be of use in designing batch reactors for treatment of water to remove heavy metal ions.

5.2.4 Complex formation with fungal chitin and chitosan

Although much of the work on metal ion adsorption of chitosan has made use of chitosan produced from crustacea exoskeletons, some work has been carried out on fungal chitin and chitosan. Muzzarelli *et al.*[44–47] have studied the metal ion complexing ability of the mixed polysaccharides obtained on treating waste mycelia with 40 wt-% NaOH to convert any chitin component to chitosan. The mycelia of *Aspergillus niger*, *Streptomyces*, *Mucor rouxii*, *Phycomyces blakesleeanus* and *Choanephora cucurbitarum* were examined and no attempt was made to isolate the chitosan from the product after the deacetylation step. Other samples of mycelia were not given a deacetylation treatment but instead were given a treatment in 0.01M NaOH to remove water-soluble materials, proteins and lipids.

The products were stirred with solutions of a number of transition metal salts at a mole ratio of approximately 40:1 for –NH$_2$: metal ion (calculated on the assumption that the total weight of product added to the metal ion solution is chitosan[0.16]). The material obtained from *Streptomyces* appeared to be particularly effective in removing the metal ions from solution, removing 100% of the Cr(III), Co(II), Ni(II), Cu(II) and Zn(II) ions,

99% of the Pb(II) ions and 95% of the Mn(II) ions. Surprisingly however its uptake of Hg(II) ions was 0% in contrast to the results obtained with crustacean-derived chitosan where Hg(II) ions are among the most readily adsorbed.[33, 38] Although the other deacetylated products adsorbed Hg(II) ions, in no case was it the most readily adsorbed species, in fact with the product from *Phycomyces blakesleeanus* it was the least readily adsorbed metal ion.

The products obtained by the milder treatment, washing with 0.01M NaOH, also adsorbed metal ions from solution but to a lesser extent in most cases. Again Hg(II) showed anomalous behaviour, the product from *Streptomyces* adsorbing 90% of the Hg(II) ions while that from *P. blakesleeanus* adsorbing 0%. It was claimed that the metal ion capacity of the mixed polysaccharide product was greater than that of chitosan itself when compared on the basis of mmol g^{-1} and this was attributed to the chitosan component having a more accessible physical structure.[44]

A very interesting study of the metal ion adsorption by *Rhizopus arrhizus* has been reported by Tsezos and Volesky.[31, 48] Biosorption of U(VI) was shown by transmission electron microscopy to take place throughout the fungal cell wall but nowhere else, and this conclusion was confirmed by X-ray energy dispersion analysis. IR spectroscopy showed new bands at 908 and 372 cm^{-1} which were assigned to UO_2^{2+} ions and to the U–N stretch respectively, this latter indicating association between the N atom of chitin and uranium ions.

At pH 4 the uranium–*R. arrhizus* biosorption system reached an initial equilibrium within 60 seconds, remaining in this pseudo-equilibrium state for about 30 minutes before reaching final equilibrium within 60 minutes. Biosorption at pH 2 was considerably slower, requiring 3–4 h, and did not exhibit an initial rapid equilibrium. The equilibrium uptake of U(VI) by chitin was found to be 6 mg g^{-1} whereas that for *R. arrhizus* was ~180 mg g^{-1}, despite only occurring in the cell wall of the mycelium which contains less than 100 wt-% chitin. This means that complexed U(VI) can only account for a small fraction of the uptake and three processes were suggested. Process A involves complex formation between dissolved uranium ionic species and chitin, the U(VI) coordinating with the N atoms of the chitin – presumably at the occasional free amine group along the chitin chain. Additional U(VI) ions are adsorbed by the chitin network in the fungal cell wall, close to that complexed by the chitin (Process B). Process C involves hydrolysis of the uranium–chitin complex formed during Process A and precipitation of the hydrolysis product [$UO_2(OH)_2$, uranyl hydroxide] in the cell wall. The freed chitin N atoms may then reform complexes with U(VI) ions until the accumulation of hydrolysis products inhibits the complexation–hydrolysis–precipitation cycle, thereby giving the final equilibria.

$$\text{Chit–N} + \text{U} \rightleftharpoons \text{Chit–NU}$$

$$\text{Chit–NU} + \text{H}_2\text{O} \rightleftharpoons \text{Chit–N} + \text{UO}_2(\text{OH})_2 \downarrow$$

The reaction of N-acetyl-D-glucosamine and UO_2^{2+} gave a fine precipitate of $UO_2(OH)_2$, considered to be the hydrolysis product of a U(VI)–N-acetyl-D-glucosamine complex. The observation that biosorption at pH 2 does not exhibit two equilibria is in agreement with the proposed mechanism since at this pH only the combined process A and process B would occur, hydrolysis of complexes not being usually associated with low solution pH values. At pH 4 processes A and B represent about 66% of the total uptake.

The second paper[48] deals with thorium uptake on *R. arrhizus* where a different pattern is observed. Thorium biosorption is very rapid, reaching maximum uptake within 60 seconds, and two processes were considered to take place. The processes, which are not necessarily sequential, are process A which involves formation of a coordination complex between Th(IV) ion and chitin that is similar to the U(VI) chitin complex – evidence of this is the appearance of a new band in the IR spectrum of the cell walls at 362 cm^{-1} due to the Th–N bond – and process B in which hydrolysed thorium ions are adsorbed by the outer layers of the cell wall. Process B accounts for ~95% of the total uptake. At pH 4 the biosorption process is dominated by process B whereas at pH 2 process A is also important. The distribution within the cell differs at the two pH values. At pH 2 the adsorbed metal ion is quite evenly distributed throughout the cell wall but at pH 4 it is concentrated at the outer layer of the cell wall because of the larger size of the hydrolysis product $Th(OH)_4$ compared with the Th(IV) ion, making it more difficult for the former to penetrate the cell wall uniformly.

5.2.5   Structure of chitosan–metal ion complexes

Chitosan is frequently referred to as a chelating polymer but the evidence for chelate formation is not conclusive if the term 'chelate' is taken to have its usual meaning of the metal ion being bound by two or more ligands from the same molecule. Thus for chitosan to form a chelate requires either the involvement of –OH or –O$^-$ groups on the D-glucosamine residues as ligands, or else two or more –NH$_2$ groups from one chain binding to the same metal ion. In chitosan derived from α-chitin, where chain folding is thought to be the origin of the antiparallel arrangement of chains in the unit cell, the two –NH$_2$ groups could come from two chain segments running in opposite directions but belonging to the same chain (Figure 5.5). Studies have mainly been carried out on chitosan–Cu(II) complexes.

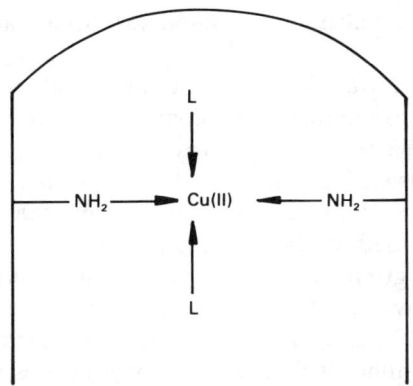

FIGURE 5.5 Possible chelate formation involving two $-NH_2$ groups from the same chitosan chain

Evidence supporting the formation of a complex in which one or more N atoms act as ligands comes mainly from ESR studies.[5, 49–51] Muzzarelli *et al.*[5] examined the ESR spectra of chitosan membranes complexed with Cu(II) ions and concluded that as the pH of complex formation increased the number of amine groups involved also increased, from 1 at pH < 5 to 3 at pH 6–8.

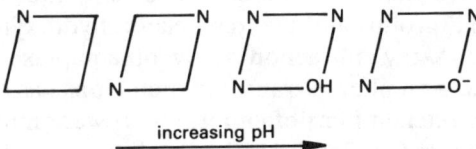

increasing pH

However it seems unlikely that three amine groups could get sufficiently close to act as ligands for one Cu(II) ion and this concept has not received support.

More recently ESR studies have been carried out on chitosan–Cu(II) complexes formed by precipitation at various pH values using either a 1:1 or a 0.01:1 mole ratio of Cu(II) : NH_2.[49, 50] The results have been interpreted in terms of two structures, designated centres I and II, the relative ratios of which depend on the pH of formation. At mole ratios of 1:1, centre I was found when the precipitate was prepared at pH 2 and centre II when it was prepared at pH 4.5, both types being found in precipitates formed at pH 2.5. Using mole ratios of 0.01:1, both centres were found in precipitates formed at pH < 4 with centre II always predominating and increasing with increase in pH of formation. Centre II was the only centre formed at pH > 4 and also in the complexes formed at any pH by chitosan oligomer.

Comparison of the ESR spectra of these chitosan–Cu(II) complexes with those of D-glucosamine–Cu(II) complexes[22] led to the suggestion that both centres involve two NH_2 groups, together with –OH groups from chitosan in the case of centre I and $–O^-$ groups with centre II. The formation of centre II at all pH values when the ligand was chitosan oligomer was explained as resulting from a difference in the acidic character of the hydroxyl groups of the oligomer, making possible their dissociation even at very low pH values. However this seems unlikely in view of the results obtained by Micera et al.[22] for complex formation between Cu(II) ions and D-glucosamine, where the hydroxyl groups of the latter showed negligible dissociation below pH 6. As both the monomer[22] and the polymer[5] show similar dissociation behaviour for their hydroxyl groups, markedly different behaviour would not be expected for the intermediate oligomer.

Micera et al.[51] have studied the ESR spectra of chitosan–Cu(II) complexes formed by heterogeneous reaction at pH values ranging from 4.6 to 11.5. They concluded that below pH 6 the complex involves only one $–NH_2$ group, together with 3 hydroxyl groups or water molecules, while over the pH range 6–7 the most likely complex involves two $–NH_2$ groups. Deprotonation of the hydroxyl groups was considered to occur over the pH range 7–9 where a third complex predominates, and this was considered to involve two $–NH_2$ groups and two dissociated hydroxyl groups.

An extensive X-ray diffraction study of complexes formed between chitosan[0.0] and a number of transition metal ions has been carried out by Ogawa et al.[52] Stretched films of chitosan[0.0] were immersed, while held at constant length, in 0.1–0.4M metal salt solution for 1 h, then washed and dried prior to carrying out the X-ray diffraction measurements. The complexes of chitosan with Co(II), Ni(II), Cu(II), Zn(II) and Hg(II) gave new diffraction patterns, the sharpness of the pattern depending on the specific metal ion and the counter ion.

The chitosan–CdSO$_4$ complex was found to have an orthorhombic unit cell with $a = 11.89$, $b = 10.22$ and $c = 44.28$ Å while that of chitosan–CdCl$_2$ complex, also orthorhombic, had $a = 11.24$, $b = 10.26$ and $c = 35.72$ Å. Calculations based on density measurements showed that in both complexes the unit cell contained eight chains. Thus complex formation increases interchain distances in both the a and c axis directions, but there is no change in the b axis direction showing that metal complex formation does not change the chitosan conformation along the fibre axis. All the reflections of each of the various chitosan–metal sulphate complexes could be indexed with the same unit cell as that of the chitosan–CdSO$_4$ complex, while the same relationship was found between the chitosan–CdCl$_2$ complex and the other chitosan–metal chloride complexes. This indicates that the base (*ac*) plane parameters of these complexes depend on the sulphate and chloride ions respectively, but are not the same as those for the chitosan sulphate and chitosan chloride salts.

Assuming a similar arrangement of chains in chitosan as in α-chitin gives a minimum N–N distance of 5.62 Å between two adjacent chains along the *a* axis, whereas the maximum length for a N–Cu(II)–N coordinated bond was calculated by Ogawa et al.[52] to be between 3.74 Å and 4.86 Å. This eliminates the possibility of chitosan–Cu(II) complexes containing two or more –NH$_2$ groups, and this also applies to the other chitosan–metal salt complexes examined since the N–Cu(II)–N bond length is the longest one of the metal ions studied. Ogawa et al.[52] also concluded that it was unlikely that the hydroxyl groups of the D-glucosamine residues were involved in the complex, since these metal ions are rarely coordinated with undissociated hydroxyl groups, and that therefore the most probable structure is one in which the metal ion is bound to an amine group of the chitosan chain in a pendant fashion. In the case of the chitosan–cadmium chloride complex the cadmium ion was considered to be coordinated with one amine group, two chloride ions and one water molecule whereas with the chitosan–cadmium sulphate complex the cadmium was considered to be coordinated with one amine group and three water molecules, the hydrated sulphate ion forming an ion pair with the complexed Cd(II) ion.

Other evidence in support of the formation of chitosan–metal ion complexes in which only one amine group bonds to each metal ion comes from maximum metal ion sorption levels such as that for Cu(II) which corresponds[34] to a metal ion:–NH$_2$ mole ratio of 0.8:1 or those for Pd(II), Au(II) and Hg(II) which correspond[38] to mole ratios of ~1:1, as is also the case[40] for UO$_2^{2+}$. These mole ratio figures would preclude any major contribution from complexes having two or more amine groups as ligands. The structure proposed by Yaku and Koshijima[21] for the soluble complex formed between chitosan oligomer and Cu(II) ions cannot apply to the analogous chitosan complex for the additional reason that it requires the destruction of the twofold helix conformation which has been shown[52] to be retained in chitosan–metal ion complexes.

Indirect evidence of the formation of chitosan–Cu(II) complexes having two amine groups bonding to one Cu(II) ion was claimed by Blair and Ho[36] who treated chitosan firms with CuSO$_4$ solutions to give samples having Cu(II) ion contents ranging from 0 to 1.02 wt-%. The treatment rendered the films more rigid and brittle than the untreated film and this was attributed to crosslinking through formation of N–Cu(II)–N interchain bonds. The rate of diffusion of an anionic dye (C.I. Acid Orange 10) through the films decreased with increase in the copper content and this was also attributed to crosslinking. This is not the sole possible explanation since the presence of a pendant complex, as proposed by Ogawa et al.,[52] would hinder diffusion of the dye through the film, both from the purely mechanical effect of introducing obstacles in the path of diffusing dye ions and, more importantly, because of the introduction of groups capable of interacting with the dye either through electrostatic interaction between $-NH_2-Cu^{2+}(L_3)$ and $Dye-SO_3^-$ or by complex formation with the dye acting as a bidentate ligand (5.1).

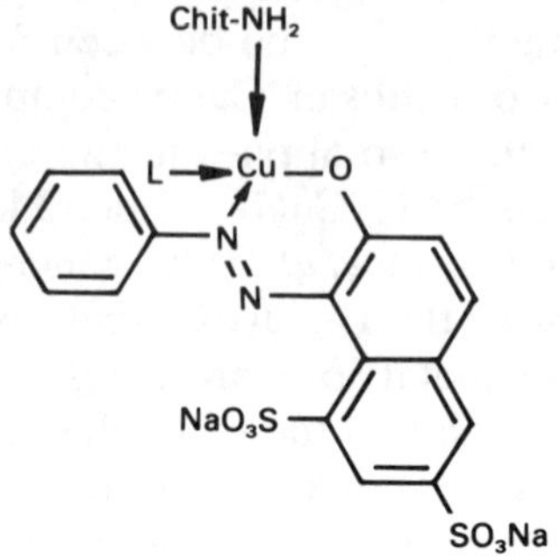

5.1

Whether or not the C(3)OH group is involved in complex formation is not definitely established, although both from studies on model systems in solution[22] and from ESR spectroscopic studies of chitosan complexes[5, 49-51] it seems clear that at pH values above 7 the dissociated C(3)O$^-$ group acts as a ligand, thereby forming a true chelate. This behaviour is similar to that of both the dissociated C(2)OH and C(3)OH groups of cellulose in forming chelate structures with copper ions in cuprammonium hydroxide solution. Whether undissociated C(3)OH groups are also involved is not so clear. From a study of the interaction in solution between chitosan[0.0] and Cu(II) ions, using potentiometric and CD measurements, Domard[8] concluded that only one complex was formed under the conditions used and assigned it structure (5.2). This is similar to that proposed by Ogawa *et al.*[52] except that two OH$^-$ groups are included as ligands. The fourth group was considered to be either a water molecule or the C(3)OH group. Micera

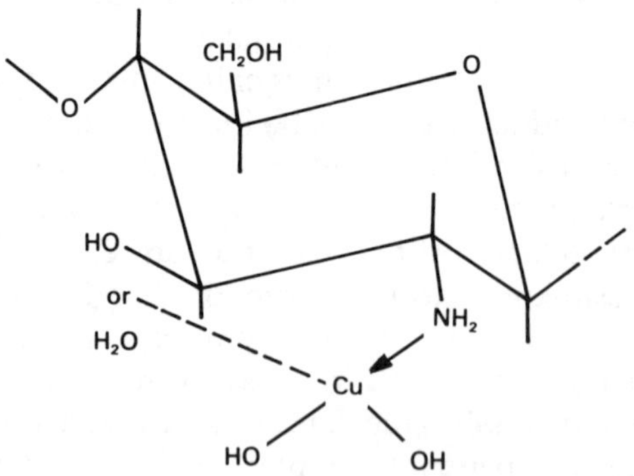

5.2

et al.[22] calculated the stability constants for several of their proposed D-glucosamine–Cu(II) complexes and based on the logarithm of the stability constant (log β_{102}) for the CuL$_2$ complex, which is about one order higher (8.76) than that for the corresponding monodentate-bonded species in the Cu(II)–NH$_3$ system (log β_{102} = 7.6) and considerably lower than that found for the (N,O) chelate coordination in CuL$_2$ complexes with amino acids, concluded that there was only a slight, if any, involvement of the hydroxyl groups of D-glucosamine in the metal coordination of the CuL$_2$ species. How applicable this conclusion is to chitosan–Cu(II) complexes is debatable since in depicting the various D-glucosamine–Cu(II) complexes the authors involved the anomeric hydroxyl group, or its dissociated form, and not the C(3)OH group which is the only one of the two available in chitosan. Furthermore about half of the sugar molecules were drawn as D-glucosamine in the $^1C_4$ conformation, in which the amine and hydroxyl groups are axial, whereas the anhydro-D-glucosamine residues in chitosan have the $^4C_1$ conformation with these groups equatorial. The remainder of the sugar molecules were drawn as L-glucosamine with the $^4C_1$ conformation, which again has the amine and hydroxyl groups axial.

On balance the evidence currently available suggests that, under heterogeneous conditions and at pH $<{\sim}6$, chitosan behaves as a poly(mononodentate ligand), while at higher pH values it behaves as a poly(bidentate ligand) forming chelates. Complex formation in solution, while allowing the formation of complexes having two amine groups bound to each metal ion, may also require alkaline pH conditions to form true chelates through involvement of the C(3)O$^-$ groups, since the two amine groups may be from different polymer chains. If they are from the same chain they are most likely to be from different segments so forming a macrocyclic chelate ring.

5.2.6 Complex formation by derivatives of chitin and chitosan

Complex formation by chitin and chitosan phosphates

The use of chitin and chitosan phosphate as complexing agents for metal ions has been reported[30, 53–58] but once again there is disagreement between the results with different ranking orders for uptake of M(II) ions reported by different workers. Furthermore Sakaguchi and Nakajima[56] state that phosphorylated derivatives of chitin and chitosan can adsorb large amounts of heavy metals such as uranium, copper and cadmium, compared with non-phosphorylated chitin and chitosan, but Nishi *et al.*[30, 58] found that the adsorption of Cd(II), Zn(II), Cu(II) and Ni(II) was better on chitosan than on the phosphorylated derivatives (Table 5.2) while Hirano *et al.*[40] found chitosan complexes readily with UO$_2^{2+}$ ions, adsorbing one mole for every mole of amine group in the chitosan. These contrary reports are difficult to explain even allowing for differences in the procedures used.

TABLE 5.2   *Adsorption of metal ions on chitin and chitosan and their crosslinked phosphate derivatives*[30, 58]

Substrate	Metal ion adsorption $\times 10^3$ (mol g^{-1})									
	Mg(II)	Ca(II)	Sr(II)	Ba(II)	Mn(II)	Ni(II)	Cu(II)	Zn(II)	Cd(II)
Chitin	0.4	0.4	0.4	0.3	0.5	0.3	0.4	0.4	0.6
Chitin phosphate	1.1	1.4	1.4	1.3	1.1	1.6	1.0	2.1	0.9
Chitosan[0.55]	0.3	0.8	1.5	1.1	1.1	3.5	5.3	5.5	6.5
Chitosan[0.55] phosphate	1.5	1.9	2.4	1.9	2.2	0.4	2.2	1.7	2.0
Chitosan[0.03]	0.5	0.4	0.6	0.8	0.5	2.3	4.8	3.2	4.9
Chitosan[0.03] phosphate	1.3	1.5	1.7	2.0	1.8	2.3	2.6	2.7	3.0

Sakaguchi and Nakajima studied the adsorption of uranium on phosphates of chitin and chitosan in some detail.[56] Both derivatives show maximum adsorption of UO_2^{2+} ions at pH 5, during the initial rapid adsorption period, with a rapid decrease above and below this pH. The adsorption increased with increase in temperature over the range 20–80°C, the enthalpy of adsorption being calculated to be 17.9 kJ mol^{-1} indicating the adsorption process to be endothermic. The adsorbed uranyl ions could be readily desorbed with dilute $NaHCO_3$ solution but nevertheless uranium could be readily adsorbed from solutions having $NaHCO_3$ concentrations of up to 0.3×10^{-3}M.

Nishi et al.[57] examined the adsorption of Cu(II) ions by water-soluble chitin[0.95] phosphate ($DS = 1.0$). The results indicate that one Cu(II) ion is complexing with two phosphate groups. The phosphate derivatives of chitin and chitosan were crosslinked by reaction with adipoyl chloride in methanesulphonic acid,[30, 58] giving insoluble materials having characteristic metal-binding properties. These are given in Table 5.2 and show that the alkaline earth metal ions and Mn(II) ions are adsorbed more readily on the crosslinked phosphate derivatives than on the starting chitin and chitosans, while the reverse is found with Ni(II), Cu(II), Zn(II) and Cd(II) for the chitosans. In general the order of complex formation for the alkaline earth metal ions is

chitin phosphate < chitosan[0.03] phosphate < chitosan[0.55] phosphate

while that for the transition metal ions is

chitin phosphate < chitosan[0.55] phosphate < chitosan[0.03] phosphate

but these orders are not absolute.

The pH dependence of Ca(II) ion binding by the phosphorylated derivatives – the capacity of crosslinked chitosan[0.55] phosphate increases about threefold on raising the pH from 4 to 9 – indicates that this binding depends primarily on the phosphate groups. Furthermore since the phosphate content of the chitosan[0.55] phosphate is 4.5×10^{-3} mol g^{-1}, and it binds 3×10^{-3} mol g^{-1} of Ca(II) ions at pH 9, the complex formed cannot involve two phosphate groups per Ca(II) ion. The authors proposed[30, 58] that either the N-acetyl group and hydroxyl groups cooperate in binding the Ca(II) ions, or that a particular balance of amine to phosphate group is required for binding the Ca(II) ions, in order to account for the greater binding by chitosan[0.55] phosphate compared with chitosan[0.03] phosphate. This difference in binding capacity increases with increase in pH and amounts to approximately 1×10^{-3} mol g^{-1} at pH 9.[30]

Complex formation by O-carboxymethyl chitin

Tokura et al.[59] have examined the interaction between 6-O-carboxymethyl chitin[0.95] and alkaline earth metal ions. The carboxymethyl chitins had $DS = 0.25$–0.28 and hence were not water-soluble.

TABLE 5.3   *Selectivity coefficients for adsorption of alkaline earth metal ions by O-carboxymethyl chitin and carboxymethyl cellulose[59]*

	Selectivity coefficient			
Substrate	Mg(II)	Ca(II)	Sr(II)	Ba(II)
CM–chitin (H$^+$)	0.98	6.89	0.44	0.62
CM–chitin (Na$^+$)	0.91	45.60		
Methyl chitin ($DS = 1.2$)	0.0	0.0		
CM–cellulose (H$^+$)	trace	0.03		
CM–cellulose (Na$^+$)		2.63		

Although Ca(II) ions were adsorbed by the carboxymethyl chitins in both the H$^+$ and Na$^+$ forms there was a large difference in the adsorption capacity of the two forms, with the Na$^+$-form adsorbing Ca(II) ions 6.6 times more strongly than the H$^+$-form, the selectivity coefficients (K^M) being 45.6 and 6.89 respectively. They also found a large difference in the selectivity coefficients shown by 6-*O*-carboxymethyl chitin for the different alkaline earth metal ions and in the selectivity coefficients for Ca(II) shown by *O*-carboxymethyl derivatives of chitin and cellulose (Table 5.3).

The high affinity of 6-*O*-carboxymethyl chitin towards Ca(II) ions suggests that a chelated complex is formed and this was supported by differential IR spectroscopy which showed that adsorption bands at 3350 cm^{-1} (–NH–), 1660 cm^{-1} (amide I), 1555 cm^{-1} (amide II) and 1050 and 1110 cm^{-1} (OH) underwent extensive changes on the adsorption of Ca(II) ions, indicating involvement of –NHCOCH$_3$ and –OH groups in addition to –COOH groups. The authors proposed structure **5.3** for the Ca(II) complex.

5.3

Uraki *et al.*[60] examined the binding behaviour of water-soluble 6-*O*-carboxymethyl chitin, $DS = 0.9$, and 3,6-di-*O*-carboxymethyl chitin with the alkaline earth metal ions and a number of heavy metal ions, Mn(II),

Ni(II), Cu(II), Cd(II) and Pb(II). The complexes formed with 6-O-carboxymethyl chitin were all soluble, with Ca(II) ions being the most strongly bound. Indeed bound Ca(II) ions could only be completely released by the addition of chelating agents such as EDTA, and the involvement of the –NHCOCH$_3$, C(3)OH and C(6)OH groups in complex formation with Ca(II) ions was demonstrated by FTIR spectroscopy. With 3,6-O-carboxymethyl chitin Ba(II) ions were the most strongly bound, closely followed by Cd(II) and Pb(II) ions. However the complexes formed with these three ions were insoluble and precipitated out of solution, as did those formed with Ni(II) and Cu(II), while those formed with Mg(II), Ca(II), Sr(II) and Mn(II) remained in solution.

The results obtained with 3,6-O-carboxymethyl chitin and Ba(II), Cd(II) and Pb(II) ions are difficult to explain since the carboxyl content of 3,6-O-carboxymethyl chitin having $DS = 2.0$ is 6.27×10^{-3} eq g^{-1}, while the stated concentrations of the adsorbed ions range from 8.5×10^{-3} to 10×10^{-3} eq g^{-1} for these three species. Thus the mechanism of binding cannot be complex formation, as was proposed for Ca(II) ions,[59] since in that case the maximum uptake would be 3.14×10^{-3} eq g^{-1}, nor can binding of each divalent ion to a single carboxyl group be the sole mechanism. The adsorption values given for Ca(II), Sr(II), Mn(II) and Cu(II) are in keeping with the structure previously proposed for the 6-O-carboxymethyl chitin–Ca(II) complex[59] while the values for Mg(II) and Ni(II) are slightly high.

Complex formation by N-(carboxyalkyl)chitosans

Complex formation between N-(carboxyalkyl)chitosans and metal ions has been extensively studied by Muzzarelli et al.[61–66] Chitosan derivatives examined include N-(carboxymethyl) chitosan (glycine glucan, **5.4a**), N-(carboxybutyl) chitosan (**5.4b**), N-(o-carboxybenzyl)chitosan (**5.4c**), glutamate glucan (**5.4d**), aminogluconate glucan (**5.4e**), and N-[(dihydroxyethyl)tetrahydrofuryl]chitosan (**5.4f**) (Figure 5.6) (section 4.3.3).

N-(Carboxymethyl)chitosan and N-(o-carboxybenzyl)chitosan readily form insoluble precipitates with Co(II), Cu(II), Zn(II), Ni(II), Cd(II), Cr(III), Hg(II), Pb(II) and UO$_2^{2+}$. With **5.4a** maximum insolubilisation occurs at pH 5.5 with UO$_2^{2+}$, 7.5 with Ni(II) and Cd(II), and at pH 6–7 with the remainder,[61] while with **5.4c** Co(II) and Ni(II) have maximum insolubilisation at pH 8.5 and Cu(II), Zn(II), Cr(III), Hg(II) and Pb(II) have it at pH 6–7. Precipitates are formed even in relatively dilute solutions ~0.01 mM. N-(Carboxymethyl)chitosan may be crosslinked by treating in suspension in IPA with 40 wt-% NaOH, followed by epichlorohydrin in aqueous dioxane to give rigid gel-like beads which showed good adsorption characteristics with Co(II) and Cu(II) ions. The adsorption behaviour was attributed[62] to the totally amorphous structure of the crosslinked material.

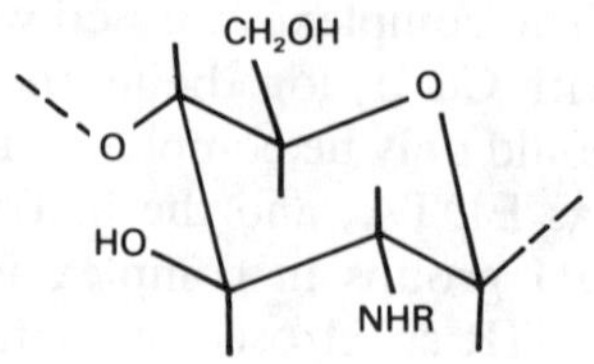

5.4

R

a: $-CH_2COOH$

b: $-CH(CH_3)CH_2CH_2COOH$

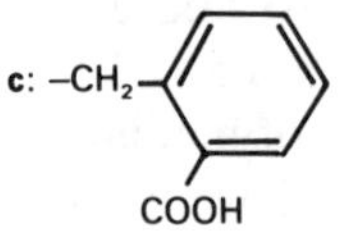

R

d: $-CH(COOH)-CH_2CH_2COOH$

e: $-CH(CH_2OH)CHOH.CHOH.CHOH.COOH$

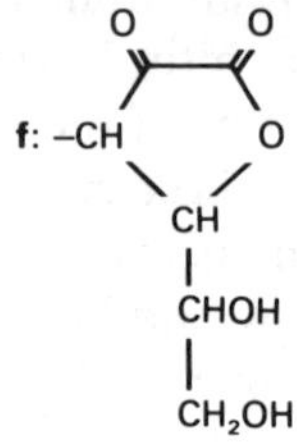

FIGURE 5.6 Structures of representative *N*-(carboxyalkyl)chitosan derivatives

Competitive adsorption studies with a solution containing a mixture of transition metal ions showed[63] that with glutamate glucan (**5.4d**) the order of affinity at pH 4.9 is Cu(II) > Cr(III) > Zn(II) > Ni(II) > Co(II) while with aminogluconate glucan (**5.4e**) at pH 5.1 it is Cu(II) > Cr(III) > Zn(II) > Co(II) > Ni(II). The adsorption behaviour is pH dependent and at pH 3.0 only Cu(II) and Cr(III), from the five metal ion species listed above, are still adsorbed. In a second series the order with glutamate glucan at pH 5.1 was Fe(II) > Cd(II) > Co(II) > Hg(II) – Mn(II) showed zero adsorption – and with aminogluconate glucan at pH 5.2, Cd(II) > Fe(II) >> Co(II), with both Mn(II) and Hg(II) showing zero adsorption.[63] Unlike *N*-(carboxymethyl)chitosan, *N*-(carboxybutyl)chitosan (**5.4b**) does not give a precipitate with Co(II) or Cr(III) but does so with Fe(II), Ni(II), Cu(II), Zn(II), Cd(II) and Ag(I).[64]

The product from the reductive alkylation of chitosan with dehydroascorbic acid, *N*-[(dihydroxyethyl)tetrahydrofuryl]chitosan (**5.4f**), shows a very high adsorption capacity for uranium[65] that is greatly in excess of that shown by *Rhizopus arrhizus*[31] (section 5.2.4). Under conditions where *R. arrhizus* collects 3.8×10^{-4} eq g^{-1}, the chitosan derivative **5.4f** collects 2.9×10^{-3} eq g^{-1}, while the amount collected by **5.4f** from a solution of

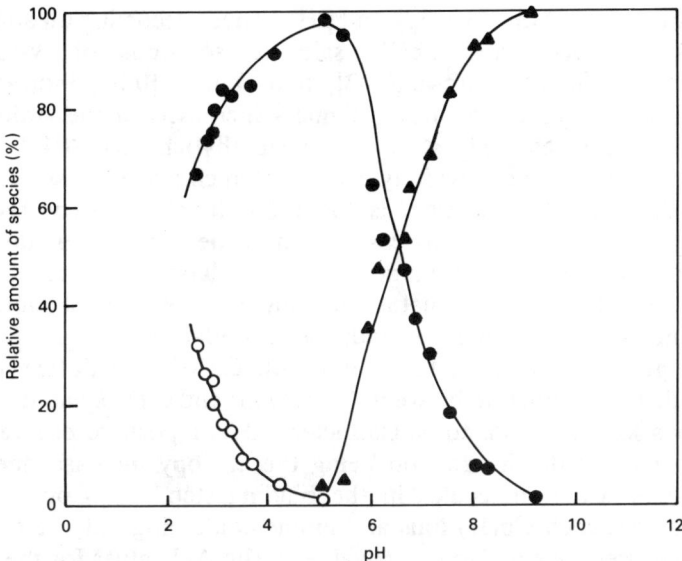

FIGURE 5.7   Variation in the extent of dissociation as a function of pH for a 4.8×10^{-3}M solution of N-(carboxymethyl)chitosan: $\bigcirc$, fully protonated form; $\bullet$, monodissociated form; $\blacktriangle$, fully dissociated form[66]

uranium in brine, 8.4×10^{-4} eq g^{-1}, is still more than twice that collected by *R. arrhizus* in the absence of NaCl.[65] In general the N-(carboxy-alkyl)chitosans have greater adsorption capacities and selectivity than unmodified chitosan, particularly under conditions of acid pH or high neutral electrolyte concentration.

Delben *et al.*[66] have made a detailed study of the solution behaviour of N-(carboxymethyl)chitosan. The potentiometric titration curve shows two inflection points corresponding to the dissociation of the –COOH and the $-\overset{+}{N}H_2R$ groups. The pK_a for the carboxylic acid group lies between 2 and 3, approximately, for α values of 0.7–1.0, while that for the secondary amine lies between 6.0 and 7.5, approximately, for α values of 1.1–2.0. The approximate relative proportions of the three forms – fully protonated, $-\overset{+}{N}H_2$–CH$_2$COOH; fully dissociated, –NH–CH$_2$COO$^-$; and monodissociated, –NH–CH$_2$–COOH $\rightleftharpoons$ $-\overset{+}{N}H_2$–CH$_2$–COO$^-$ – were calculated as a function of pH. The results show (Figure 5.7) that the fully protonated form only exists at pH < ~5 and that only the fully dissociated form exists at pH > ~9. Between these two pH values the polymer exists as a mixture of the monodissociated and fully dissociated forms with the latter increasing with increasing pH.

Measurements of the enthalpy and pH changes occurring on addition of solutions of Cu(II) and Pb(II) salts to solutions of N-(carboxymethyl)chitosan[0.42], chitosan[0.42], and chitosan[0.03] indicated that the binding ability of the three polymers increased in the order chitosan[0.03] < chitosan[0.42] < N-(carboxymethyl)chitosan[0.42]. It is of interest that the more or less fully deacetylated chitosan is less effective at binding these M(II) ions than is the more highly N-acetylated chitosan[0.42] sample, since in this case it cannot be regarded as due to the increased accessibility of the amine group in the latter as was suggested[23, 29] for heterogeneous chitosan–metal ion complex formation (section 5.2.2).

Evidence for involvement of both the secondary amine group and the carboxyl group in complex formation with Cu(II) and Pb(II) ions was presented. The interaction between Cu(II) ions and carboxylated polymers in aqueous solution tends to be characterised by a positive ΔH value, the driving force for the interaction being the entropy increase due to the liberation of water molecules in the binding step.[67, 68] Conversely the interaction between Cu(II) ions and amine-containing polymers tends to be characterised by a negative ΔH value.[69] The ΔH values for the interaction of N-(carboxymethyl)chitosan with both Cu(II) and Pb(II) were negative, as they were for the interactions of Cu(II) and Pb(II) ions with both chitosan[0.42] and chitosan[0.03], indicating the participation of the amine groups in the complex formation. Dilatometric measurements showed that an increase in volume occurred on interactions of Cu(II) and Pb(II) ions with N-(carboxymethyl)chitosan (ΔV was positive). This increase in volume may be considered as arising from the desolvation of the interacting ionic species during the binding step, indicating the involvement of the carboxyl group in the binding. In agreement with this interpretation, both chitosan[0.42] and chitosan[0.03] gave ΔV values of zero indicating non-ionic interactions.

Adsorption by N-permethylated chitosans

This has been studied briefly by Muzzarelli and Tanfani.[70] CD measurements showed interaction in solution between the polymer and a number of transition metal ions, the interaction being particularly strong in the case of Cu(II) and Hg(II) ions. The extent of metal ion adsorption under heterogeneous conditions was also examined and this showed that UO_2^{2+}, Cr(III) and Cu(II) ions are all readily adsorbed. The uptake of these ions increased linearly with increase in the concentration of metal ion initially present in solution, the percentage uptake being between 90% and 100% of the total amount of metal ion in the system for all values of the initial concentration. In the case of Ni(II) ion adsorption the percent uptake increased with increase in the initial concentration of metal ion, being 0% for a 2.5×10^{-4}M solution, 10% for a 5×10^{-4}M solution, and 70% for a

2×10^{-3}M solution. The behaviour of Mn(II) and Co(II) ions was the reverse of this, the percent uptake decreasing with increase in the initial concentration. On going from 2.5×10^{-4}M to 2×10^{-3}M the percent uptake decreased from 22% to 6% for Mn(II) and from 31% to 11% for Co(II).

In all cases in this study the capacity of the N-permethylated chitosan, which had an N-methylated amine group concentration of 3×10^{-3} mol g^{-1}, was considerably in excess of the amount of metal ion available for adsorption; this varied from 6.25×10^{-5} to 5×10^{-4} mol g^{-1} polymer.

5.3   ADSORPTION ON CHITIN AND CHITOSAN

5.3.1   Adsorption of dyes

Giles and co-workers[71, 72] were the first to study the uptake of dyes on chitin. They found that the amount of dye adsorbed increases with decrease in pH, levelling off over the pH 6–4 region, then rising again as the pH is decreased further. The level of dye uptake represented by the plateau, and the pH at which the plateau begins, increase with increase in the molecular weight of the dye.

Although the authors did not discuss these results they are in fact very similar to those obtained by Peters with nylon 6.6[73] and may be explained on the same basis with reference to the idealised curve of Figure 5.8.

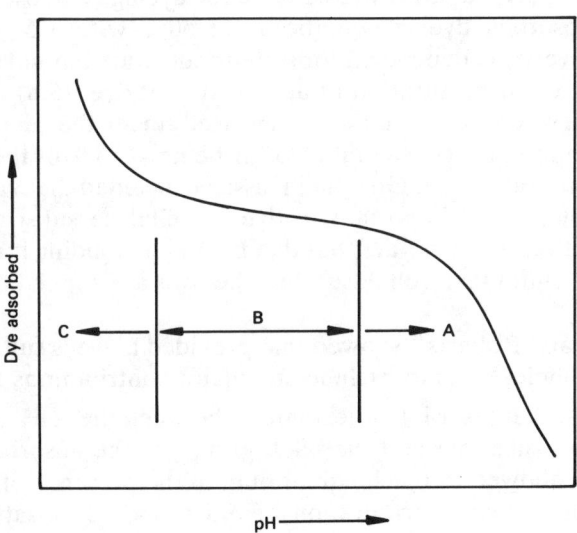

FIGURE 5.8   Idealised curve of pH dependence of anionic dyes on chitin

Region A represents the gradual titration of the amine groups present in the chitin as the pH is reduced (section 5.1), B represents the region where all the amine groups have been protonated, and C represents the region where protonation of the C(2)NHCOCH$_3$ groups occurs. On the assumption that the principal mechanism of dye uptake is ion exchange

$$\text{Chit–}\overset{+}{\text{N}}\text{H}_3\text{X}^- + \text{Dye–SO}_3^- \rightleftharpoons \text{Chit–}\overset{+}{\text{N}}\text{H}_3\text{O}_3\text{S–Dye} + \text{X}^-$$

then A shows that as the pH is lowered the proportion of amine groups protonated increases, creating more sites at which ion exchange can take place, while at B the concentration of these sites is fixed at their maximum value over the pH range of the plateau. In region C the protonation of amide groups enables additional dye to be taken up by ion exchange. Even highly N-acetylated samples can adsorb appreciable quantities of dye by ion exchange compared with textile fibres. The amine group concentration of a chitin[0.99] sample is approximately equivalent to that of commercial nylon 6.6 fibres while the adsorption levels obtained by Giles et al.[71] with dyes **5.5** and **5.6** (Figure 5.9) indicate that their sample was chitin[0.96].

Additional dye ions may be adsorbed through other attractive forces – van der Waals' forces, hydrophobic bonding and hydrogen bonding, depending on the nature of the dye ion – so that the plateau may be in excess of that expected from an ion exchange process alone. Also the adsorption level of a very water-soluble, low-molecular-weight dye may be below the plateau if the equilibrium for the ion exchange lies too far over towards the left-hand side. Giles and Hassan showed[72] that with Congo Red (C.I. Direct Red 28, **5.7**), a direct dye suitable for dyeing cellulose, the concentration of adsorbed dye in equilibrium at 50°C with a 5×10^{-4}M dye solution was very pH dependent for a chitin substrate but not for cellulose. Furthermore the concentration of adsorbed 'half-dye' (**5.8**) on chitin was similar, if slightly less, to that of Congo Red under the same conditions, while on cellulose it is considerably lower being $\leq 25\%$ of the Congo Red concentration (Table 5.4). Giles and Hassan concluded that van der Waals' forces, although very important in dyeing cellulose substrates, are not important with chitin substrates but that hydrogen bonding is more important in dyeing chitin than cellulose when the dyes are applied from aqueous solution.

Maghami and Roberts[74] showed that provided a dye is sufficiently small and water-soluble, so as to exclude any major contributions from van der Waals' forces, there is a 1:1 stoichiometry between the $-\overset{+}{\text{N}}\text{H}_3$ groups in the chitin/chitosan substrate and the $-\text{SO}_3^-$ groups of the adsorbed dye when the system is allowed to reach equilibrium in the presence of an excess of dye. Thus the molar concentration of **5.5** adsorbed at saturation by a chitosan sample is three times the molar quantity of dye **5.9**, and this stoichiometry enables dye adsorption measurements under acid conditions to be used to determine the degree of deacetylation of chitin and chitosan

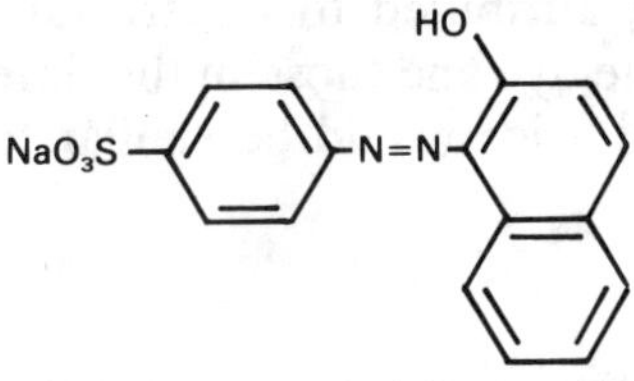

5.5

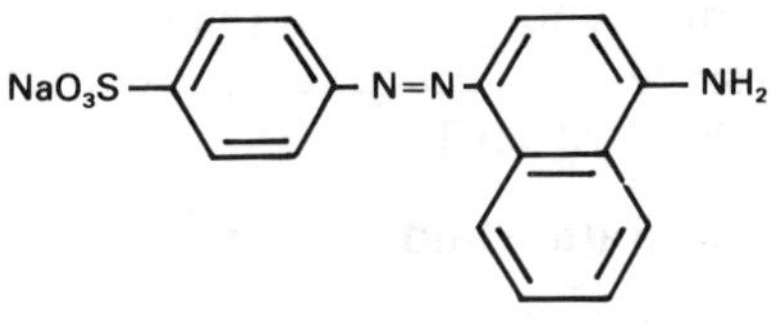

5.6

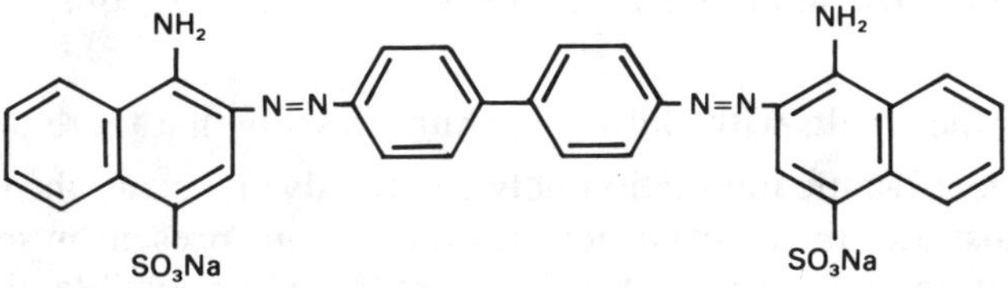

5.7

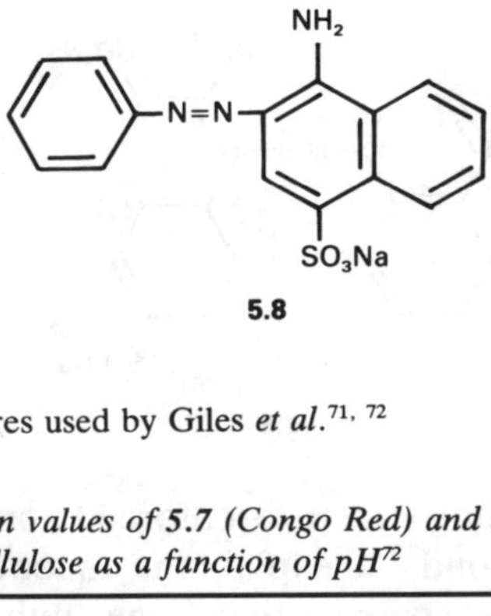

5.8

FIGURE 5.9   Dye structures used by Giles et al.[71, 72]

TABLE 5.4   *Adsorption values of 5.7 (Congo Red) and its 'half-structure' 5.8 on chitin and on cellulose as a function of pH[72]*

Substrate	Dye adsorption $\times 10^3$ (mol kg^{-1})			
	Neutral		pH 9	
	5.7	5.8	5.7	5.8
Chitin	47	45	18	15
Cellulose	33	6.5	31	5

samples (section 3.2.3). This stoichiometry has been confirmed using picric acid as the dye[75] (section 3.2.3).

The forces giving rise to the affinity between dye ions and oppositely charged substrates is still a matter of debate but the consensus is that specific electrostatic attraction between dye ions and charged sites does not occur. Thus Peters[76] states:

Clearly the affinity of an anion cannot be attributed to electrostatic interaction between the charged groups in the dye and those in the fibre since the adsorption properties of the [dye] acids would be similar to those of HCl.

while Rattee and Breuer[77] argue that:

Clearly the whole question of coulombic interaction is of considerable importance in relation to dye sorption. However it should be remembered that the factors involved depend upon the charge and potential effects alone. No distinction is made between ions of the same charge although these might be as simple as chloride or as complex as [an azo dye].

However the results with chitin and chitosan show that there must be some specific electrostatic interaction between the dye ions and the $-\overset{+}{N}H_3$ groups in the substrate, to the exclusion of other anions present in much greater concentration in the system. Furthermore the other possible binding forces – hydrogen bonding, van der Waals' and hydrophobic bonding – cannot be of importance with dyes such as **5.5** and **5.9** since otherwise dye would be taken up in excess of the 1:1 stoichiometry found.[74] The most probable

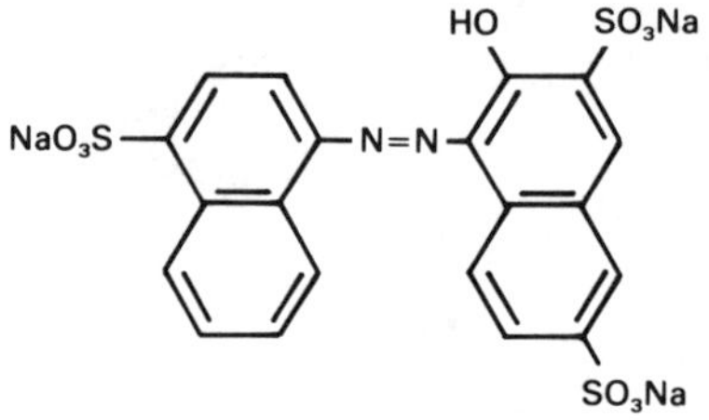

5.9

mechanism is that of water-structure–enforced ion pairing which was initially proposed by Diamond[78] to explain the aqueous solution behaviour of large organic ions. This assumes that large, unhydrated organic ions cause an increase in the degree of order of the water molecules immediately surrounding them and that, if there are both cations and anions of this type together in the solution, the presence of this structured water sheath tends to force the cation and anion into a single cavity or sheath of highly ordered water molecules, thereby giving an increase in entropy through release of water molecules. This association is facilitated by the opposite charges on the two species, the pairing lowering the ionic free energy of the associated pair. The ions of low-molecular-weight acids and electrolytes will be hydrated and hence not subject to this type of ion pair formation, thereby enabling selective electrostatic attraction between dye ions and charged sites on substrates.

Seo et al.[79] have examined the adsorption of C.I. Acid Orange 20 (Methyl Orange, **5.10**) on chitosan[0.15] and its N-octanoyl-, N-dodecanoyl-, and N-octadecanoyl-derivatives having various levels of N-acylation. Partial N-octanoylation over the DS range 0.1–0.6 gives a considerable increase in the uptake of dye at a pH of ~6.3 compared with that obtained either by chitosan[0.15] or by the more highly N-octanoylated samples. There is very little difference between the adsorption isotherms for N-octanoylchitosan[0.15/0.3] and N-octanoylchitosan [0.15/0.6], or between those for N-octanoylchitosan[0.15/0.3], N-dodecanoylchitosan[0.15/0.3] and N-octadecanoylchitosan[0.15/0.3], but in all cases the concentration of dye ions adsorbed is considerably less than the amine group concentration of the chitosan derivatives. This is partly due to the relatively high pH used, ~6.3, which causes a reduction in the fraction of amine groups protonated and therefore capable of functioning as cationic dye sites, and partly to limited accessibility of these dyesites. The increase in uptake on partial N-acylation may be attributed to increased accessibility of the charged sites, being similar to the increase in metal ion adsorption observed by Kurita et al.[27-29] on partial N-nonoylation of chitosan [0.12].

Lowering the pH from 6.4 to 5.9 causes a 10-fold increase, approximately, in the adsorption of **5.10** on chitosan[0.15], while a decrease in pH from 6.3 to 5.8 causes an even greater increase in adsorption on N-octanoylchitosan[0.15/0.3], indicating the importance of electrostatic interactions. Increasing the pH from 6.3 or 6.4 to 8.0 has only a limited effect on the adsorption, although the decrease in uptake appears to be somewhat greater for chitosan[0.15] than for N-octanoylchitosan[0.15/0.3] and N-octanoylchitosan[0.15/0.6]. This would suggest that the extent of protonation of the amine groups is low at pH 6.3.

The adsorption isotherms were analysed using a model involving two sorption mechanisms, site adsorption (Langmuir) and diffuse adsorption (partition). The results showed that the saturation value S, which represents the concentration of charged sites in the Langmuir model, attained in all cases was considerably lower than the amine group content. Saturation values were only obtained at pH 6.3 or 6.4 and 8.0 and there was very little difference between the S values for a given sample when determined at these two pH values, indeed they were identical for chitosan[0.15] while for N-octanoylchitosan[0.15/0.3] the value for S obtained at pH 6.3 was smaller than that determined at pH 8.0.

The study was subsequently extended to the adsorption behaviour of a number of dyes, including **5.5**, **5.6** and **5.10–5.13** (Figure 5.10), on the same substrates and the results were analysed using the same model involving two sorption mechanisms.[80] The saturation value S for **5.13** was the lowest of all the dyes on each of the substrates to which it was applied, while its partition coefficient K, which applies to the diffuse adsorption

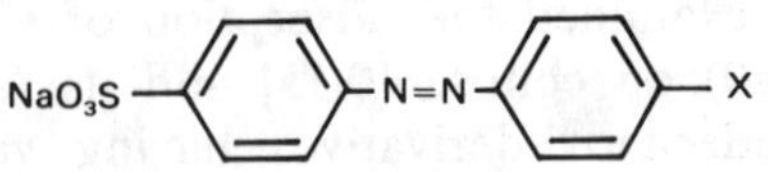

5.10: X = N(CH₃)₂
5.11: X = NH₂
5.12: X = OH
5.13: X = N(C₄H₉)₂

FIGURE 5.10   Dye structures used by Seo et al.[79, 80]

mechanism, was the largest. Dye **5.11** showed the reverse behaviour, having in general the highest S values and the lowest K values. It was concluded that the affinity of dye **5.13** for the chitosan substrates is primarily due to hydrophobic bonding forces while that for **5.11** is primarily electrostatic in nature. Support for this came from measurements of the thermodynamic parameters which show that, for the adsorption of **5.13** on fully N-acylated N-octanoylchitosan, the value of the entropy change, ΔS, is positive.

Knorr[81] has also reported a study of the adsorption of an anionic dye, **5.14**, on chitin and chitosan. However the dyeing conditions used, 30 minutes at room temperature, are insufficient to reach equilibrium; with dye **5.5** a dyeing time of 2 h at 100°C is required.[74] Furthermore the quantities of dye used were such that there was a very large excess of amine

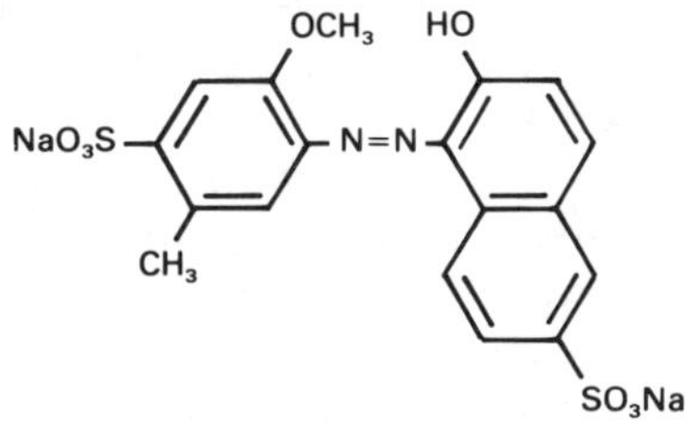

5.14

groups to dye ions. Even assuming that the dye was pure, which it would not be since a commercial sample was used, and that the substrates were chitin[0.95] and chitosan[0.15], the most concentrated dye solution gave $-\overset{+}{N}H_3:-SO_3^-$ ratios of approximately 40:1 for chitin and 800:1 for chitosan. Hence no definite conclusions can be drawn from the results.

Because of its high capacity for anionic dyes, chitin has been proposed as an adsorbent for treatment of dyehouse effluent and a detailed study of

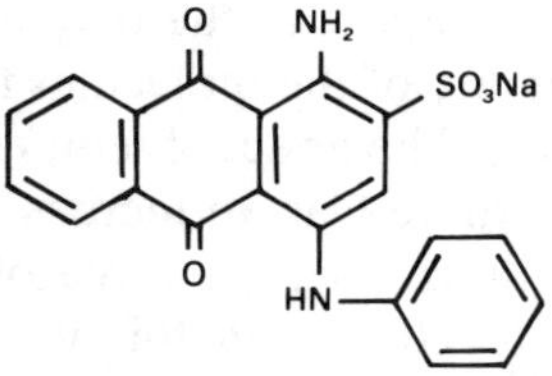

5.15

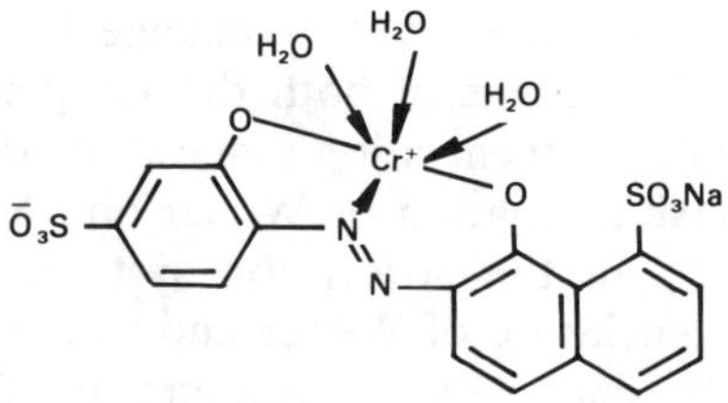

5.16

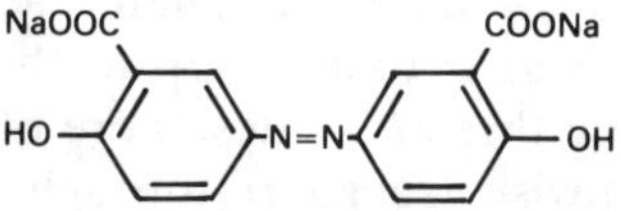

5.17

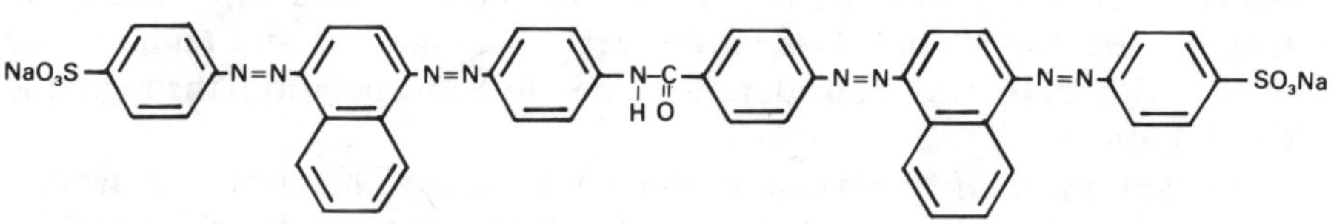

5.18

FIGURE 5.11 Dyes used by McKay *et al.*[82-85]

the chemical engineering aspects of this has been reported.[82-85] The equilibrium adsorption isotherms were determined for the four dyes C.I. Acid Blue 25 (**5.15**), C.I. Acid Blue 158 (**5.16**), C.I. Mordant Yellow 5 (**5.17**), and C.I. Direct Red 84 (**5.18**), and the effects of several variables examined (Figure 5.11). Particle size was found to have little or no influence on the ability of chitin to adsorb three of the four dyes; the fourth dye, **5.18**, showed a particle size effect with more dye being adsorbed the smaller the particles. Since **5.18** is the largest of the dye molecules studied this effect was attributed to the inability of the molecules to penetrate all the internal pore structure of the chitin. The effect of temperature on the adsorption of **5.18** was peculiar in that there was an increase in equilibrium adsorption with increase in temperature, with a threefold increase on raising the temperature from 20°C to 60°C, and this was thought to be due to an increase in dye mobility and a temperature-induced swelling effect within the internal structure of the chitin, allowing the large dye ions to penetrate into the particles. The behaviour shown by **5.17** was also peculiar as its equilibrium adsorption was unaffected by change in temperature.[82] The data could be expressed in terms of both the Langmuir and Freundlich equations over parts of the concentration range studied but the best fit was provided by the composite equation of Weber and Matthews.[86] Despite variations in behaviour, all the adsorption systems were found to be 'favourable' in the nomenclature of Weber and Chakravorti.[87]

The film mass transfer coefficients were determined in a second paper and found to be independent of the initial dye concentration, mass of chitin, particle size, and temperature, although some dependence on agitation was observed.[83] Subsequently the intraparticle diffusion processes were examined using as a rate parameter, k, the slope of the plot of amount of dye adsorbed *versus* time$^{0.5}$. Up to three separate linear portions of these plots, giving three k values, were obtained. This was interpreted as indicating stepwise transport through the different pore size ranges: micropores having radii of 1–2 nm, transitional pores with radii of 2–100 nm, and macropores with radii of 100–200 nm. The k values were independent of the mass of chitin, agitation, and the initial concentration, but decreased with increase in particle size and increased with increase in temperature for all the dyes except **5.16**. The lack of any temperature effect with **5.16** was considered to be due to chelation through the Cr(III) ion.[84]

Venkatrao *et al.*[88] have also looked at some chemical engineering aspects of dye adsorption, but on chitosan rather than chitin, using C.I. Direct Red 31 (**5.19**) and C.I. Reactive Red 73. These workers also used, as the rate constant k, the slope of the dye uptake *versus* time$^{0.5}$ plot. They observed that the time required to reach equilibrium increased with increase in the initial dye concentration, while the values of k were almost constant, in the case of the direct dye. No such dependence of time to equilibrium on initial dye concentration was observed with the reactive dye, the k values for which were found to increase with increase in initial dye concentration. Both dyes gave a correlation between k and the square root, approximately, of the particle diameter, indicating that intraparticle diffusion is the main rate-determining factor.

5.19

The rate parameter k was found to increase with increase in temperature for both dyes, as would be expected for a diffusion-controlled process, but the activation energy for the adsorption rate for the reactive dye was almost double that for the direct dye. This is difficult to explain since on cellulose substrates the activation energy of diffusion of reactive dyes as a group is considerably lower than that for direct dyes. The authors attempted to explain the greater activation energy as arising from aggregation of the reactive dye in solution, with the diffusing species being the aggregate. This is unlikely since reactive dyes are designed as relatively small, highly sulphonated structures and are therefore much less likely to aggregate than are the larger, more hydrophobic direct dyes such as **5.19**.

Both this paper[88] and those of McKay et al.[82–85] are open to criticism regarding the method used for obtaining the rate constant. The rate of adsorption depends on the diffusion coefficient and also on the concentration gradient which will be dependent, initially, upon the surface concentration. Thus the plot should be dye uptake/equilibrium dye uptake *versus* time$^{0.5}$ if the slope is to be proportional to $D^{0.5}$. Unless this is done, any change in k with change in a parameter, such as temperature or initial dye concentration, will only be meaningful provided the equilibrium uptake does not change.

The diffusion of an anionic dye (C.I. Acid Orange 10) in both chitosan film and chitosan film pretreated with $CuSO_4$ has been studied[89] using the Sekido roll technique.[90] The diffusion behaviour of the dye in the chitosan film was unusual, the diffusion coefficient, D, decreasing with increase in the relative concentration, whereas on nylon 6.6 it increases with increase in the relative concentration.[91] A tentative explanation was that the change in D was due to decreasing solubility of the chitosan owing to its interaction with the dye ions, leading to a more compact and less permeable structure. The diffusion coefficient for the dye in the $CuSO_4$-treated film (0.45 wt-% copper) was approximately 2 orders of magnitude less than in the chitosan film itself and this was attributed to crosslinking (section 5.2.5). However the distribution of Cu(II) ions within the film was not determined but it may be reasonably assumed, bearing in mind Averbach's results[35] (section 5.2.3) and the fact that Guthrie et al.[89] treated the chitosan film in $CuSO_4$ solution for only 24 h, that the Cu(II) ion adsorption is not uniform across the film thickness and that if there is any interaction between dye ions and Cu(II) ions, this non-uniform distribution would make analysis of the results extremely complex.

5.3.2 Adsorption of proteins

Hackman was the first to demonstrate that chitin can adsorb proteins from aqueous solution.[92] His results indicated that the maximum adsorption occurs at the isoelectric point of the protein and that, at a given pH, the extent of adsorption decreases with increase in the ionic strength of the solution. The adsorbed protein could be released on raising the pH to 9.

Agrawal and Scheller[93] studied the adsorption of arylphorin, which may be involved in sclerotisation, onto chitin from a 0.05M phosphate buffer at pH 5.8. Treatment of the chitin–protein complex with 7M urea at 4°C released a substantial amount of protein but after two such treatments no more protein was released. The protein removed by these treatments was considered to be non-covalently bound. Extraction with 1M NaOH at 4°C released covalently bound protein but after two extractions the chitin still retained ~25% of the protein adsorbed initially. Similar results were obtained using both crustacean and blowfly chitin. The authors proposed that the chitin first forms covalent bonds with the protein, yielding a glycoprotein which then adsorbs more protein by non-covalent bonding.

The ability of chitin to adsorb proteins has been made use of in affinity chromatography. Bloch and Burger[94] purified wheat germ agglutinin on a chitin column, the agglutinin being eluted with 0.05M HCl after a non-binding protein, and protein bound non-specifically, had been eluted with a 0.01M solution of tris(hydroxymethyl)aminomethane hydrochloride adjusted to pH 8.5 and containing 1M NaCl. The purified product, which amounted to less than 0.5% of the crude protein, had a specific activity of 1800 units mg^{-1} compared with 11.4 units mg^{-1} for the starting material. Chitin has also been used as the adsorbent for the purification, by affinity chromatography, of lectin of the Japanese horseshoe crab.[95] In this case elution of the purified product was carried out using a solution of N-acetyl-D-glucosamine.

Baba et al.[96] examined the adsorption of phosphorylases on columns of N-acylchitosans. Rabbit muscle phosphorylase a was retarded on N-acetylchitosan and adsorbed on N-propionylchitosan, N-butyrylchitosan, N-pentanoylchitosan and N-hexanoylchitosan. It could be eluted in over 82% yield from N-propionylchitosan with 1M NaCl, but adsorption increased with increase in the length of the hydrocarbon chain and the phosphorylase a could not be eluted from N-hexanoylchitosan with 0.2M acetic acid, 50 vol.-% ethylene glycol, Triton X-100 (5 vol.-%) or 0.5M imidazole–HCl (pH 5.6). They concluded that N-propionylchitosan was the most suitable of the series for purifying phosphorylase a but that

N-acetylchitosan was better for maize phosphorylase. This difference was attributed to structural differences between the two types of phosphorylase.

Although immobilisation of enzymes on chitin and chitosan normally requires a covalent bond between the enzyme and the support polymer (section 4.4), in some cases the binding forces giving rise to physical adsorption are sufficient to bring about immobilisation under the conditions needed for the enzyme to function. One of the first reports was that of Muzzarelli et al.[97] who immobilised α-chymotrypsin (EC 3.4.21.1) and acid phosphatase (EC 3.1.3.2) on chitosan particles by adsorption at pH 8.5 and 6.5 respectively, using in the latter case chitosan preconditioned with 0.05M H_2SO_4. Immobilisation of 90% of the α-chymotrypsin and 100% of the acid phosphatase was achieved with a contact time of 60 minutes. Acid phosphatase could be eluted from the column at $4 > pH > 9$ and α-chymotrypsin at $5 > pH > 11$. Maximum activity was found to occur at pH 8.5 for α-chymotrypsin and pH 5.0 for acid phosphatase. Because the optimum pH of operation for the immobilised acid phosphatase was 5.0, immobilisation was also carried out at this pH although the efficiency is less than at pH 6.5. However this ensured that no enzyme was desorbed during the subsequent activity studies, which showed that the activities of these enzymes were much higher when immobilised by physical adsorption than when immobilised by reaction with glutaraldehyde.

Amylase has been immobilised on chitin by adsorption from buffer solution at pH 3.8. The chitin-immobilised amylase showed almost the same digestion rate for raw corn starch and gelatinised potato starch as did the free enzyme, although it showed a narrower pH range for activity and lower pH stability and thermostability compared with the free enzyme. Recycling several times caused only a slight decrease in the activity and no amylase activity could be observed in the hydrolysate, indicating that the amylase was bound tightly to the chitin making the system suitable for continuous use.[98-100]

The activities of immobilised enzymes on krill chitin decrease with decrease in the purity of the sample, that is with increase in the amount of residual protein or oganic material,[101] and increase with increase in the free amine group content of the chitin,[102] the rate increasing by up to 230% on changing the adsorbent from chitin[0.96] to chitin[0.92]. Comparison of the activities of α-amylase (EC 3.2.1.1), β-amylase (EC 3.2.1.2) and amyloglucosidase (EC 3.2.1.3) immobilised by different techniques showed that crosslinking with glutaraldehyde reduces the activity to as little as 60% of that for the enzyme immobilised solely by physical adsorption.

Ohtakara et al.[103] examined the immobilisation of two acidic proteins, α-galactosidase (EC 3.2.1.22) and amyloglucosidase from Aspergillus sp. K-27, and a basic protein, amyloglucosidase from R. niveus, on chitosan beads. The beads were of four types: A, uncrosslinked; B, crosslinked with hexamethylene diisocyanate; C, crosslinked with 4,4'-diphenylmethane

TABLE 5.5   *Immobilisation of α-galactosidase and two amyloglucosidases on various types of chitosan beads[103]*

Chitosan beads[b]	Activity yield[a] (%)			
	α-Galactosidase	Amyloglucosidase		
		Asp.sp.	K-27	R. niveus
A	8.1			
B	30.0	24.8	8.6[c]	4.4
C	27.9	31.6	26.7[c]	41.8
D		13.9	2.4[c]	2.6

[a] The activity yield is the immobilised activity expressed as a percentage of the added activity.

[b] A, uncrosslinked beads; B, crosslinked with hexamethylene diisocyante; C, crosslinked with 4,4'-diphenylmethane diisocyanate; D, crosslinked with a quaternary ammonium compound.

[c] Activity yield after incubation in 0.05M acetate buffer containing 1.0M NaCl for 1 h at 37°C.

diisocyanate; D, crosslinked and containing a quaternary ammonium compound.

The α-galactosidase was immobilised on bead types A–C but the activity yield on A, the non-crosslinked beads, was very low in the absence of glutaraldehyde pretreatment while there was no noticeable difference in the activity yields between untreated and glutaraldehyde-treated beads with the crosslinked beads B and C (Table 5.5). Treatment with 0.2M NaCl at pH 5 removed the activity of enzyme immobilised on A and B but the activity of α-galactosidase immobilised on C was retained even on treating in 2.0M NaCl.

The amyloglucosidase from *R. niveus* was only immobilised to any significant extent on bead type C, while that from *Asp.* sp. K-27 showed similar activity yields to those of α-galactosidase on B and C, and considerably more than amyloglucosidase *R. niveus* on D. The activity yield for both amyloglucosidases on C increased with increase in the extent of deacetylation of the chitosan used. The amyloglucosidase from *R. niveus* immobilised on C retained about 84% of the initial activity after shaking in 0.05M acetate buffer containing 1M NaCl, and a similar result was obtained for that from *Asp.* sp. K-27. However the retention of activity of this latter amyloglucosidase on D and B after similar treatment was only 17.4% and 34.6% respectively (Table 5.5). The authors concluded that the acidic proteins, α-galactosidase and amyloglucosidase *Asp.* sp. K-27, are bound to beads A and B through weak ion exchange forces which are very sensitive to changes in ionic strength. The basic protein amyloglucosidase *R. niveus* is not adsorbed on these beads to any extent owing to electrostatic repulsion but is adsorbed strongly onto C, as are the other two enzymes, through hydrophobic bonding with the aromatic groups in the crosslinks, hence the low sensitivity to added electrolyte.

Immobilised amyloglucosidase *R. niveus*, on C-type beads without glutaraldehyde, was used in column form for the hydrolysis of liquefied starch; stable, continuous production of D-glucose was maintained for 20 days, after which the rate gradually decreased.

Since chitin-degrading enzymes will bind selectively to chitin they may be purified by affinity chromatography on chitin columns and this approach has been used quite extensively with lysozyme. Cherkasov and Kravchenco[104] found that lysozyme binds tightly on chitin at pH 7–9 and low ionic strength, and could be eluted from the column with 0.2M acetic acid. This was confirmed by Hayashi *et al.*[105] who determined that maximum adsorption occurs at pH 9, although adsorption from a 0.5M NaCl solution is independent of pH over the pH range 2–8.

Jensen and Kleppe[106] have used chitin columns to both concentrate and purify T4 lysozyme, finding that the column could be used repeatedly for up to 1 month without any noticeable change in its capacity. Different binding behaviour on chitin, particularly in respect of optimum pH and the effects of added electrolyte, were shown by this lysozyme and by lysozyme from egg-white. Cherkasov and Kravchenco[107] determined the binding constant (K_b) for lysozyme on chitin using a column of powdered chitin. The values of K_b were 1.15×10^5 M^{-1} at pH 4.7 and 8.95×10^5 M^{-1} at pH 8.4, but heat-denatured lysozyme was found not to adsorb on chitin.

The study of adsorption behaviour of chitinases on chitin may be complicated by the gradual production of lower-molecular-weight species that will have different binding characteristics. Skujins *et al.* found that the chitinase from *Streptomyces* sp. was adsorbed on to chitin very rapidly, the adsorption being complete in a matter of seconds. Although isotherms could be obtained for adsorption at 0°C, this was much more difficult at 25°C because of the rapid rate of hydrolysis of the chitin substrate by the chitinase.[108]

5.3.3 Adsorption of aromatic hydrocarbons

Giles and Hassan[72] found that phenol and resorcinol are readily adsorbed on chitin from aqueous solution, in contrast to their non-adsorption on cellulose. However both cellulose and chitin behaved similarly in their adsorption of these solutes from non-aqueous solution. The authors concluded that adsorption from aqueous solution involves hydrogen bond formation between the phenolic group and the C(2)NHCOCH$_3$ group, the

$$\text{Ar–OH} \ldots \text{NHC–CH}_3 \quad (\overset{O}{\underset{\|}{})}$$

proposed bond being Ar–OH . . . NHC–CH$_3$. However the carbonyl group would be expected to be a better acceptor group and since phenol is also adsorbed from aqueous solution at pH 9, when it will be mainly dissociated and can act only as the acceptor molecule, hydrogen bonds such as $Ar-O^- \ldots HN-\overset{\overset{\displaystyle O}{\displaystyle \|}}{C}-CH_3$ may also contribute to the adsorption from neutral solution.

Aromatic sulphonic acids were also found to be adsorbed; benzene sulphonic acid, naphthalene-2-sulphonic acid, 2-naphthol-6-sulphonic acid, anthraquinone-2-sulphonic acid and anthracene-1-sulphonic acid are all adsorbed to approximately the same extent. Their isotherms indicate a saturation level of $\leqslant 0.3$ mol kg^{-1}, which would be the saturation level expected for specific site adsorption on protonated amine groups for a chitin[0.94] sample, suggesting that with these solutes the bulk of the adsorption occurs through an ion exchange mechanism.

The adsorption of tannins on chitin was examined by Nakabayashi and Makita.[109] High-molecular-weight tannins were adsorbed preferentially and the adsorption rate was increased by the addition of NaCl or by decrease in the particle size, but decreased by the addition of ethanol. The effects of the two solutes may be explained in terms of changes in the partition coefficient for the tannins, but more surprisingly the adsorption rate was inversely related to the temperature. The use of chitin or of chitosan, which had a higher adsorption, to remove tannins from fruit and vegetable juices to improve the taste was suggested by the authors.

A more recent study[110] has examined the adsorption of polychlorinated biphenyls (PCBs) on chitosan, which was found to be a more efficient adsorbent than chitin, activated charcoal or sand. Complete removal of PCBs from contaminated water could be achieved by cycling the water in a closed-loop system.

5.3.4 Adsorption of other chemical substances

Giles and Hassan reported[72] that chitin adsorbs methanol from benzene, although to a lesser extent than does cellulose, and Austin has described the formation of complexes from chitin and alcohols, aldehydes or ketones.[111] The general method of preparation involves treating water-swollen or solvent-swollen chitin with the alcohol, aldehyde or ketone, so as to displace the water, the treatment being repeated several times with intervening filtration steps. The mixture is then air-dried to yield the complex. Evidence cited in support of complex formation includes the abrupt change in the rate of solvent evaporation during the drying process, the curves obtained by DTA which show strong endothermic peaks, and the greater solubility of the products compared with the starting chitin. Complexes were formed between chitin and *inter alia* methanol, ethanol, acetone, methyl ethyl ketone, chloral hydrate and camphor, with mole ratios of complexing agent:N-acetyl-D-glucosamine residue of from 5:1 to

1:6. With low boiling substances such as methanol or acetone the complex has a high vapour pressure of the alcohol or carbonyl component, and hence will break down completely unless stored in a closed container in the presence of a small excess of the complexing agent.

Complexes between chitosan and halogens have also been reported,[112–115] both bromine and iodine being adsorbed from aqueous solution to give red and dark purple complexes respectively, the latter reaction being the basis of the well-known van Wisselingh test for detecting the presence of chitosan, or of chitin after deacetylation, in biological specimens. The nature of these complexes was first studied by Shigeno et al.[112–114] who showed that adsorption of iodine was accompanied by some destruction of the crystal structure.[113] These workers also showed that when the halogen was present in excess above a mole ratio of 1:1 for halogen:$-NH_2$, the amount of halogen adsorbed was linearly proportional to the amount of chitosan present,[112, 114] indicating that the adsorption process was one of chemisorption, a charge-transfer complex being formed between the halogen and the amine groups of chitosan.

Adsorption of bromine onto chitosan was found to reach saturation in 6 h at 30°C although the characteristic red colour could be observed after only a few minutes because of the initial reaction at the surface. The amount of bromine adsorbed was dependent on the reaction medium and increased in the order

chloroform << ethanol < water < methanol

The reaction has also been investigated by Hirano et al.[115] who found that the amount adsorbed decreased with increase in the degree of N-acetylation (Table 5.6). The value of zero for the adsorption by chitin[1.0] agrees with the observation of Shigeno et al.[114] that the amount adsorbed by chitin was very small and could be attributed to the presence of a limited number of amine groups in the chitin. The amount adsorbed decreased with increase in the molecular weight of the N-acyl group, being 0.37 for N-octanoylchitosan[0.0/0.5], but increased on introduction of an aromatic

TABLE 5.6 *Adsorption of bromine on chitosan and chitin as a function of the degree of N-acetylation[115]*

Adsorbent	Br/N ratio[a]
Chitosan[0.0]	1.11
Chitosan[0.25]	0.61
Chitosan[0.50]	0.55
Chitin[0.75]	0.13
Chitin[1.0]	0.0

[a] Ratio of atoms of bromine to atoms of nitrogen.

N-acyl group.[115] Similar trends were observed for iodine adsorption and no selectivity was observed for adsorption from a solution containing both bromine and iodine.

The complexes were stable to 0.5M HCl and 0.5M acetic acid, and while bromine was released from the chitosan–bromine complex by treatment with 0.5M $(NH_4)_2CO_3$, the chitosan–iodine complex required treatment with 0.5M NaOH in order to release iodine. The proposed structure (5.20) for the complexes, shown in Figure 5.12, is in agreement with the IR spectra. Treatment with 0.5M NaOH gave complete release of the halogens, together with the formation of water-soluble, colourless products which were fractionated, hydrolysed with 1M H_2SO_4 and reduced with $NaBH_4$ before acetylating. Analysis of the alditol acetates by GLC showed the presence of D-arabinitol, xylitol, D-mannitol, galactitol, D-glucitol, and an unknown sugar alcohol, most probably D-altritol. They concluded that the chitosan–halogen complex breaks down by oxidative deamination of the chitosan, and proposed a reaction mechanism, Figure 5.12, analogous to that for the degradation of cellulose by Cl_2–NaOH mixtures.

5.3.5 The use of chitin and chitosan in chromatography

Because of their behaviour as adsorbents, chitin and chitosan have been examined for use in chromatography. Chitosan in powder form has been used in TLC for the separation of a number of food dyes and of dyes commonly used in histology.[116] This latter group contains both anionic and cationic dyes and in general the separations achieved on chitosan thin layers were superior to those obtained using O-diethylaminoethyl cellulose. Several of the dyes, which on DEAE cellulose were either immobile or gave long tails, were resolved into several components. A particular example is Eosine Yellowish (C.I. Acid Red 87) which was separated into eight components with R_f values of from 0.51 to 0.96 on chitosan, but which was immobile on DEAE cellulose. The food dyes examined, all anionic, mainly gave a single spot or at most two with no tailing.

Chitin powder has been used in TLC for the separation of amino acids.[117, 118] Basic amino acids had greater R_f values on chitin than on polyamide or silica gel while acidic amino acids gave smaller values, and it was suggested[118] that this behaviour might be due to partial deacetylation of the chitin giving rise to a basic powder. Two-dimensional TLC of casein hydrolysates gave good resolution on chitin using a 3:1:1 mixture of 1-butanol:acetic acid:water as the first solvent and a 67:33 mixture of 1-propanol:34 vol.-% ammonia as the second.

Chitin powder was found to give better separation of nucleic acid bases, nucleosides and nucleotides than can be achieved on cellulose powder, the R_f values being larger on the former material.[118, 119] Similarly, phenols were separated more satisfactorily on chitin than on polyamide or silica gel

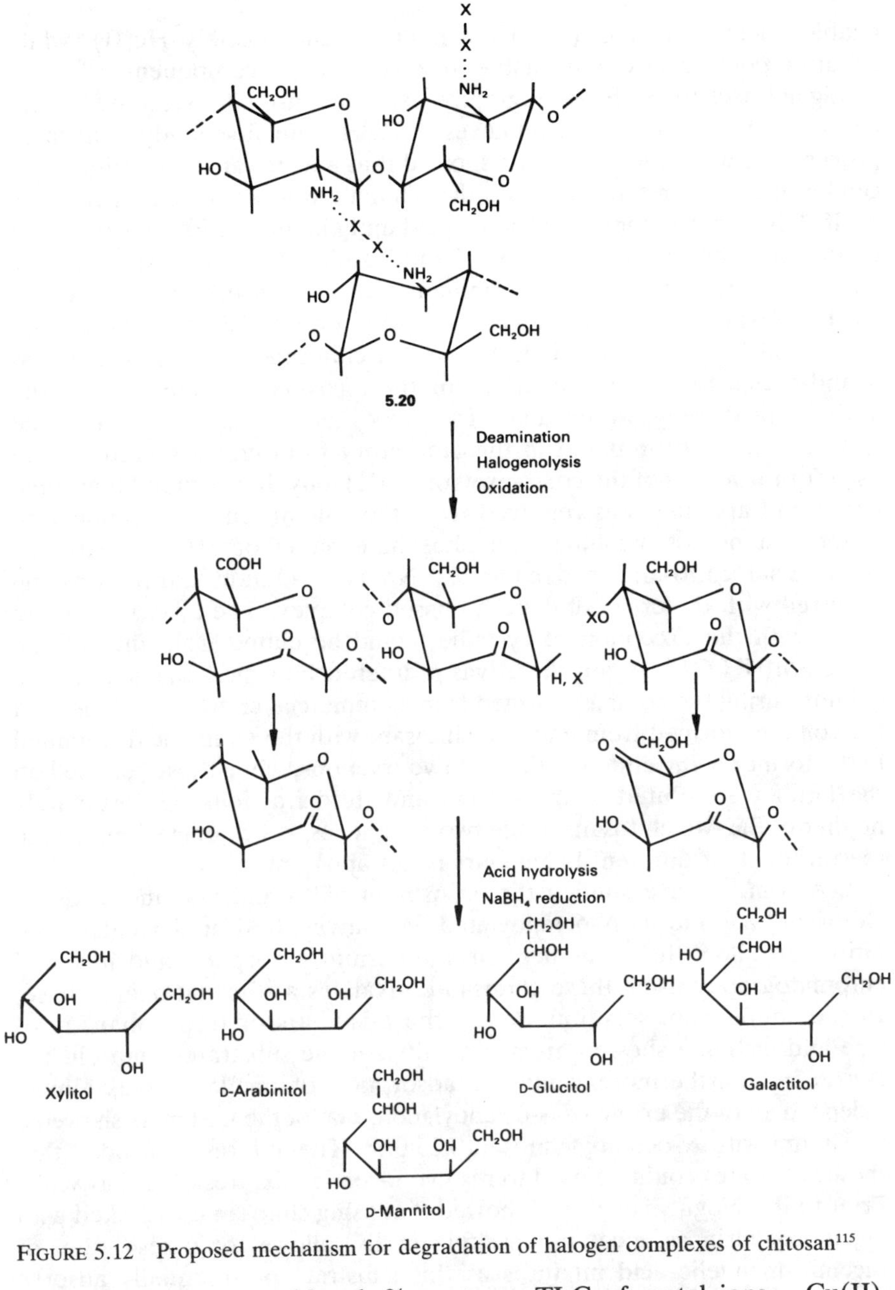

FIGURE 5.12 Proposed mechanism for degradation of halogen complexes of chitosan[115]

using either water or 20 vol.-% acetone. TLC of metal ions – Cu(II), Cd(II), Hg(II), Pb(II) and Bi(III) – gave good resolution using a 20:4:1 mixture of 1-butanol:1.5M HCl:acetylacetone, the separation particularly of Cu(II) and Pb(II) being better than on cellulose. The use of chitin for TLC of the metals in the hydrolysates of oysters and viscera of cuttlefish enabled detection of Cu(II), Cd(II), Pb(II) and possibly Hg(II), while cellulose powder TLC was unable to separate these components.[118]

Ligand-exchange chromatography has been used to separate amino acids on chitosan–Cu(II) complexes.[116, 120] In the first study[116] a batch process was used and the results showed that a wide range of amino acids can be collected on the complex although none were collected on chitosan itself. Glycine, tryptophan, aspartic acid and glutamic acid were the most readily adsorbed of those examined and histidine the least. Although the majority of the amino acids examined were also adsorbed on chitosan–Cu(II)–NH₃ complexes, the levels of adsorption were lower in all cases.

No and Meyers[120] examined the use of chitosan–Cu(II) complexes as ligand-exchange column materials for the recovery of amino acids from seafood processing wastewater. Two types of chitosan, a commercial chitosan and one prepared in the laboratory from crawfish chitin, were used for formation of the complex with Cu(II) ions. It was found that some lysine and argenine was removed from the column during washing with water, but not on washing with phosphate buffer at pH 8.2, with the commercial chitosan-based complex. No such elution during washing occurred with the crawfish chitosan-based complex. The adsorbed amino acids, with the exception of cysteine, could be eluted from the column using 3M NH₄OH. Again there was a difference in the behaviour of the columns using the complex formed from commercial chitosan and those of the complex formed from crawfish chitosan, with the amino acids retained in the former being eluted in the first two fractions, while those retained on the latter were eluted in the second and third fractions. Unfortunately neither of the two chitosans or the two complexes was characterised, so the reason for their different behaviours is not apparent.

Seo et al.[121] have studied the adsorption of D- and L-amino acids on chitosan[0.23] and its N-octanoylated derivatives, both under batch conditions to allow the attainment of equilibrium adsorption and in liquid chromatography using these chitosan derivatives as the stationary phase. In the equilibrium sorption studies the L-isomers of tryptophan, tyrosine and histidine showed greater affinity for the substrates than did the D-isomers. Furthermore while the adsorption of the latter was almost independent of the extent of N-octanoylation, that of the L-isomers showed a maximum with N-octanoylchitosan[0.23/0.3]. The authors concluded that these substrates could be used to resolve racemic mixtures of amino acids. Prior to this Noguchi et al.[3, 122] showed that using chitosan crosslinked with epichlorohydrin as a column packing in LC allows some resolution of racemic mandelic acid mixtures as this substrate preferentially adsorbs D-mandelic acid, thereby enriching the effluent with L-mandelic acid and the eluent with the D-isomer.

In the studies of LC[121] three stationary phases were used – chitosan[0.23], N-octanoylchitosan[0.23/0.3] and N-octanoylchitosan[0.23/0.7] – and the separation of the L-isomers of tryptophan, tyrosine and histidine examined. The elution volumes of these three amino acids are different, indicating the possibility of effecting separation, and the sharpness of the peaks in the chromatogram increased in the order N-octanoylchitosan[0.23/0.3] < chitosan[0.23] < N-octanoylchitosan[0.23/0.7]. The improved performance of the latter substrate was attributed to its more compact gel structure, and the separation on it to hydrophobic interactions.

Hirano et al. have examined the use of several chitosan derivatives as the gel medium in GPC. The first of these derivatives was N-methylidenechitosan,[123, 124] prepared by the reaction between chitosan and formaldehyde, and the others were N-acyl-N-carboxyacylchitosans[124] prepared by a two-stage homogeneous N-acylation process. In the first stage chitosan[0.0] was N-acylated with a cyclic anhydride – malonic, phthalic or succinic anhydride – at a mole ratio of anhydride:anhydro-D-glucosamine residues of 0.4:1. After completion of this reaction, acetic anhydride or octadecanoic anhydride was added to the mixture to acylate a proportion of the remaining amine groups. The ion-exchange capacities were determined by titration (five of the six products were ampholites since complete N-acylation was not attained) and in all cases these values were less than those calculated from the results of elemental analysis. This was attributed[124] to some of the functional groups being inaccessible during titration, but it is more likely that the structures assigned on the basis of elemental analysis are in error. Indeed in this particular study the values for N-carboxyacylation reported for the intermediate products at the end of the first stage reaction are higher than the values reported for the final products, in some examples by a factor of almost 2. Since de-N-acylation is unlikely to be taking place concurrently with N-acylation, these results suggest that the values for ion-exchange capacity obtained by titration are more likely to be correct.

Of the derivatives examined only N-methylidenechitosan, N-acetyl-N-(3'-carboxypropionyl)chitosan and N-octadecanoyl-N-(3'-carboxypropionyl)-chitosan were suitable for use in GPC, as shown by their total bed volume (V_t) and water regain (W_r) (Table 5.7). On calibration of these substrates in GPC columns N-octadecanoyl-N-(3'-carboxypropionyl)chitosan showed a rectilinear relationship between elution volume and log (molecular weight) over the molecular weight range of 2×10^4–6×10^5 and N-methylidenechitosan over the range 2.9×10^3–6×10^4. The recovery of the standards was 80–100% on both gel media.

Various N-acylated chitosans have been examined as packing material for gas chromatography columns.[125, 126] Muzzarelli et al.[125] were the first to study GC on chitin/chitosan columns, investigating the behaviour of water and a number of amines and alcohols. Using a column temperature of 125°C they obtained retention times for methanol of 153 s on a chitin sample and of 48 s on chitosan. They also obtained a linear relationship

TABLE 5.7   *Properties of N-substituted chitosan derivatives suitable for use in gel chromatography*[123, 124]

Chitosan derivative	Bed volume $(cm^3\ g^{-1})$	Water regain $(cm^3\ g^{-1})$	Ion-exchange capacity $(meq\ g^{-1})$
N-Acetyl-N-(3'-carboxypropionyl)–	80–90	40–50	0.8
N-Octadecanoyl-N-(3'-carboxypropionyl)–	50–70	30–40	1.1
N-Methylidene–	30–35	20–30	—

TABLE 5.8   *Retention times and column efficiency for a chitin bead packing material*[126]

Compound	Retention time[a] (s)	Column efficiency	
		Number of theoretical plates (N)	Height equivalent to a theoretical plate (H)
Pentane	10	256	1.76 mm
Hexane	18	207	2.10 mm
Xylene	28	348	1.29 mm
Methyl acetate	13	220	2.04 mm
Ethyl acetate	20	400	1.13 mm

[a] Column length, 45 cm; column temperature, 40°C; N_2 flow rate, 30 cm^3 min^{-1}.

between the retention time and the degree of N-acetylation and found that in the alcohol series the retention time decreased with increase in molecular weight of the alcohol, from 153 s for methanol to 29 s for 1-butanol. These results are in disagreement with more recent work[126] in which the use of uniform spherical beads of chitin[127] and its N-acyl analogues was examined. A number of hydrocarbons, esters and alcohols were used as test compounds and in all cases the retention times were very small for column temperatures in excess of 100°C, although at 40°C they were sufficient to enable preliminary calculations of the column efficiency to be made (Table 5.8). The resolution for a 1:1 mixture of pentane and xylene was calculated to be 0.86, which is below the minimum value of 1.0 required for quantitative separation, although qualitative separation may be carried out at values as low as 0.6.

Contrary to the findings of Muzzarelli *et al.*,[125] little retention of alcohols was observed and the use of higher-molecular-weight N-acyl groups, for example N-hexanoylchitosan, gave no increase. At 50°C the retention times on chitin[0.9] beads for ethanol, 1-butanol and 1-hexanol were 7 s, 10 s and 28 s respectively, which is the reverse of the inverse relationship between retention time and alcohol molecular weight found previously.[125] This trend suggests that the retention time for methanol at 50°C would be < 7 s, whereas according to the reported relationship between % N-acetylation and retention time,[125] the retention time on chitin[0.9] should be ~300 s at 125°C. The difference in column lengths or the possible differences in surface areas of the two packing materials[127] cannot account for these widely differing results but it would appear that chitin and its N-acyl analogues are not suitable materials for use in GC, a major factor being the relatively low surface areas obtained under the operating conditions.

5.4 DEGRADATION OF CHITIN AND CHITOSAN

5.4.1 Acid hydrolysis

Both chitin and chitosan, being poly(glycosides), are hydrolysed by acids, but they differ considerably in their respective susceptibilities to acid hydrolysis. Irvine and Hynd[128] found that a methyl 2-amino-2-deoxy-D-glucopyranoside showed a greater resistance to acid hydrolysis than did either methyl α-D-glucopyranoside or the β enantiomer. This observation, together with the lability of a methyl 2-acetamido-2-deoxy-D-gluco-pyranoside compared with either methyl D-glucopyranoside or with a methyl 2-amino-2-deoxy-D-glucopyranoside, led Neuberger and co-workers[129, 130] to propose that the latter is hydrolysed very slowly because under acid conditions the amine group is protonated and the $C(2)-\overset{+}{N}H_3$ group shields the glycosidic oxygen from protonation, which is the first step in the acid hydrolysis reaction. Confirmation of this effect was reported by Onodera and Komano[131] who found that the dependence of the rates of hydrolysis of methyl β-D-glucopyranosides on the nature of the C(2)–substituent was in the order $-NHCOCH_3 < -OH << -NH_2$, the times to 50% hydrolysis being 4.0, 240 and 540 minutes respectively. The same order was found for the methyl α-D-glucopyranoside series.

Additional evidence comes from studies on the polymers themselves by Nagasawa and Tanoura[132] who found that on hydrolysing chitosans in 96 wt-% H_2SO_4 the yield of non-dialysable material decreased with increase in the extent of N-acetylation over the range chitosan[0.17] to chitosan[0.55], while Barker et al.[133] reported that a sample of chitosan was not completely hydrolysed on boiling in 3.3M HCl for 3 days.

The acid hydrolysis of chitin was first studied by Meyer and Wehrli[134] who determined the energy of activation of hydrolysis of chitin in 50 wt-% HNO_3 to be 123.9 kJ mol^{-1}, similar to the value of 125.2 kJ mol^{-1} found previously[135] for the hydrolysis of cellulose in 51 wt-% H_2SO_4. Rupley[136] carried out a study of the hydrolysis of chitin in solution in concentrated HCl for the purpose of preparing oligomeric substrates for investigating the action of lysozyme. Solutions of chitin were prepared at 0°C in 5.5, 7, 9 and 11M HCl, the extent of hydrolysis occurring during dissolution being very limited since no free reducing groups could be detected in the solutions as prepared. Rates of both hydrolysis and de-N-acetylation were determined over the temperature range of 0–80°C, the rate of increase in reducing group concentration being approximately ten times the rate of increase in amine group concentration. A detailed study of the results showed that the hydrolysis follows first-order kinetics from about 10% to 90% hydrolysis but that the initial hydrolysis occurs at a slower rate, an effect which becomes more noticeable with decrease in the HCl concentration. First-order rate constants calculated for the linear portions of rate of hydrolysis curves were found to vary with the third to the fifth power of the HCl concentration. The slower initial hydrolysis rate was explained by analogy to a previous suggestion for the hydrolysis of cellulose,[135] namely that all the interior glucoside bonds have the same reactivity whilst the terminal ones are more reactive.

The energy of activation for hydrolysis in 11M HCl was found to be 94.1 kJ mol^{-1} which is lower than the value obtained earlier.[134] It is also lower than the value of 113.2 kJ mol^{-1} obtained by Senju[137] for the hydrolysis of O-hydroxyethyl chitin in 0.5M H_2SO_4. However Meyer and Wehrli used only two temperatures for their determination, as did Senju, whereas Rupley measured the rate of hydrolysis at seven temperatures, hence his data should be more reliable. It is also in agreement with the observed greater susceptibility to hydrolysis of the monosaccharide model compounds having a $C(2)NHCOCH_3$ group in place of a $C(2)OH$ group.[132] The monomer, dimer and trimer were found to be the predominant species throughout the duration of the hydrolysis and this does not fit the pattern expected for random splitting of the chain. Rupley suggested therefore that the chitin chains are aggregated in solution and that the rate-limiting step is a disaggregation step, followed by rapid hydrolysis. A hydrolysis mechanism in which the physical arrangement of the chitin chains is important in determining the kinetic behaviour is in agreement with the observed high reaction order with respect to acid concentration.

The hydrolysis of chitin and chitosan by H_2SO_4 has been studied by Nagasawa et al.[132, 138, 139] This shows considerable differences from the behaviour on hydrolysis with HCl, the most obvious being that hydrolysis is accompanied by O- and N-sulphation of chitosan and by O-sulphation of chitin, with no evidence of concurrent de-N-acetylation of chitin. The effect of the reaction time and temperature on the hydrolysis of chitin was interpreted in terms of a two-stage process in which the greatest hydrolysis occurs at the time of dissolution, the subsequent hydrolysis in solution being slower.[138] This is unlike the behaviour with HCl where very little hydrolysis took place during the solubilisation step.[136] Another difference in behaviour compared with hydrolysis in HCl is that the chain scission appears to occur at random; after 4 h hydrolysis at 5°C in 96 wt-% H_2SO_4 the number average DP of the dialysable fraction, which amounted to 94%

of the product, was 16, while all the products for hydrolysis times of 1–6 h at −20°C were non-dialysable. Thus, unlike hydrolysis by HCl, the predominant species are not the monomer, dimer and trimer.

Hydrolysis of chitosan also gave products which were predominantly non-dialysable, indicating random scission of the chains. Surprisingly, the temperature of the ether used to precipitate the products at the end of the hydrolysis stage has a considerable effect on the number average DP, lowering the ether temperature from −10°C to −60°C caused an increase in the DP from 42 to 66. The non-dialysable fraction also increased with increase in the concentration of H_2SO_4 used.[132]

Very little work has been carried out on hydrolysis in organic acids. Lee[140] observed a steady decrease with time in the LVN of chitin solutions held at 2°C and at 25°C, the rate being considerably greater at 25°C. A complicating factor in such studies may be the formylation of chitin which occurs on standing in formic acid solution.[141] Kaifu et al.[142] concluded that very little hydrolysis of chitin takes place in solution in methanesulphonic acid at 0°C, while Austin et al.[143] report that solutions of chitosan in acetic acid have a useful life of more than 1 month.

One use of HCl hydrolysis of chitin and chitosan has been for the preparation of oligomers of the two polymers. Barker et al.[133] hydrolysed chitosan using 3.3M HCl at 100°C for 32 h, followed by selective N-acetylation prior to separation on a charcoal–Celite column to give the first seven members of the chitin oligomer series. Subsequently Rupley[136] and Capon and Foster[144] carried out hydrolysis of chitin at 40°C in concentrated HCl obtaining, respectively, the first five and the first six members of the series. While Rupley used a charcoal–Celite column Capon and Foster used gel filtration chromatography, which they claimed was superior, being more rapid and more suitable for dealing with reasonably large quantities.

Horowitz et al.[145] hydrolysed chitosan then separated the oligomer mixture without prior N-acetylation, thereby obtaining D-glucosamine together with five other fractions, the first of which was characterised as chitobiose. Much more recently the preparation and characterisation of the D-glucosamine oligomer series has been studied by Domard and Cartier[146, 147] who also used gel filtration chromatography in the separation step. Hydrolysis was carried out using a modified version of the procedure of Horowitz et al.,[145] the conditions used being 12M HCl at 72°C and a sample of chitosan[0.0]. Consideration of the oligosaccharide distribution as a function of hydrolysis time led to the selection of 75 minutes as the optimum reaction time. By comparison, Horowitz et al. used 48 h at 53°C. Domard and Cartier obtained chromatographically pure samples of each of the oligomers up to DP 15 and, in addition, four other fractions having DP values of 20, 28, 32 and 37 and polydispersities < 1.01 were also obtained. The oligomers were characterised by MS and ^{13}C NMR spectroscopy.

Acetolysis, the simultaneous hydrolysis and acetylation using H_2SO_4–acetic anhydride mixtures, has also been used to prepare oligomers but only the dimer[148–151] and trimer,[149, 150] chitobiose octaacetate and chitotriose undecaacetate, have been obtained by this method.

Acid hydrolysis is also used for the production of microcrystalline chitin and chitosan, formation of which involves controlled mineral acid hydrolysis followed by subjection to high shear forces while suspended in water to give gel-like thixotropic dispersions. Microcrystalline chitin was first described by Dunn and Farr[152] who used boiling 2.0M HCl as the hydrolytic medium, hydrolysis times of 5 minutes being sufficient. It was claimed that the properties of the microcrystalline chitin were superior to those of a commercial microcrystalline cellulose.

Austin and Brine subsequently claimed a more effective process for preparing microcrystalline chitin using 85 wt-% H_3PO_4 as the hydrolytic agent and IPA as the reaction medium, treatment times of 2 h being required.[153] This process was claimed to give greater control over the molecular weight of the microcrystalline chitin but the material contained low concentrations of phosphorus (0.4–0.8%) and had limited solubility. Deschamps and Castle[154] showed that all of the phosphorus, which is present as the phosphate salt of the small concentration of free amine groups in the chitin, could be removed by washing with dilute aqueous NaOH solution. However dispersions formed from the phosphate-free microcrystalline chitin are not as stable as those formed from the phosphate-containing material, unless a dispersing agent is added. Deschamps and Castle concluded that the phosphate salts act as self-dispersing agents.

Microcrystalline chitosan has been prepared by hydrolysis of chitosan in solution followed by precipitation by addition of NaOH solution while subjecting the mixture to rapid stirring[155] – the so-called aggregation technique. The product had a crystallinity index (CrI) of up to 95% compared with a value of 58.9% for the starting material.[155]

Two unusual observations concerning acid degradation of chitin have been reported in the literature. Giles et al.[15] found that the nitrogen content of a chitin sample fell from 7.1% to 3.7% on heating for 24 h at 60°C in 2M HCl and concluded that the acetamido group is removed, ammonia and acetic acid being lost from the system. However the work of Rupley[136] and of Barker et al.,[133] among others, points to no loss of amine groups although some deacetylation may occur, and no similar results to those of Giles et al. have been reported.

Kandaswamy[156] examined the effects of fuming HNO_3 and of 6M HCl on α- and β-chitin, treatment conditions being 4 h at room temperature. The results showed that both treatments reduced the % N-acetyl content of the β-chitin samples from approximately 9.5% to approximately 7.5% but that no such decrease occurred with α-chitin, the change being only from

7.5% to 7.25%. In all cases the nitrogen contents were relatively unchanged. Measurements on chitosans produced under standard conditions from the chitins showed that there was very little difference in the viscosities of samples prepared from α-chitin after acid treatment compared with the viscosity of the chitosan prepared from the control sample of α-chitin, which had not been given an acid treatment. In contrast to this, the viscosities of the chitosans prepared from the two acid-treated β-chitin samples were only about 50% of the viscosity of the control sample.

Kandaswamy concluded that the two forms exhibit different stabilities to these acids, and this could be related to the greater degree of swelling shown by β-chitin compared with α-chitin. However the values for the % N-acetyl content of the chitin samples are very low, indeed the highest value (9.64%) represents a chitin[0.4] while the α-chitin sample (7.53%) represents chitin[0.3]. Obviously a considerable degree of deacetylation must have occurred during the isolation stage and it is difficult to discuss the results in terms of chitin. One possible explanation for the greater chain cleavage obtained with the 'β-chitin' samples is the lower degree of shielding afforded the glycosidic links because of their greater degree of N-acetylation.

The action of anhydrous HF on chitin and chitosan has been studied by Gadelle and coworkers,[157–159] following the reaction by ^{13}C NMR spectroscopy. Chitin dissolves readily in HF at 20°C, a homogeneous solution being obtained in 10 minutes, but degradation is slower than degradation of cellulose.[160] The initial reaction is chain cleavage and after 4 h the predominant species is the dimeric glucopyranosyl oxazolinium ion, **5.21** (Figure 5.13), which is the expected initial product from the HF-scission of (1→4)-linked oligomers. After 16 h both this and the glucofuranosyl oxazolinium ion **5.22**, formed from **5.21** by ring contraction, are present; while **5.22** is the predominant species after 24 h. This latter product was also obtained from a solution of 2-acetamido-2-deoxy-D-glucose in HF. Precipitation in ether, followed by dissolving in water, gave a mixture of chitin oligosaccharides, **5.23**, with any glycosyl fluoride residues, **5.24**, being hydrolysed. Neither migration nor loss of the N-acetyl groups occurred, unlike the case of mineral acid hydrolysis where some deacetylation occurs.[136]

Hydrolysis of chitosan was more difficult[159] although the formation of a mixture of oligomers with an average DP of 4 was achieved after 19 h at 20°C. However obtaining anomerically unsubstituted chitosan oligomers was hindered by the presence of anomeric fluorides which, unlike the chitin oligomers, were not hydrolysed even in boiling water. However they could be removed by heating in aqueous $HClO_4$ at 60°C for 12 h, or by treatment with trifluoracetic anhydride followed by refluxing the N,O-trifluoracetamido derivatives in aqueous methanol under reflux.

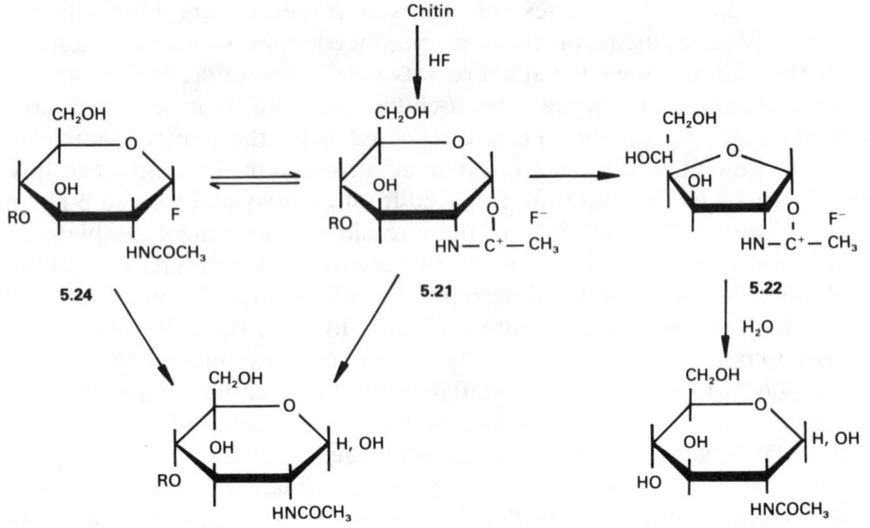

FIGURE 5.13　Proposed mechanism for the fluorohydrolysis of chitin[158, 159]

5.4.2　Alkaline degradation

Alkaline degradation of chitin is a well known phenomenon, occurring most noticeably during the deacetylation of chitin to produce chitosan. A number of procedures have been proposed for reducing the deleterious effects of alkali during this process and these include the use of an inert atmosphere such as nitrogen[161–166] or argon,[167] the addition of an oxygen scavenger such as thiophenol,[164–166] or the use of a reducing agent such as NaBH$_4$.[164] However very little fundamental work has been done and no mechanistic studies appear to have been carried out.

Two different aspects need to be considered, the first being the reaction between unmodified chitin, or chitosan, and alkali. Batista[166] has proposed mechanisms for two degradative reactions of both unmodified chitin and unmodified chitosan, the two reactions being random scission of the glycosidic link and end-group peeling, or unzipping, from the reducing terminal unit. The mechanisms, which are based on the analogous mechanisms for cellulose, are given in Figures 5.14 and 5.15. However that for chitosan chain scission is doubtful since formation of a 1,2-epimine, as depicted in Figure 5.14b would require either a better leaving group on C(1) or formation of a C(2)NH$^-$ ion which can then attack the anomeric centre in an analogous way to that proposed for the C(2)O$^-$ group in cellulose degradation.[168] Two alternatives are that degradation of chitosan

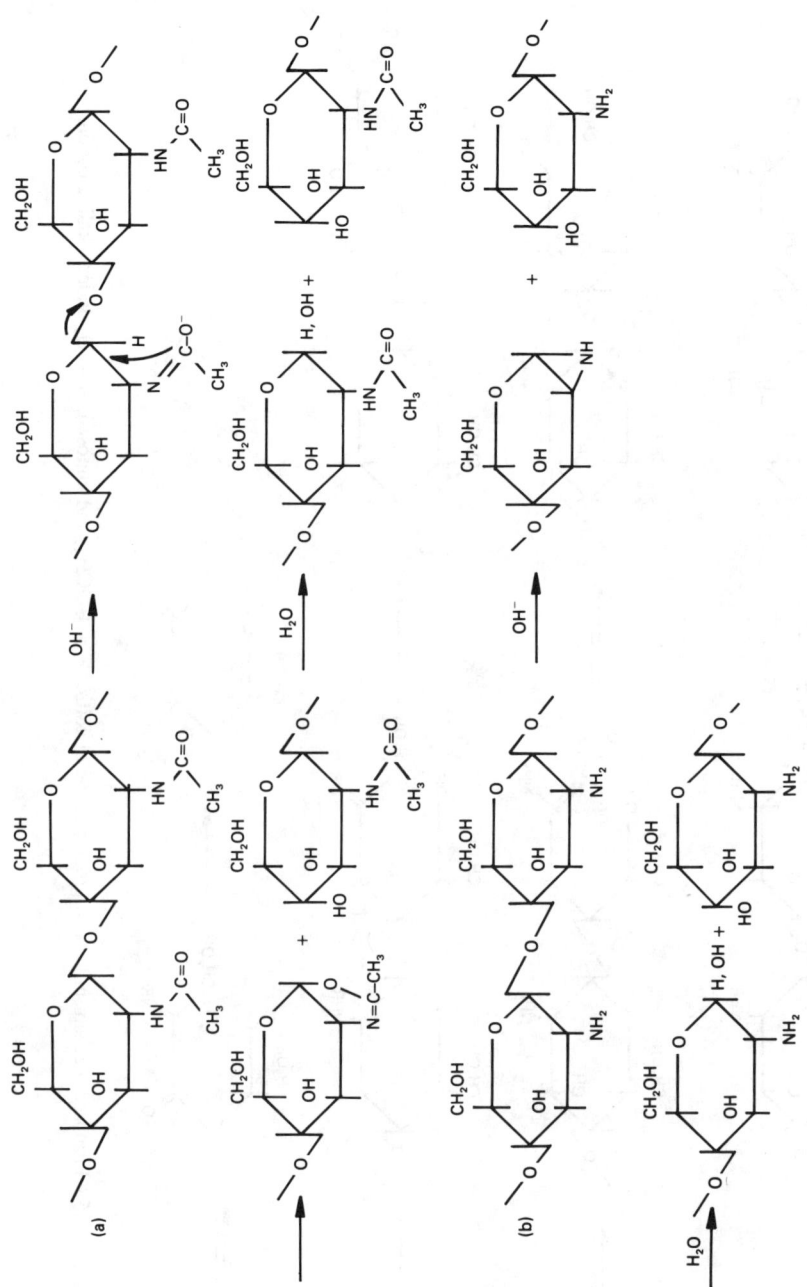

FIGURE 5.14 Proposed mechanisms for alkaline chain cleavage of (a) chitin and (b) chitosan[166]

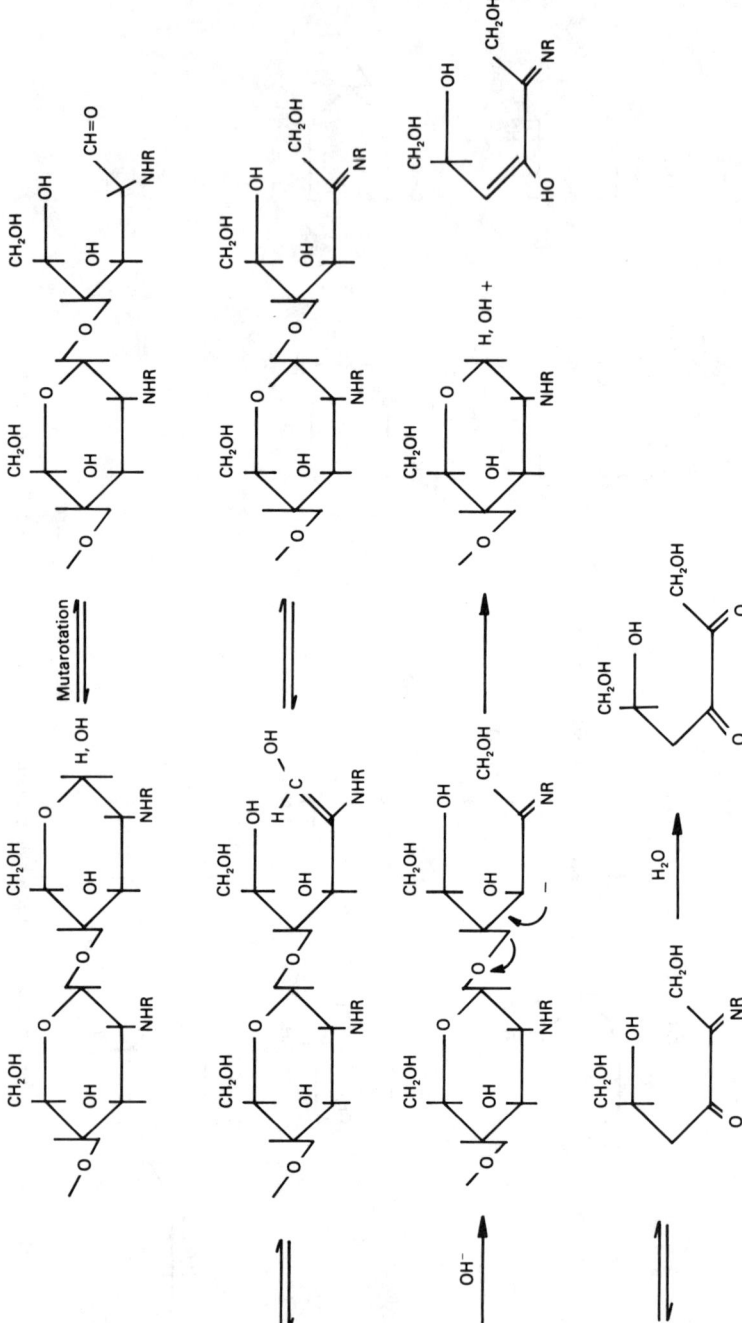

FIGURE 5.15   Proposed mechanism for end-group peeling of chitin, R = COCH₃, and chitosan, R = H, under alkaline conditions[166]

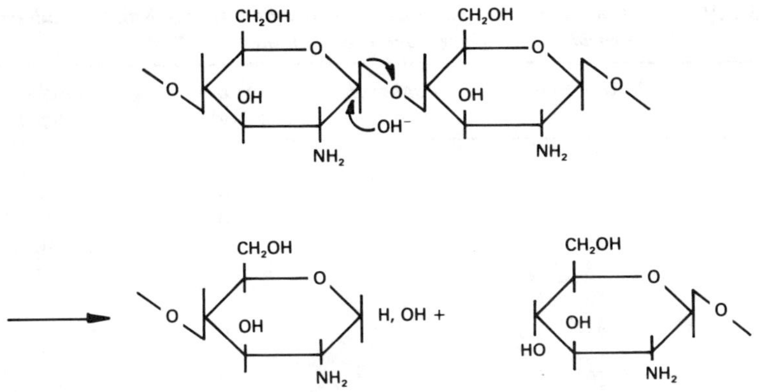

FIGURE 5.16   Possible mechanism for alkaline chain cleavage of chitosan

involves chain scission at one of the remaining N-acetyl-D-glucosamine residues as depicted in Figure 5.14a, or that there is a direct attack by an OH⁻ ion at the anomeric centre (Figure 5.16). Such a mechanism has been proposed for cellulose degradation[169] and it should be possible to distinguish between these two alternatives by a study of the susceptibility of chitosan[0.0] to non-oxidative alkaline degradation.

Oxidation under alkaline conditions could give rise to carbonyl groups at either the C(6) or C(3) positions in chitin and chitosan, and possibly also at C(2) in chitosan. According to the β-alkoxy carbonyl elimination mechanism proposed by Isbell,[170] the presence of such groups would make the adjacent glycosidic links very susceptible to alkaline hydrolysis. The presence of $NaBH_4$ could reduce the extent of degradation either by preventing the formation of these carbonyl groups or by reducing them once formed. This is a better explanation for the observed effects of $NaBH_4$ during chitin deacetylation than the prevention of end-group peeling of the chain.[164] Although peeling may result in a considerable weight loss it would not be expected to give rise to any appreciable decrease in molecular weight, hence no noticeable improvement in the latter should occur on stabilising the terminal unit by reduction to the hexitol (Table 5.9).

5.4.3 Deamination and chain scission with nitrous acid

The depolymerisation of chitosan by reaction with nitrous acid was first reported by Ambrecht[171] and subsequently by Meyer and Wehrli[134] who noted a rapid decrease in the viscosity of a solution of chitosan hydrochloride on the addition of $AgNO_2$. Other studies include that of Karrer and White,[172] who determined the amine group concentration of chitosan by

TABLE 5.9   *Effect on the product viscosity of the addition of thiophenol or sodium borohydride to the deacetylation reaction mixture*[a] [164, 166]

Run	NaOH conc. (wt-%)	Thiophenol (cm^3)	$NaBH_4$ (g)	Viscosity[b] (cps)
1[c]	50	0	0	52.5
2[c]	50	1	0	104
3	45	0	0	46
4	45	1	0	80
5	45	0	1	52
6	45	1	1	85
7	42.5	0	0	79
8	42.5	0	1	765
9[c, d]	50	0	0	412
10[c, d]	50	1	0	650

[a] Except where otherwise indicated deacetylation was carried out at 80°C for 3 h using 10 g chitin:160 cm^3 IPA:15 cm^3 NaOH solution.
[b] Viscosities were measured using 10 g dm^{-3} solutions in 1 vol.-% acetic acid.
[c] Reaction time of 2 h.
[d] Deacetylation carried out using 10 g chitin:150 cm^3 of 50 wt-% NaOH and no IPA.

measuring the volume of N_2 released on deamination, and that of Foster et al.[173] who used the differences in the rates of deamination of methyl 2-amino-2-deoxy-α-D-glucopyranoside and of the β-D-anomer to determine the glycoside link configuration in chitosan, which was deaminated at the same rate as the methyl 2-amino-2-deoxy-β-D-glucopyranoside.

The reaction of HNO_2 with aliphatic amines is very fast, even at room temperature,[174] and comparison of the rates of increase in the reducing power of chitosan showed that treatment with refluxing 3M HCl required a reaction time of 160 minutes to obtain the same extent of depolymerisation as could be achieved by treatment with HNO_2 for 5 minutes at room temperature.[175] Although the reaction is conventionally described as the nitrous acid deamination of chitosan, nitrous acid itself is not directly involved, being of too low activity. Instead the active species will be the nitrous acidium ion, $ON.\overset{+}{O}H_2$, or nitrous anhydride, N_2O_3, the former being more important at low pH values. Another possible active species is NOCl.[176]

The reaction path involves formation of a diazonium ion which decomposes to give a carbonium ion which undergoes nucleophilic addition (Figure 5.17). The chain scission induced by this reaction results in one unmodified chain of lower DP and one chain, also of lower DP, in which the terminal unit is a 2,5-anhydro-D-mannose unit, and a study by Hirano et al.[177] showed that no other groups were introduced. In this work a sample of chitosan[0.48] in 20 vol.-% acetic acid was treated with $NaNO_2$ (mole ratio of $NaNO_2$:$-NH_2$ of 5.45:1) followed by acetylation with acetic

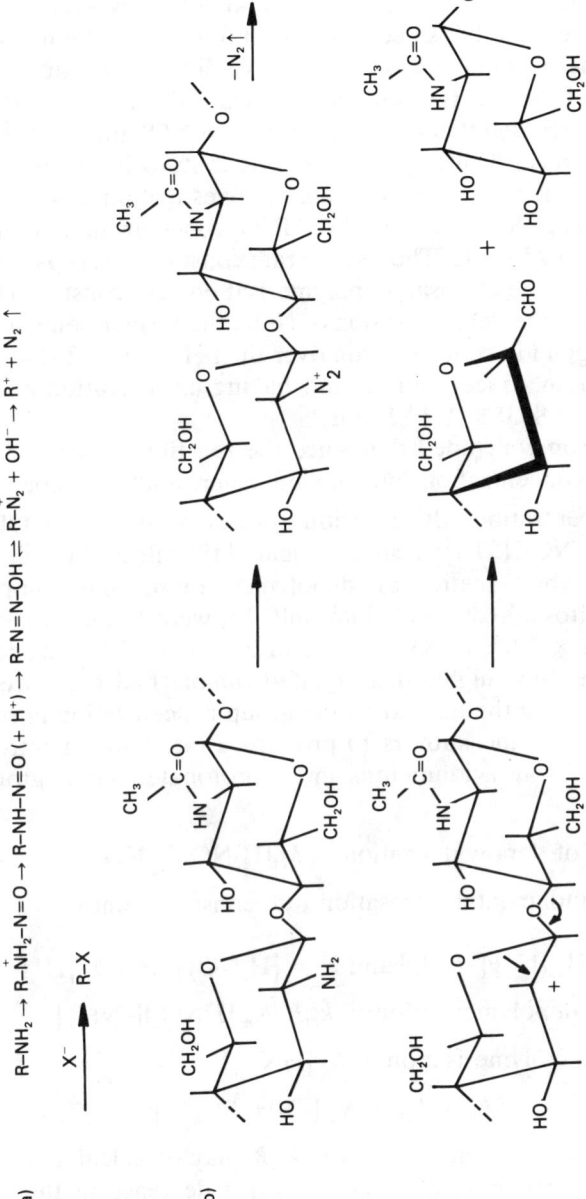

(a) $R-NH_2 \rightarrow R-NH_2-N=O \rightarrow R-NH-N=O \ (+ H^+) \rightarrow R-N=N-OH \rightleftharpoons R-\overset{+}{N_2} + OH^- \rightarrow R^+ + N_2 \uparrow$

FIGURE 5.17   Nitrous acid deamination of chitosan: (a) general reaction; (b) chain scission in chitosan

anhydride–pyridine and fractionation by column chromatography. Six products were obtained, one of which was 3,4,6-tri-O-acetyl-2,5-anhydro-D-mannose. Hydrolysis of the remaining five products, after de-O-acetylation, yielded only D-glucosamine and 2,5-anhydro-D-mannose, as shown by GLC and TLC, while the hydrolysates of the $NaBH_4$-reduced oligomers contained only D-glucosamine and 2,5-anhydro-D-mannitol. This specificity is remarkable since it has been found that the nitrous acid deamination of methylamine yields at least five different products.[178]

A brief study of the reaction kinetics was carried out by Peniston and Johnson[179] who concluded that the consumption of NO_2^- follows first-order kinetics, but more recently a very detailed kinetic study has been reported by Allan and Peyron.[180] This comprehensive investigation confirmed that the kinetics are first order with respect to HNO_2 concentration at temperatures in the range of 25–45°C. The overall rate constant was separated into its two components, the chitosan depolymerisation rate constant (k_c) and the rate constant for the decomposition of NO_2^-, the former being independent of the hydrogen ion concentration over the pH range 1.2–2.4, dependent on the chitosan concentration, and having an activation energy of depolymerisation of 86.0 ± 1.7 kJ mol^{-1}.

Allan and Peyron concluded that since the reaction is first order with respect to HNO_2 concentration, not second order, and is independent of chloride ion concentration, the reaction species must be $ON.\overset{+}{O}H_2$ and not either N_2O_3 or $NOCl$. They also investigated the effect of the degree of deacetylation on the kinetics of depolymerisation, and samples of chitosan[0.42], chitosan[0.30] and chitosan[0.12] were found to have rate constants of 1.038×10^{-3}, 2.083×10^{-3} and 0.75×10^{-3} s^{-1} respectively, indicating that the most highly deacetylated sample had the lowest rate constant despite having the highest amine group concentration in solution (Table 5.10). This led the authors to propose a rate-limiting nitrosation reaction between nitrous acidium ions and unprotonated amine groups, so that

$$\text{rate of depolymerisation} = k_{cf}[H_2NO_2^+][-NH_2]$$

where k_{cf} is the fundamental nitrosation rate constant. Since

$$K_a = [-NH_2][H^+]/[-NH_3^+] \text{ and } K = [H_2NO_2^+]/[HNO_2][H^+]$$

$$\text{rate of depolymerisation} = k_{cf}K K_a [HNO_2][-NH_3^+]$$

and since rate of depolymerisation = $k_c [HNO_2]$

$$k_c = k_{cf} K K_a [-NH_3^+]$$

from which values of the composite term $k_{cf}K$ may be calculated.

The values obtained indicate (Table 5.10) a decrease in the relative reactivity of the amine group towards the nitrous acidium ion with increase

TABLE 5.10   *Effect of the extent of deacetylation on the reactivity of chitosan toward nitrous acid*[180]

Sample	$[-NH_2]$ (mol dm^{-3})	pK_a^a	k_c (s^{-1})	$k_{cf} K$ ($s^{-1} M^{-2}$)
Chitosan[0.42]	1.62×10^{-2}	6.65	1.038×10^{-3}	2.87×10^5
Chitosan[0.30]	3.91×10^{-2}	6.65	2.083×10^{-3}	2.38×10^5
Chitosan[0.12]	6.33×10^{-2}	6.08	0.75×10^{-3}	0.14×10^5

[a] Calculated from Domard[8] assuming all amine groups protonated.

decrease in pK_a with increase in the extent of deacetylation (section 5.1). The nitrosation rates of aliphatic amines have been correlated with their basicity,[181] their reactivity increasing by a factor of 50 as the pK_a increased from 9.2 to 10.8.[182] This is in agreement with an increase in reactivity by a factor of about 20 for an increase in the pK_a value of chitosan from 6.08 to 6.65 (Table 5.10).

5.4.4 Biodegradation

Although the biology of chitin and chitosan is outside the scope of this book, a brief account of their biodegradation is appropriate at this point. The earliest report of enzymatic degradation of chitin is that of Beneke[183] in 1905, but the first specific information came from the studies of Karrer *et al.*[184, 185] who obtained *N*-acetyl-D-glucosamine in high yield from enzymatic hydrolysis of both crustacean and fungal chitin, thereby establishing the chemical similarity of the chitin from both sources (section 1.3.2). Karrer *et al.* used a chitinase obtained from the snail *Helix pomatia* and subsequently Zechmeister and Tóth[186, 187] demonstrated that both this chitinase, and also that obtained from emulsin, after being freed from β-glucosidase and α-galactosidase, hydrolysed phenyl 2-acetamido-2-deoxy-β-D-glucopyranoside but not the α-isomer, proving that the 1→4 glycoside link in chitin has the β-configuration. The crude chitinases could each be separated into two components, a chitinase that breaks down chitin into oligomers, including di-*N*-acetylchitobiose, and a di-*N*-acetyl-chitobiase that hydrolyses this latter substrate and the other lower oligomers to *N*-acetyl-D-glucosamine.[188, 189] In most instances the complete enzymatic hydrolysis of chitin to *N*-acetyl-D-glucosamine requires such a dual system of chitinase (EC 3.2.1.14) and di-*N*-acetylchitobiase or *N*-acetyl-β-glucosaminidase (EC 3.2.1.30), although a chitinase from *Verticillium albo-atrium* which hydrolyses chitin, tetra-*N*-acetylchitotetrose, tri-*N*-acetylchitotriose, and di-*N*-acetylchitobiose to *N*-acetyl-D-glucosamine, without the need for the addition of any di-*N*-acetylchitobiase, has been reported.[190]

Chitinolytic bacteria and fungi occur widely in nature while chitinolytic enzymes have also been found in invertebrates such as snails,[184-189] crustacea and insects[191] and in protozoa, nematods, earthworms, coelenterates and polychaete worms.[192] They have also been found in the digestive tracts of some vertebrate species such as birds and lizards that are insectivorous.[192, 193] It has also been shown[194-196] that chitinases are separate and distinct from the lysozymes which also hydrolyse chitin.[197, 198] Methods have been published for the isolation and purification of chitinases from *Serratia marcescens*,[199] *Pycnoporus cinnabarinus*,[200] *Neurospora crassa*,[201] *Verticillium albo-atrium*,[190] *Phaseolus vulgaris*,[202] *Lycopersicon esculentum*,[203] *Calvatia cyathiformis*,[204] soybean seeds[204] and wheat germ.[205]

The presence of chitosan in living systems was first demonstrated in the cell walls and sporangiophores of *Phycomyces blakesleeanus* by Kreger[206] and subsequently in the cell walls of *Mucor rouxii* by Bartnicki-Garcia and Nickerson,[207] and it has since been identified in a wide range of species. Chitosan-degrading enzymes were first identified by Monaghan *et al*.[208] and Ramirez-Leon and Ruiz-Herrera[209] and they are known to be extremely common in soil from a number of environments; garden, forest, salt marsh and agricultural land.[210] Based on reported studies on the specificity of chitosanases from *Penicillium islandicum*, *Streptomyces* No. 6, Bacterium No. 8, Bacterium K-1, *Myxobacter* A1-1, and *Bacillus* R4, Fenton *et al*.[210] proposed that these enzymes should be divided into two classes: those specific for chitosan, and those which attack both chitosan and carboxymethyl cellulose. No decision was reached as to whether or not this latter group should be classified as cellulases rather than chitosanases.

The available evidence to date indicates that chitosanases exhibit maximum activity towards substrates intermediate between chitin[1.0] and chitosan[0.0]. Thus a chitosanase from *Penicillium islandicum* showed greatest activity towards chitosan[0.3]–chitosan[0.6] substrates,[210] while that of a chitosanase produced by *Bacillus circulans* MH-K1 was greatest towards a chitosan[0.2] substrate, decreasing sharply with further increase in the extent of deacetylation.[211] This may be related to the fact that naturally occurring chitosans are normally partially N-acetylated. Studies with the *P. islandicum* chitosanase indicated that the enzyme shows high specificity towards hydrolysis of the glycosidic link between C(1) of an N-acetyl-D-glucosamine residue and C(4) of a D-glucosamine residue.[210] Methods have been published for the isolation and purification of chitosanases from *Bacillus* sp. No. 7-M[212] and *Streptomyces griseus*.[213]

Chitosan occurring naturally is produced by enzymatic deacetylation of chitin within the living system[214-216] and the action of chitin deacetylase obtained from *Mucor rouxii* has been studied by Araki and Ito[214, 215] using O-hydroxyethyl chitin, colloidal chitin and chitin oligomers as substrates. They demonstrated its specificity towards chitin and chitin oligomers, although it was found to be inactive towards N-acetyl-D-glucosamine and di-N-acetylchitobiose. The enzyme has an optimum pH of 5.5, does not require the presence of any metal ion, and is strongly inhibited by the presence of acetic acid. It was suggested that it is more active towards nascent chitin than towards preformed chitin, and that its action is inhibited by substitution at the C(3)OH group. Chitin deacetylase activity has also been found[217] in the culture filtrate of *Colletotrichum lindemuthianum* and in *Mucor miehe*, and in the cuticle of the abdomen of the physogastric queen of the termite *Macrotermes estherae*.[218] Methods have been published for the isolation and purification of chitin deacetylase from *Mucor rouxii* AHU 6019[219] and *Colletotrichum lindemuthianum*.[220]

Davis and Eveleigh[221] proposed a scheme for chitin biodegradation (Figure 5.18) analogous to that for cellulose biodegradation by the fungal cellulase complex. They admitted that no exochitinases or exochitosanases had been reported and thus their inclusion in the scheme was speculative, but stated that such an absence was surprising since an exo-splitting di-N-acetylchitobiohydrolase or chitobiohydrolase would be expected in view of the twofold helical structure of the chains. However prior to this Ohtakara and Mitsutomi had reported[222] a study of the chitinase obtained from *Pycnoporus cinnabarinus*, where the major product obtained from the oligomeric substrates tri-N-acetylchitotriose to hexa-N-acetylchitohexose, and their reduced analogues, was di-N-acetylchitobiose obtained by splitting off a disaccharide unit from the non-reducing end. In subsequent articles the authors state[200, 223] that the predominant mode of action is 'splitting the second β-N-acetylglucosaminide bond from the non-reducing end,' but despite this they classified it as an endochitinase. However from their results it would appear to be an exo-splitting di-N-acetylchitobiohydrolase as indicated in the biodegradation scheme of Davis and Eveleigh.[221]

Bade and Hickey[224] have proposed a test system to determine the mode of action of chitinases. The test makes use of two forms of regenerated chitin: partially esterified fibrous material obtained by regeneration from solution in H_2SO_4, and a more particulate form obtained by precipitation from solution in HCl. Exochitinases rapidly attack the former material, owing to the ready accessibility of the chain ends, but show very little or no activity with the latter substrate since the chain ends are not readily available. In a mixture of the two forms an exochitinase will hydrolyse all the fibrous material while leaving the particulate material relatively untouched, but an endochitinase will attack both substrates. Thus rapid production of di-N-acetylchitobiose, determined as N-acetyl-D-glucosamine after treatment of the supernatent with N-acetyl-β-glucosaminidase, from the fibrous substrate indicates the presence of an exochitinase, while the presence of an endochitinase is indicated by the rate at which the turbidity of a suspension of particulate chitin is reduced. Using these tests with two chitinases isolated from the medium of chitin-adapted

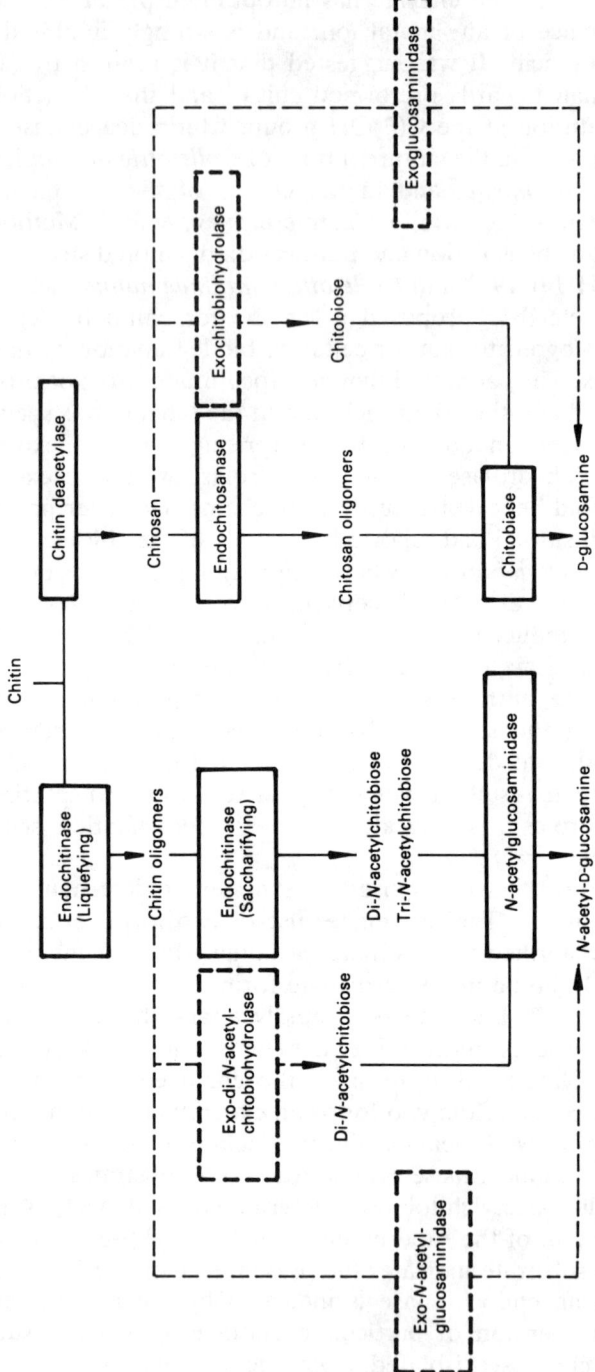

FIGURE 5.18 Potential routes for enzymatic degradation of chitin.[221] Confirmed hydrolytic steps (——); hypothetical hydrolytic steps (– – –)

TABLE 5.11   *Relative rates of hydrolysis of* N-*acylchitosans prepared by homogeneous* N-*acylation*[227, 228]

R	H–	CH$_3$–	CH$_3$CH$_2$–	CH$_3$(CH$_2$)$_2$–	(CH$_3$)$_2$CH–	H$_2$N–CH$_2$–	Cl–CH$_2$–
Relative rate of hydrolysis[a]	0.8	8.3	1.5	0.5	0.4	0.3	0.1

[a] Rate of hydrolysis relative to the rate for native chitin from crab shell.

Streptomyces cultures, the authors characterised one as an exochitinase and one as an endochitinase. The components of the chitinase complex from *Streptomyces* sp. ATCC 11238 (*Streptomyces olivaceovirides*) has been separated by fast protein liquid chromatography[225] to give an endochitinase (molecular weight 4.6×10^4), two exochitinases (molecular weights 7.0×10^4 and 9.2×10^4) and a di-N-acetylchitobiase (molecular weight 4.7×10^4). In all studies to date of the mode of action of chitosanases[210–213, 226] the behaviour has been found to be that of an endochitosanase.

Some studies on the effects of both chemical and physical modifications of chitin on its biodegradation have been reported. Karrer and White[172] found that the chitinase in the extract from *Helix pomatia* did not hydrolyse N-formyl-, N-propionyl- or N-butyrylchitosan. However a similar study carried out subsequently by Hirano and Yagi,[227, 228] using a chitinase from *Streptomyces griseus*, gave conflicting results. In this latter work the enzyme was found to hydrolyse a variety of N-acylated chitosans, as shown in Table 5.11. The authors suggested that the differences between their results and those of Karrer and White[172] were due to differences in enzyme source and to the lower levels of N-substitution in the derivatives used by Karrer and White. However all the modified chitins used by Hirano and Yagi were prepared by homogeneous N-acylation (section 4.2.1) leading to more accessible substrates, as shown by the fact that dried, powdered N-acetylchitosan prepared in this way is over eight times as susceptible to hydrolysis by the chitinase as is native, crystalline chitin. Karrer and White prepared their chitin analogues by heterogeneous N-acylation and these substrates would be expected to have physical structures which more closely resemble that of native chitin. Thus the different results obtained may be due to a physical effect, in which case the figures in Table 5.11 should be divided by a factor of 8 to obtain approximate values for the

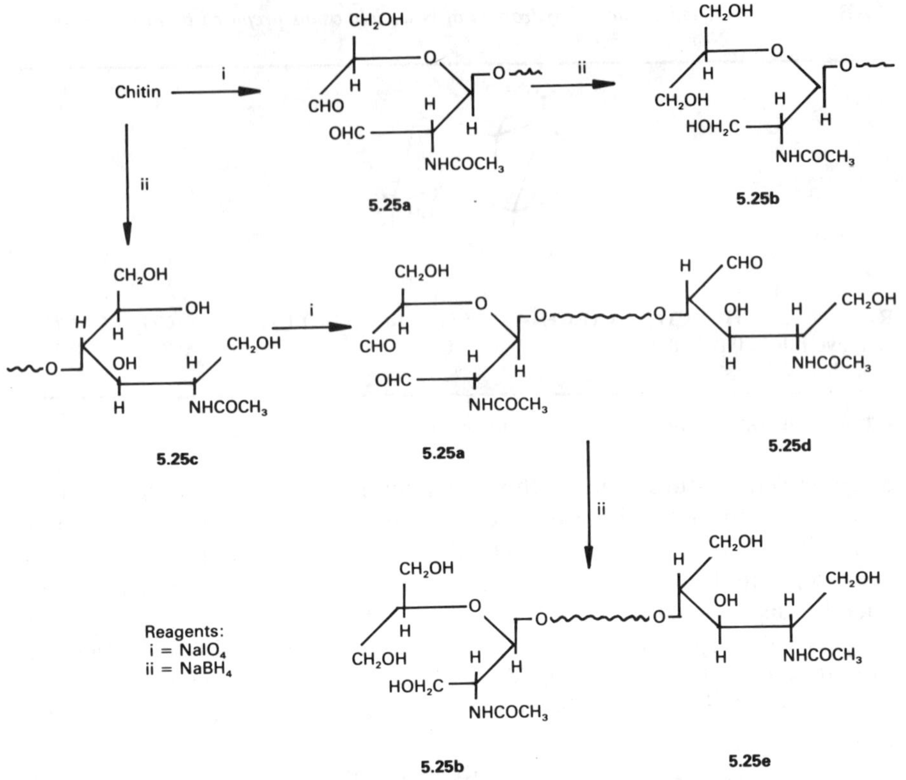

FIGURE 5.19   Chemical modifications of the terminal units in chitin by $NaIO_4$ oxidation and $NaBH_4$ reduction[229]

relative rates of hydrolysis that might be expected for the N-acylchitosans of Karrer and White. If this is done the relative rate for the most reactive analogue, N-propionylchitosan, becomes less than 20% of that of chitin itself.

Hirano and Yagi also found that N-pentanoyl-, N-hexanoyl-, N-dodecanoyl-, N-octadecanoyl-, N-benzoyl-, N-nicotinyl-, N-methylidene- and N-benzylidenechitosan, together with chitosan itself, were not hydrolysed by the chitinase or by lysozyme.

N-Acetylchitosan, as the undried gel, had a relative rate of hydrolysis of 13 compared with 8.3 for the dried material, while N-acetylchitosan oxidised with $NaIO_4$ gave a value of 15, lowered to 8.5 on reduction with $NaBH_4$.[228] In a later paper the authors examined the effect of $NaIO_4$–$NaBH_4$ treatments in greater detail[229] but of the structures given in Figure 5.19 only 5.25a gave an increase in the rate of hydrolysis with chitinase from Streptomyces griseus.

Hara *et al.*,[230] working with a chitinase produced by *Streptomyces erythraeus*, examined its action on partially *O*-methylated derivatives of di-*N*-acetylchitobiose and concluded that the C(2)NHCOCH$_3$ and C(3)OH groups of the reducing ends, and the C(6)OH of the non-reducing ends of the disaccharides, were important for binding to the active site of the enzyme, while the C(6)OH of the reducing ends and the C(3)OH of the non-reducing ends were of less importance.

Finally the effect of *O*-alkyl substitution on the susceptibility of chitin to degradation by lysozyme has been reported.[231] The order of reactivity for samples having *DS* values for etherification of ~0.4 was

$$H(\text{chitin}) < CH_3(CH_2)_3- < (CH_3)_2CHCH_2- < (CH_3)_3C- \\ << NaOOC.CH_2-$$

which follows the order of increasing swelling in water, showing the importance of accessibility.

REFERENCES

1. H. Terayama, *J. Pol. Sci.*, **8** (1952) 243.
2. J. Doczi, *US Patent 2,795,579* (1957).
3. J. Noguchi, K. Arato and T. Komai, *Kogyu Kogaku Zasshi*, **72** (1969) 796.
4. L.A. Nud'ga, E.A. Plisko and S.N. Danilov, *Zhur. Obsch. Khim.*, **43** (1973) 2756.
5. R.A.A. Muzzarelli, F. Tanfani, M. Emanuelli and S. Gentile, *J. Appl. Biochem.*, **2** (1980) 380.
6. J.W. Park, K-H. Choi and K.K. Park, *Bull. Korean Chem. Soc.*, **4** (1983) 68.
7. A. Katchalsky and P. Spitinik, *J. Pol. Sci.*, **2** (1947) 432.
8. A. Domard, *Int. J. Biol. Macromol.*, **9** (1987) 98.
9. A. Katchalsky, *J. Pol. Sci.*, **12** (1954) 159.
10. Z. Tamura, M. Miyazaki and T. Suzuki, *Chem. Pharm. Bull.*, **13** (1965) 330.
11. P. Broussignac, *Chim. Ind. Genie Chim.*, **99** (1968) 1241.
12. E.R. Hayes and D.H. Davies, in *Proceedings 1st International Conference on Chitin/Chitosan (1977)*, R.A.A. Muzzarelli and E.R. Pariser (eds), MIT Sea Grant Program Report MITSG-78-7, 1978, p. 406.
13. J.G. Domszy and G.A.F. Roberts, *Makromol. Chem.*, **186** (1985) 1671.
14. A. Domard and M. Rinaudo, *Int. J. Biol. Macromol.*, **5** (1983) 49.
15. C.H. Giles, A.S.A. Hassan, M. Laidlaw and R.V.R. Subramanian, *J. Soc. Dyers Colourists*, **74** (1958) 647.
16. R.A.A. Muzzarelli, *Natural Chelating Polymers*, Pergamon, Oxford, 1973.
17. R.A.A. Muzzarelli, *Chitin*, Pergamon, Oxford, 1977.
18. R.A.A. Muzzarelli, ref. 17, p. 142.
19. R.A.A. Muzzarelli, A. Isolati and A. Ferrero, *Ion Exchange Membr.* **1** (1974) 193.
20. T. Yoshinari and T. Subramanian, *Environ. Biogeochem.*, **2** (1976) 541.
21. F. Yaku and T. Koshijima, in ref. 12, p. 386.
22. G. Micera, S. Deiana, A. Dessi, P. Decock, B. Dubois and H. Kozlowski, *Inorg. Chim. Acta*, **107** (1985) 45.
23. K. Kurita, T. Sannan and Y. Iwakura, *J. Appl. Pol. Sci.*, **23** (1979) 511.

24. T. Sannan, K. Kurita and Y. Iwakura, *Makromol. Chem.*, **177** (1976) 3589.
25. K. Kurita, Y. Koyama and A. Taniguchi, *J. Appl. Pol. Sci.*, **31** (1986) 1169.
26. K. Kurita, Y. Koyama and A. Taniguchi, *J. Appl. Pol. Sci.*, **31** (1986) 1951.
27. K. Kurita, in *Industrial Polysaccharides: Genetic Engineering, Structure/ Property Relations and Applications*, M. Yalpani (ed.), Elsevier, Amsterdam, 1987, p. 337.
28. K. Kurita, S. Chikaoka and Y. Koyama, *Chem. Letters*, (1988) 9.
29. K. Kurita, Y. Koyama and S. Chikaoka, *Polym. J.*, **20** (1988) 1083.
30. N. Nishi, Y. Maekita, S. Nishimura, O. Hasegawa and S. Tokura, *Int. J. Biol. Macromol.*, **9** (1987) 109.
31. M. Tsezos and B. Volesky, *Biotech. Bioeng.*, **24** (1982) 385.
32. R.A.A. Muzzarelli and O. Tubertini, *Microchem. Acta*, **5** (1970) 892.
33. T. Koshijima, R. Tanaka, E. Muraki, A. Yamada and F. Yaku, *Cellulose Chem. Technol.*, **7** (1973) 197.
34. G. McKay, H.S. Blair and A. Findon, in *Chitin in Nature and Technology*, R.A.A. Muzzarelli, C. Jeuniaux and G.W. Gooday (eds), Plenum Press, New York, 1985, p. 559.
35. B.L. Averbach, in *Chitin and Chitosan*, S. Hirano and S. Tokura (eds), The Japanese Society of Chitin and Chitosan, Tottori, 1982, p. 248.
36. H.S. Blair and T-C. Ho, *J. Chem. Tech. Biotechnol.*, **31** (1980) 6.
37. H. Irving and R.J.P. Williams, *Nature*, **162** (1948) 746.
38. M.S. Masri, F.W. Reuter and M. Friedman, *J. Appl. Pol. Sci.*, **18** (1974) 675.
39. R.A.A. Muzzarelli, in ref. 17, p. 145.
40. S. Hirano, Y. Kondo and Y. Nakazawa, *Carbohydr. Res.*, **100** (1982) 431.
41. R.G. Pearson, *J. Amer. Chem. Soc.*, **85** (1963) 3533.
42. J.E. Hukeey, *Inorganic Chemistry: Principles of Structure and Reactivity*, Harper International, New York, 3rd edn, 1983.
43. H. Hauer, in ref. 12, p. 263.
44. R.A.A. Muzzarelli, *Belgian Patent 876,990* (1979).
45. R.A.A. Muzzarelli, F. Tanfani and G. Scarpini, *Biotech. Bioeng.*, **22** (1980) 885.
46. R.A.A. Muzzarelli, F. Tanfani and M. Emanuelli, *J. Appl. Biochem.*, **3** (1981) 322.
47. R.A.A. Muzzarelli and F. Tanfani, in ref. 35, p. 183.
48. M. Tsezos and B. Volesky, *Biotech. Bioeng.*, **24** (1982) 955.
49. B. Focher, A. Massoli, G. Torri, A. Gervasini and F. Morazzoni, *Makromol. Chem.*, **187** (1986) 2609.
50. B. Focher, A. Massoli, G. Torri, A. Gervasini and F. Morazzoni, in ref. 34, p. 306.
51. G. Micera, S. Deiana, A. Dessi, P. Decock, B. Dubois and H. Kozlowski, in ref. 34, p. 565.
52. K. Ogawa, K. Oka, T. Miyanishi and S. Hirano, in *Chitin, Chitosan, and Related Enzymes*, J.P. Zikakis (ed.), Academic Press, New York, 1984, p. 327.
53. T. Sakaguchi, T. Horikoshi and A. Nakajima, *J. Agric. Chem. Soc. Japan*, **53** (1979) 149.
54. T. Sakaguchi, A. Nakajima and T. Horikoshi, *J. Agric. Chem. Soc. Japan*, **53** (1979) 211.
55. T. Sakaguchi, T. Horikoshi and A. Nakajima, *Agric. Biol. Chem.*, **45** (1981) 2191.
56. T. Sakaguchi and A. Nakajima, in ref. 35, p. 177.
57. N. Nishi, A. Ebina, S. Nishimura, A. Tsutsumi, O. Hasegawa and S. Tokura, *Int. J. Biol. Macromol.*, **8** (1986) 311.

58. N. Nishi, S. Nishimura, Y. Maekita, A. Ebina, A. Tsutsumi and S. Tokura, in ref. 34, p. 297.
59. S. Tokura, S. Nishimura and N. Nishi, *Polym. J.*, **15** (1983) 597.
60. Y. Uraki, N. Nishi, S. Nishimura and S. Tokura, in *Chitin and Chitosan*, G. Skjåk-Braek, T. Anthonsen and P. Sandford (eds), Elsevier, London, 1989, 537.
61. R.A.A. Muzzarelli and F. Tanfani, in ref. 35, p. 45.
62. R.A.A. Muzzarelli, *Carbohydr. Polym.*, **8** (1988) 1.
63. R.A.A. Muzzarelli and A. Zattoni, *Int. J. Biol. Macromol.*, **8** (1986) 137.
64. R.A.A. Muzzarelli, M. Weckx, O. Filippini and C. Lough, *Carbohydr. Polym.*, **11** (1989) 307.
65. R.A.A. Muzzarelli, in ref. 34, p. 321.
66. F. Delben, R.A.A. Muzzarelli and M. Terbojevich, *Carbohydr. Polym.* **11** (1989) 205.
67. V. Crescenzi, F. Delben, S. Paoletti and J. Skerjanc, *J. Phys. Chem.*, **78** (1974) 607.
68. A. Cesàro, F. Delben and S. Paoletti, *J. Chem. Soc., Faraday Trans. I*, **84** (1988) 2573.
69. R. Barbucci, M. Casolaro, V. Barone, P. Ferruti and M. Tramontini, *Macromolecules*, **16** (1983) 1159.
70. R.A.A. Muzzarelli and F. Tanfani, *Carbohydr. Polym.*, **7** (1985) 297.
71. C.H. Giles, A.S.A. Hassan and R.V.R. Subramanian, *J. Soc. Dyers Colourists*, **74** (1958) 685.
72. C.H. Giles and A.S.A. Hassan, *J. Soc. Dyers Colourists*, **74** (1958) 846.
73. R.H. Peters, *J. Soc. Dyers Colourists*, **61** (1945) 95.
74. G.G. Maghami and G.A.F. Roberts, *Makromol. Chem.*, **189** (1988) 2239.
75. W.A. Neugebauer, E. Neugebauer and R. Brzezinski, *Carbohydr. Res.*, **189** (1989) 363.
76. R.H. Peters, *Textile Chemistry*, Elsevier, New York, vol. 3, 1975.
77. I.D. Rattee and M.M. Breuer, *The Physical Chemistry of Dye Adsorption*, Academic Press, London, 1974, p. 14.
78. R.M. Diamond, *J. Phys. Chem.* **67** (1963) 2513.
79. T. Seo, T. Kanbara and T. Iijima, *J. Appl. Pol. Sci.*, **36** (1988) 1443.
80. T. Seo, S. Hagura, T. Kanbara and T. Iijima, *J. Appl. Pol. Sci.*, **37** (1989) 3011.
81. D. Knorr, *J. Food Sci.*, **48** (1983) 36.
82. G. McKay, H.S. Blair and J.R. Gardner, *J. Appl. Pol. Sci.*, **27** (1982) 3043.
83. G. McKay, H.S. Blair and J.R. Gardner, *J. Appl. Pol. Sci.*, **27** (1982) 4251.
84. G. McKay, H.S. Blair and J.R. Gardner, *J. Appl. Pol. Sci.*, **28** (1983) 1767.
85. G. McKay, H.S. Blair and J.R. Gardner, *J. Appl. Pol. Sci.*, **33** (1987) 1249.
86. W.J. Weber and A.P. Matthews, *Assoc. Inst. Chem. Eng. Symp. Ser. No. 166*, **73** (1976) 91.
87. T.W. Weber and R.K. Chakravorti, *Assoc. Inst. Chem. Eng. J.*, **20** (1974) 228.
88. B. Venkatrao, A. Baradarajan and C.A. Sastry, in ref. 34, p. 554.
89. J. Guthrie, H.S. Blair and R.P. O'Donnell, *Polymer Commun.*, **27** (1986) 53.
90. M. Sekido, T. Iijima and A. Takahashi, *Kogyo Kogaku Zasshi*, **68** (1965) 524.
91. R. McGregor, R.H. Peters and J.H. Petropoulos, *Trans. Faraday Soc.*, **58** (1962) 1054.
92. R.H. Hackman, *Aust. J. Biol. Sci.*, **8** (1955) 530.
93. O.P. Agrawal and K. Scheller, in ref. 34, p. 316.
94. R. Bloch and M.M. Burger, *Biochem. Biophys. Res. Commun.*, **58** (1974) 13.
95. I. Matsumoto, H. Yamaguchi, N. Seno, Y. Shibata and T. Okuyama, in ref. 35, p. 165.

96. T. Baba, R. Yamaguchi, Y. Arai and T. Ito, *Carbohydr. Res.*, **86** (1980) 161.
97. R.A.A. Muzzarelli, G. Barontini and R. Rocchetti, *Biotech. Bioeng.*, **18** (1976) 1445.
98. P.Q. Flor and S. Hayashida, in ref. 35, p. 153.
99. S. Hayashida and P.Q. Flor, *Agric. Biol. Chem.*, **46** (1982) 1639.
100. P.Q. Flor and S. Hayashida, *Biotech. Bioeng.*, **25** (1983) 1973.
101. J. Synowiccki, Z.E. Sikorski and M. Naczk, *Biotech. Bioeng.*, **23** (1981) 2211.
102. J. Synowiecki, in ref. 34, p. 417.
103. A. Ohtakara, G. Mukerjee and M. Mitsutomi, in ref. 60, p. 643.
104. I.A. Cherkasov and N.A. Kravchenco, *Biokhimiya*, **33** (1968) 761.
105. K. Hayashi, Y. Hamasu, Y. Doi and M. Funatsu, *J. Fac. Agric. Kyushu Univ.*, **17** (1973) 327.
106. H.B. Jensen and K. Kleppe, *Eur. J. Biochem.*, **26** (1972) 305.
107. I.A. Cherkasov and N.A. Kravchenco, *Biochim. Biophy. Acta*, **206** (1970) 289.
108. J. Skujins, A. Pukite and A.D. McLaren, *Molec. Cell. Biochem.*, **2** (1973) 221.
109. T. Nakabayashi and T. Makita, *Nippon Shokuhin Kogyu Gakkai-shi*, **19** (1972) 111.
110. J.P. Thomé and Y. Van Daela, in ref. 34, p. 551.
111. P.R. Austin, *US Patent 4,063,016* (1977).
112. Y. Shigeno, K. Kondo and K. Takemoto, *J. Appl. Pol. Sci.*, **25** (1980) 731.
113. Y. Shigeno, K. Kondo and K. Takemoto, *Angew. Makromol. Chem.*, **90** (1980) 55.
114. Y. Shigeno, K. Kondo and K. Takemoto, *Angew. Makromol. Chem.*, **90** (1980) 211.
115. S. Hirano, Y. Kondo, M. Kuketa and A. Yamashita, in ref. 35, p. 57.
116. R.A.A. Muzzarelli, in ref. 12, p. 335.
117. M. Takeda and T. Tomida, *J. Shimonoseki Univ. Fish.*, **20** (1972) 107 (quoted in ref. 118).
118. M. Takeda, in ref. 12, p. 355.
119. M. Takeda and T. Tomida, *J. Shimonoseki Univ. Fish.*, **17** (1969) 37 (quoted in ref. 118).
120. H.K. No and S.P. Meyers, *J. Food Sci.*, **54** (1989) 60.
121. T. Seo, Y. AiGan, T. Kanbara and T. Iijima, *J. Appl. Pol. Sci.*, **38** (1989) 997.
122. J. Noguchi, S. Tokura, M. Inomata and C. Asano, *Kogyo Kogaku Zassi*, **68** (1965) 904.
123. S. Hirano, N. Matsuda, O. Miura and T. Tanaka, *Carbohydr. Res.*, **71** (1979) 344.
124. S. Hirano and Y. Nishiguchi, *Carbohydr. Polym.*, **5** (1985) 13.
125. R.A.A. Muzzarelli, F. Tanfani, G. Scarpini and G. Laterza, *J. Biochem. Biophys. Methods*, **2** (1980) 299.
126. K.E. Taylor, Ph.D. Thesis (CNAA), Nottingham Polytechnic, UK 1990.
127. G.A.F. Roberts and K.E. Taylor, in ref. 60, p. 577.
128. J.C. Irvine and A. Hynd, *J. Chem. Soc.*, **101** (1912) 1128; **105** (1914) 698.
129. R.C.G. Moggridge and A. Neuberger, *J. Chem. Soc.*, (1938) 745.
130. A. Neuberger and R. Pitt Rivers, *J. Chem. Soc.*, (1939) 122.
131. K. Onodera and T. Komano, *Agric. Biol. Chem.*, **25** (1961) 932.
132. K. Nagasawa and N. Tanoura, *Chem. Pharm. Bull.*, **20** (1972) 157.
133. S.A. Barker, A.B. Foster, M. Stacey and J.M. Webber, *J. Chem. Soc.*, (1958) 2218.

134. K.H. Meyer and H. Wehrli, *Helv. Chim. Acta*, **20** (1937) 353.
135. K. Freudenberg and G. Blomquist, *Ber.*, **68** (1935) 2070.
136. J.A. Rupley, *Biochim. Biophys. Acta*, **83** (1964) 245.
137. R. Senju, *J. Agric. Chem. Soc. Japan*, **25** (1951/52) 301 (through *Chem. Abs.*, **47** (1953) 7873).
138. K. Nagasawa, Y. Tohira, Y. Inoue and N. Tanoura, *Carbohydr. Res.*, **18** (1971) 95.
139. K. Nagasawa and Y. Inoue, *Chem. Pharm. Bull.*, **19** (1971) 2617.
140. V.F. Lee, *University Microfilms (Ann Arbor)*, 74/29446, (1974).
141. D. Gagnaire, J. Saint-Germain and M. Vincendon, *Makromol. Chem.*, **183** (1982) 593.
142. K. Kaifu, N. Nishi, T. Komai, S. Tokura and O. Somorin, *Polym. J.*, **13** (1981) 241.
143. P.R. Austin, G.A. Reed and J.R. Deschamps, in ref. 35. p. 99.
144. B. Capon and R.L. Foster, *J. Chem. Soc. C*, (1970) 1654.
145. S.T. Horowitz, S. Roseman and H.J. Blumenthal, *J. Amer. Chem. Soc.*, **79** (1957) 5046.
146. A. Domard and N. Cartier, in ref. 60, p. 383.
147. A. Domard and N. Cartier, *Int. J. Biol. Macromol.*, **11** (1989) 297.
148. M. Bergmann, L. Zervas and E. Silberkweit, *Naturwiss.*, **19** (1931) 20.
149. L. Zechmeister and G. Tóth, *Ber.*, **64** (1931) 2028.
150. L. Zechmeister and G. Tóth, *Ber.*, **65** (1932) 161.
151. F. Zilliken, G.A. Braun, C.S. Rose and P. Gyorgy, *J. Amer. Chem. Soc.*, **77** (1955) 1296.
152. H.J. Dunn and M.P. Farr, *US Patent 4,034, 121* (1977).
153. P.R. Austin and C.J. Brine, *US Patent 4,286,087* (1981).
154. J.R. Deschamps and J.E. Castle, in ref. 35, p. 63.
155. H. Struszczyk, *J. Appl. Pol. Sci.*, **33** (1987) 177.
156. C.K. Kandaswamy, in ref. 12, p. 517.
157. A. Domard and A. Gadelle, in ref. 34, p. 295.
158. C. Bosso, J. Defaye, A. Domard, A. Gadelle and C. Pedersen, *Carbohydr. Res.*, **156** (1986) 57.
159. J. Defaye, A. Gadelle and C. Pedersen, in ref. 60, p. 415.
160. J. Defaye, A. Gadelle and C. Pedersen, *Carbohydr. Res.*, **110** (1982) 217.
161. G.W. Rigby, *US Patent 2,040,879* (1934).
162. A.C.M. Wu and W.A. Bough, in ref. 12, p. 88.
163. C.V. Lusena and R.C. Rose, *J. Fish. Res. Board Can.*, **10** (1953) 521.
164. I. Batista and G.A.F. Roberts, *Makromol. Chem.*, **191** (1990) 429.
165. A. Domard and M. Rinaudo, *Int. J. Biol. Macromol.*, **5** (1983) 49.
166. I. Batista, M.Sc. Thesis, New University of Lisbon, 1986.
167. L.A. Nud'ga, E.A. Plisko and S.N. Danilov, *Zhur. Obsh. Khim.*, **41** (1971) 2555.
168. B. Lindberg, *Svensk Papperstidn.*, **59** (1956) 531.
169. R.J. Ferrier, W.G. Overend and A.E. Ryan, *J. Chem. Soc.*, (1965) 3484.
170. H.S. Isbell, *J. Res. Natl. Bur. Stand.*, **32** (1944) 45.
171. W. Ambrecht, *Biochem. Zeit.*, **95** (1919) 108.
172. P. Karrer and S.M. White, *Helv. Chim. Acta*, **13** (1930) 1105.
173. A.B. Foster, E.F. Martlew and M. Stacey, *Chem. Ind.*, (1953) 825.
174. J.M. Williams, *Adv. Carbohydr. Chem. Biochem.*, **31** (1975) 9.
175. Y. Matsushima and N. Fujii, *Bull. Chem. Soc. Japan*, **30** (1957) 48.
176. H. Zollinger, *Diazo and Azo Chemistry*, Interscience, New York, 1961, p. 27.
177. S. Hirano, Y. Kondo and K. Fujii, in ref. 34, p. 299.
178. A.T. Austin, *Nature*, **188** (1960) 1086.

179. Q.P. Peniston and E.L. Johnson, *US Patent 3,922,260* (1975).
180. G.G. Allan and M. Peyron, in ref. 60, p. 443.
181. M.R. Crampton, J.T. Thompson and D.L.H. Williams, *J. Chem. Soc., Perkin Trans.*, **2** (1979) 18.
182. D.L.H. Williams, *Adv. Phys. Org. Chem.*, **19** (1983) 381.
183. W. Beneke, *Botan. Zeit. Abt. I*, **63** (1905) 227.
184. P. Karrer and A. Hofmann, *Helv. Chim. Acta*, **12** (1929) 616.
185. P. Karrer and G. von Francois, *Helv. Chim. Acta*, **12** (1929) 986.
186. L. Zechmeister and G. Tóth, *Naturwiss.*, **27** (1939) 367.
187. L. Zechmeister and G. Tóth, *Enzymologia*, **7** (1939) 170.
188. L. Zechmeister and G. Tóth, *Enzymologia*, **7** (1939) 165.
189. L. Zechmeister, G. Tóth and M. Balint, *Enzymologia*, **5** (1938) 302.
190. G.F. Pegg, in *Methods in Enzymology*, W.A. Wood and S.T. Kellogg (eds), Academic Press, New York, 1988, vol. 161, p. 435.
191. D.F. Waterhouse, R.H. Hackman and J.W. McKellar, *J. Insect Physiol.*, **6** (1961) 96.
192. C. Jeuniaux, *Nature*, **192** (1961) 135.
193. C. Jeuniaux, *Arch. Int. Physiol. Biochem.*, **69** (1961) 384.
194. R. Fänge, G. Lunblad and J. Lind, *Marine Biol.*, **36** (1976) 277.
195. C. Cornelius, G. Dandifosse and C. Jeuniaux, *Int. J. Biochem.*, **7** (1976) 445.
196. C. Cornelius and G. Dandifosse, *Biochem. System. Ecol.*, **5** (1977) 53.
197. L.R. Berger and R.S. Weiser, *Biochim. Biophys. Acta*, **26** (1957) 517.
198. D. Charlemagne and P. Jolles, *FEB Letters*, **23** (1972) 275.
199. E. Cabib, in ref. 190, p. 460.
200. A. Ohtakara, in ref. 190, p. 462.
201. A. Arroyo-Begovich, in ref. 190, p. 471.
202. T. Boller, A. Gehri, F. Marech and U. Vögelli, in ref. 190, p. 479.
203. G.F. Pegg, in ref. 190, p. 484.
204. J.P. Zikakis and J.E. Castle, in ref. 190, p. 490.
205. E. Cabib, in ref. 190, p. 498.
206. D.R. Kreger, *Biochim. Biophys. Acta*, **13** (1954) 1.
207. S. Bartnicki-Garcia and W.J. Nickerson, *Biochim. Biophys. Acta*, **58** (1962) 102.
208. R.L. Monaghan, D.E. Eveleigh, E.T. Reese and R.P. Tewari, *Nature New Biology*, **245** (1972) 78.
209. I.F. Ramirez-Leon and J. Ruiz-Herrera, *J. Gen. Microbiol.*, **72** (1972) 281.
210. D. Fenton, B. Davis, C. Rotgers and D.E. Eveleigh, in ref. 12, p. 525.
211. M. Yabuki, in ref. 60, p. 197.
212. Y. Uchida and A. Ohtakara, in ref. 190, p. 501.
213. A. Ohtakara, in ref. 190, p. 505.
214. Y. Araki and E. Ito, *Biochem. Biophys. Res. Commun.*, **56** (1974) 669.
215. Y. Araki and E. Ito, *Eur. J. Biochem.*, **55** (1975) 71.
216. L.L. Davis and S. Bartnicki-Garcia, *Biochem.*, **23** (1984) 1065.
217. H. Kauss, W. Jeblick and D.H. Young, *Plant Sci. Letters*, **28** (1982/83) 231.
218. G. Sundara Rajulu, M. Aruchami and N. Gowri, in ref. 35, p. 140.
219. Y. Araki and E. Ito, in ref. 190, p. 510.
220. H. Kauss and B. Bauch, in ref. 190, p. 518.
221. B. Davis and D.E. Eveleigh, in ref. 52, p. 161.
222. A. Ohtakara and M. Mitsutomi, in ref. 35, p. 117.
223. A. Ohtakara and M. Mitsutomi, in ref. 190, p. 453.
224. M.L. Bade and K. Hickey, in ref. 60, p. 179.
225. H. Diekmann, A. Tschech and H. Plattner, in ref. 60, p. 207.

226. Y. Uchida, M. Izume and A. Ohtakara, in ref. 60, p. 373.
227. S. Hirano and Y. Yagi, *Carbohydr. Res.*, **83** (1980) 103.
228. S. Hirano and Y. Yagi, *Agric. Biol. Chem.*, **44** (1980) 963.
229. S. Hirano and Y. Yagi, *Carbohydr. Res.*, **92** (1981) 319.
230. S. Hara, Y. Yamamura, Y. Fujii and T. Ikenaka, in ref. 35, p. 125.
231. S. Tokura, N. Nishi, S. Nishimura and Y. Ikeuchi, in ref. 52, p. 303.

6

Solubility and Solution Behaviour of Chitin and Chitosan

6.1 SOLUBILITY

6.1.1 Solubility of chitin

Chitin is a relatively intractable polymer and despite its structural similarity to cellulose it is insoluble in typical cellulose solvents such as cuprammonium hydroxide (Schweizer's reagent), cupriethylene diamine and Cadoxen. However there are a limited number of solvent systems which may be grouped together in three classes.

Aqueous solutions of neutral salts

Von Weimarn reported[1] that both α- and β-chitin can be dissolved by treatment with hot, concentrated solutions of neutral salts capable of a high degree of hydration, the order of effectiveness being LiCNS > Ca(CNS)$_2$ > CaI$_2$ > CaBr$_2$ > CaCl$_2$. The addition of ethanol was found to cause the formation of a translucent gelatinous mass. The rate of solubilisation of β-chitin was greater than that of α-chitin.

Clark and Smith,[2] using a LiCNS solution saturated at 60°C as solvent, found that chitin was readily dissolved at 95°C, the non-degradative nature of the solvent being demonstrated by the similarity in the X-ray diffraction patterns of starting material and precipitated chitin that had been held in solution for several months. Acetone was found to be a better precipitant than ethanol.[2] Hackman and Goldberg were able to carry out light scattering studies on a solution of α-chitin in an aqueous LiCNS solution containing sufficient LiCNS to give a saturated solution at room temperature. A temperature of 95°C was used to bring about dissolution of the chitin, after which the solution was cooled to room temperature and diluted with water to give a 5.5M LiCNS solution.[3] Despite these reports Lee[4] was unable to dissolve chitin in LiCNS solutions.

Acid solvents

Chitin is soluble in concentrated HCl, H_2SO_4, and H_3PO_4, though not concentrated HNO_3, but degradation accompanies solution in these mineral acids although Capozza states[5] that the extent of chain hydrolysis in H_3PO_4 is considerably less than in either HCl or H_2SO_4. Furthermore the use of H_2SO_4 as a solvent for chitin causes O-sulphation in addition to chain hydrolysis[6-8] (section 4.1.3). Clark and Smith[2] had previously demonstrated, by X-ray diffraction studies of chitin reprecipitated from solution in HCl at room temperature, that there is a decrease in chain length with increase in the length of time the sample is retained in solution. They also stated that the minimum concentration for HCl is 8.5M if it is to be an effective solvent for chitin, while with H_2SO_4 the minimum normality required is considerably higher.

Chitin is soluble in a number of organic carboxylic acids: formic acid, dichloroacetic acid (DCA) and trichloroacetic acid (TCA). According to Lee[4] β-chitin is soluble in HCOOH but α-chitin is not. This has been discussed in some detail by Austin[9] who has pointed out the critical nature of the water content of the formic acid, particularly in the case of α-chitin. Thus 88–90 wt-% HCOOH is a solvent for β-chitin but not for α-chitin, the latter being soluble in HCOOH of 98 wt-% concentration and above. However dissolution is slow, requiring up to three weeks,[9] and Gagnaire et al. have shown that esterification occurs under these conditions so that the solute is O-formylchitin rather than chitin itself.[10] Austin has also investigated the use of cosolvents such as 2-chloroethanol or dichloromethane in conjunction with formic acid, the cosolvent being added to the solution of chitin in HCOOH to reduce the solution viscosity.[9, 11] Noguchi et al.[12] have reported the dissolution of α-chitin in HCOOH after several repetitions of a freeze–melt cycle.

Formic acid has been shown to be a degradative solvent,[4] the LVN decreasing approximately 39% on standing for 14 days at 2°C, and by approximately 69% on standing for 3 days at 25°C.

Austin was the first to report the use of DCA and TCA as solvents for chitin.[13] Of the two acids DCA is the more convenient to use, being a liquid at room temperature, but it is a less effective solvent and gives viscous solutions at relatively low concentrations of chitin. Although TCA is a better solvent for chitin it is a solid at room temperature so its use requires the presence of a cosolvent and solutions containing 20–50 wt-% TCA in formic acid have been claimed[13] to be particularly useful, giving chitin solutions having relatively low viscosities. Hydrolysis in both DCA and TCA–HCOOH is reported to be slow, chitin solutions in these solvent systems retaining substantial viscosity after standing for 1 month at room temperature.[12] Although DCA can be used on its own as a solvent its use in admixture with HCOOH has been reported,[12, 13] although in neither case was the mixed solvent system used for dissolution. Instead DCA was added to a solution of chitin in HCOOH in the first example[12] and HCOOH to a solution of chitin in DCA in the second.[13] The mixed solvent system TCA–dichloromethane has been used as the reaction medium for homogeneous *O*-acylation of chitin.[14]

The use of methanesulphonic acid as a solvent and reaction medium for the homogeneous *O*-acylation of chitin has been studied extensively[12, 15–19] (section 4.2.2). Solution is aided by the use of low temperatures, such as standing overnight at −20°C to 0°C, and the extent of chain hydrolysis has been found to be limited under these conditions.[17] However in most of this work there is simultaneous solubilisation and derivatisation of the chitin so that the eventual solute is not chitin itself, although methanesulphonic acid is a solvent for underivatised chitin.[12]

Organic solvents

The earliest report of the use of organic solvents for chitin is by Allan *et al.*[20] who investigated the uses of DMF–N_2O_4 mixtures as solvents for a number of polysaccharides. It was found that chitin exhibited limited solubility using a DMF:N_2O_4:chitin ratio of 50:3:1, the maximum solution concentration obtainable being approximately 1 g dm^{-3}. The absence of any chemical modification to the chitin during the solution process was shown by a comparison of the IR spectra of initial and recovered materials. Following this, Capozza[5] disclosed the use of hexafluoro-2-propanol and hexafluoroacetone sesquihydrate as solvents for chitin, and the use of films cast from these solvents as biodegradable carriers for drugs. The CD spectrum of chitin film cast from solution in hexafluoro-2-propanol has been reported.[21] However both solvents have been criticised on the grounds that they are irritants and toxic.[22]

The use of DMAc–LiCl and *N*-methylpyrrolidone–LiCl mixtures as solvents for chitin was disclosed by Austin,[23] the LiCl concentration being 50 g dm^{-3} in all cases described. A more detailed report of these solvent systems has been given by Rutherford and Austin[24] and in this it was stated that DMAc–LiCl and NMP–LiCl were the best solvents for chitin of the more than 200 potential solvent systems examined. Furthermore both are non-degrading, a solution of chitin in DMAc–LiCl retaining its viscosity on standing at room temperature for 48 days. The percentage soluble material ranged from 30% to 92% depending on the source of the chitin, all of the samples being α-chitin. The solubility was not related to molecular weight

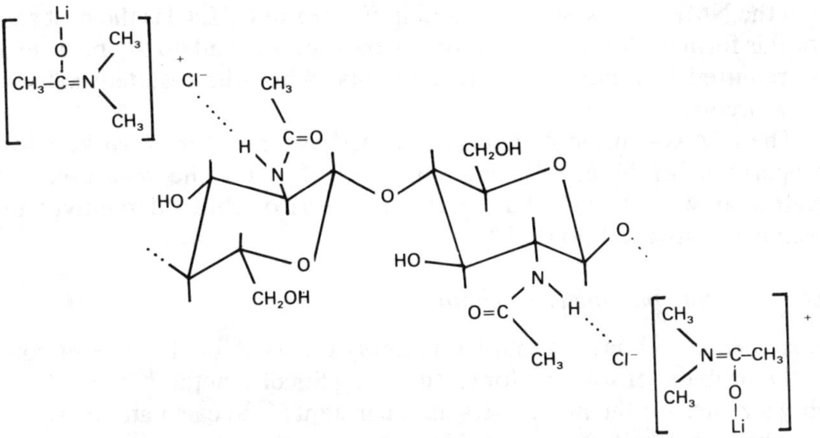

FIGURE 6.1 Proposed mechanism of solution of chitin in DMAc–LiCl, based on the mechanism of Paner and Beste[27] for the solution of poly(1,4-benzamide)

but did appear to depend on the degree of *N*-acetylation, the three most soluble chitins being the three most highly *N*-acetylated samples, while a chitosan sample prepared from one of these was completely insoluble in DMAc–LiCl. This demonstrates the importance of the C(2)NHCOCH₃ group in the solution process, while the importance of the LiCl is shown by the fact that the pure liquids alone only swell chitin and do not dissolve it.

Earlier studies using low-molecular-weight amides showed that they form complexes with LiCl in which the Li$^+$ ion is coordinated with the carbonyl oxygens of four amide groups while the Cl$^-$ ion is hydrogen-bonded to an –NH– group.[25, 26] However in the DMAc–LiCl–chitin system both the solute and the solvent contain amide groups which are potential sites for complex formation with the Li$^+$ ions. Paner and Beste concluded[27] from an NMR and IR spectroscopic study of the DMAc–LiCl–poly(1,4-benzamide) system that the Li$^+$ ion is associated with the carbonyl oxygen of DMAc, forming a weak complex that solvates the 'polyelectrolyte' formed by association between the Cl$^-$ ions and the –NH– groups of the polyamide chain (Figure 6.1). A similar mechanism was assumed to operate in the DMAc–LiCl–chitin system.[24] However Aharoni and Wasserman have concluded that the Li$^+$ ions bond strongly with the carbonyl oxygens of polyamides[28] while Vincendon,[29] in a NMR study of DMAc–LiCl–methyl di-*N*-acetylchitobioside, found that solution takes place through the interaction of LiCl molecules with all the available groups containing labile protons, OH as well as –NHCOCH₃. Thus the precise structure of the complex giving rise to dissolution of chitin in these tertiary amide–LiCl systems is not known.

Austin *et al*. have extended their studies to β-chitin and have reported[11] that the NMP–LiCl system containing 50–70 g dm^{-3} LiCl is the best solvent for this form of chitin. Dissolution is slow however and up to one week may be required to attain complete solubility, while the resultant solution is very viscous.

The DMAc–LiCl system has been used by a number of workers for the preparation of fibres[30-33] and films,[24, 30, 34-36] for the measurement of molecular weight,[37, 38] and for the synthesis of chitin derivatives under homogeneous conditions.[38-41]

The solubility parameter of chitin

Austin *et al.* [9, 11] have attempted to apply the solubility parameter concept to the problem of solvents for chitin, using Small's method,[42] together with Hoy's values for the molar attraction constants,[43] to calculate the solubility parameter value (δ) for chitin. However the solubility parameter approach is of doubtful validity if specific forces of interaction are involved, as distinct from dispersion, dipole–dipole and induction energies, or if the polymer is crystalline.[44] Furthermore the solubility parameter/cohesive energy density concept has no practical utility if ionic interactions are involved in the solution process.[44] Hence the mechanism for the dissolution of chitin in DMAc–LiCl proposed by Rutherford and Austin,[24] and the suggestion that protonation of free amine groups is involved in the dissolution of chitin in acid solvents such as formic acid and dichloroacetic acid,[11] would appear to invalidate such an approach with these solvents. Furthermore the solubility parameter values calculated for chitin,[9] $\delta = 12.1$–12.5, or for the chitin–LiCl complex,[9] $\delta = 11.2$, seem high in view of its solubility in hexafluoro-2-propanol and hexafluoroacetone sesquihydrate.[5, 21] Although there appear to be no δ values listed in the literature for these two solvents, a comparison of the δ values of other fluorinated compounds with those of their non-fluorinated analogues shows that fluorination tends to reduce the δ value.[45] Hence these two fluorinated solvents would be expected to have lower δ values than 2-propanol ($\delta = 11.5$) and acetone ($\delta = 10.0$) respectively, suggesting a lower δ value for chitin than that obtained by calculation or by use of acidic or complex solvents.

6.1.2 Solubility of chitosan

Chitosan, being a base, forms salts with acids (section 5.1) so producing polyelectrolytes whose solubilities will depend on the nature of the anion involved. Salt formation may take place as a separate step, followed by addition of the chitosan salt to water,[46] but the more common practice is to add acid to chitosan suspended in water so that salt formation and dissolution occur concurrently. In those cases where the chitosan salt has limited

TABLE 6.1   *Conflicting reported solubilities of chitosan salts*

Chitosan salt	Reported solubility[a]			
	Ref. 50	Ref. 53	Ref. 54	Ref. 56
Dodecanate	I	SS	SS	
Oxalate	SS	SS	SS	S
Decandioate	I	S	S	
Tartrate	I	S	S	S
Citrate	I	S	S	S
Benzoate	S	S	S	SS
Salicylate	S	SS	SS	SS
Cinnamate	I	SS	SS	

[a] S = soluble; SS = slightly soluble; I = insoluble.

aqueous solubility other factors such as the molecular weight and degree of
N-acetylation of the chitosan, the total amount of acid present and the
temperature, become important and it is not surprising that there are some
conflicting reports in the literature in respect of the solubilities of some
chitosan salts.

Solubility in mineral acids

Chitosan is soluble in dilute HCl, HBr, HI, HNO_3, and $HClO_4$ but may be
precipitated out of solution in HCl or HBr by increasing the concentration
of acid in the system[47-49] and it is likely that similar behaviour would be
observed with the other acids listed. Surprisingly Gross *et al.* report that a
sample of chitosan[0.29] neutralised with HCl was insoluble.[50] Chitosan is
also slightly soluble in dilute H_3PO_4[51] but is insoluble in dilute H_2SO_4 at
room temperature, although chitosan sulphate dissolves in water on heating and reforms on cooling.[52] Chitosan is soluble in concentrated H_2SO_4
but dissolution is accompanied by sulphation and extensive hydrolysis of
the polymer chain.[8]

Solubility in organic acids

Chitosan forms water-soluble salts with a large number of organic acids.
The earliest list of chitosan salts and their aqueous solubilities was com-
piled by Rigby[53, 54] and covered 68 salts. A similar list, containing the
results for several additional salts, was given by Muzzarelli.[55] Other less
extensive lists have been given by Hayes *et al.*[56] and by Gross *et al.*,[50] and
while the results are generally in agreement this is not always the case
(Table 6.1).
    The salts of monocarboxylic acids up to hexanoic are readily soluble in
water, as are the salts formed with their halogeno- and hydroxy-

derivatives, but chitosan heptanoate has been reported to be insoluble.[50] In the dicarboxylic acid series the salts formed from malonic acid up to nonanedioic acid (azelaic acid) give water-soluble chitosan salts, as does malic acid.

The chitosan salts formed with aromatic carboxylic acids have also been investigated. Chitosan benzoate has been found to be readily soluble[50, 53-55] although Hayes et al. found that chitosan was only slightly soluble in an aqueous solution of benzoic acid.[56] These authors suggested that this was due more to the low aqueous solubility of the acid, 0.041M being the maximum benzoic acid concentration obtained, rather than to an inherent low solubility of chitosan benzoate. Chitosan o-aminobenzoate[53-55] (chitosan anthranilate) and chitosan p-aminobenzoate[50, 53-55] are readily soluble, as are chitosan phenylacetate, 4-methoxyphenylacetate, 3,4-dimethoxyphenylacetate, and phenylglycolate, but chitosan 3-phenylpropionate (chitosan hydrocinnamate) has poor solubility and the p-methoxycinnamate is insoluble.[50]

Benzenesulphonic acid, p-aminobenzenesulphonic acid (sulphanilic acid) and sulphosalicylic acid yield water-soluble chitosan salts[50, 53-55] as do methanesulphonic acid and butanesulphonic acid,[50] while the methanedisulphonic acid salt is insoluble.[50]

Austin and Sennett[46] have described a process for preparing water-soluble chitosan salts as dry, free-flowing solids by treatment of chitosan dispersed in an organic medium with a suitable carboxylic acid, the preferred acids being formic, pyruvic and lactic acids. Suitable organic liquids include THF, ethyl acetate and 1,2-dichloroethane. Prior to this work Wolfrom et al.[57] prepared the perchlorate and nitrate salts of chitosan[0.15] by addition of 60 wt-% $HClO_4$ and 70 wt-% HNO_3, respectively, to suspensions of chitosan in glacial acetic acid. Austin and Sennett found that with formic and acetic acid the salts obtained contained acid in excess of a 1:1 stoichiometry and that these acid solvates or complexes gradually lost acid on standing, although the formate salt remained water-soluble for up to a year. The stoichiometric relationships given throughout the paper are open to question, being based on an average value of 169 for the repeat unit of chitosan[0.2], rather than a value of 212 for the equivalent weight of the amine group. Since it is this group which is involved in salt formation it would be more appropriate to use this latter figure unless it is assumed that the amide groups are also protonated.

Solubility in organic solvents

Chitosan dissolves readily in DMF–N_2O_4 mixtures,[20] using a N_2O_4: chitosan ratio of 3:1, to give relatively low viscosity solutions. No chemical modification of the chitosan occurs but the medium appears to be acidic enough to allow esterification with acyl anhydrides without the need to add a catalyst. This is the only organic solvent for chitosan reported to date although 1 vol.-% acetic acid in glycerol functions as a near solvent in the absence of water, causing extensive swelling. No study has been made of the solubility of chitosan in the chitin solvents hexafluoro-2-propanol and hexafluoroacetone sesquihydrate.

Aqueous solubility

Homogeneous deacetylation of chitin[58, 59] to approximately chitosan[0.5], or homogeneous *N*-acetylation to the same level,[60, 61] gives a water-soluble polymer (section 2.5.7). Although the discoverers of this material refer to it as water-soluble chitin it should more properly be classified as chitosan, being soluble in aqueous acetic acid.

6.1.3 Solvent compatibility

Despite the very limited organosolubility of chitosan, solutions of chitosan in aqueous acetic acid can tolerate the addition of large volumes of polar solvents without causing precipitation of the polymer. This tolerance is made use of in the method of gel formation developed by Hirano (section 6.3.1). Up to 70 vol.-% of alcohols from methanol to butanol, ethylene glycol, diethylene glycol, triethylene glycol, acetone, and formamide can be tolerated,[62] and also up to 40 vol.-% 2-propanol and 80 vol.-% glycerol.[51] In these studies the organic solvent was added to a solution of chitosan in dilute acetic acid rather than dissolving the chitosan in the final aqueous acid–solvent mixture. The presence of organic solvents appears to have very little effect on the solution viscosity, except in the case of polyols which cause a considerable increase.[51]

6.1.4 Electrolyte tolerance

Precipitation by low-molecular-weight electrolytes

As previously stated in section 6.1.2, chitosan may be precipitated from solution in the form of the hydrohalide salt by addition of concentrated HCl or HBr.[47–49] That the precipitation is due to the large increase in ionic strength, rather than the accompanying decrease in pH, is shown by the precipitation of chitosan hydrochloride, hydrobromide, and hydroiodide by addition of a saturated solution of the appropriate sodium halide to a solution of chitosan in aqueous acetic acid.[48] Chitosan may also be precipitated from solution in dilute acid by addition of H_2SO_4 or its salts but, unlike the hydrohalide salts which are water-soluble, chitosan sulphate is insoluble in water at room temperature although it dissolves on heating.[52] It may also be precipitated as the phosphate salt by addition of either polyphosphoric acid or its salts, and is readily precipitated by complex ions

TABLE 6.2   *Compatible electrolytes for chitosan solutions*[51]

	Sodium salts[a]	Nitrate salts[a]
Acetate	Formate	Aluminium
Bromide	Nitrate	Calcium
Chloride	Nitrite	Chromium
Citrate	Phosphate (dibasic)	Cupric
		Ferric

[a] Compatible at a concentration of 10 wt-%.

such as molybdate, tungstate, phosphomolybdate, phosphotungstate, ferrocyanide and ferricyanide.[63]

In the above studies, isolation of a chitosan salt was frequently a major aim and concentrated solutions of the precipitating agents may have been required in some cases. Filar and Wirick have studied the tolerance of chitosan solutions to a range of electrolytes at a standard concentration.[51] Under these conditions Na_2SO_4 was the only electrolyte of those examined to cause precipitation and a number of salts were found to be compatible at the selected concentration level of 10 wt-% (Table 6.2). In considering these results it is surprising that no mention was made of any reaction occurring on addition of an acidic solution of chitosan to a solution of $NaNO_2$.

Organic anions which precipitate chitosan include the salicylate,[64] picrate and tannate,[55] naphthalene sulphonic acids[65] and anionic dyes.[66] On the available evidence it is reasonable to assume that any large organic anion will act as a precipitant, particularly if it is present in excess of the stoichiometric requirements.

The precipitating action of selected eletrolytes has been used in a number of areas. Precipitation by sodium triphosphate has been used in ionotropic gel formation for cell entrapment[67-69] and significantly better performances were obtained with cells from *Amaranthu tricolor* and *Asclepias syriaca* L. entrapped in chitosan polyphosphate gels compared with the same cells entrapped in calcium alginate gels.[69]

Doczi[64] has described a process for purifying and fractionating chitosan which involves adding sodium salicylate to a solution of chitosan, warming to obtain a clear solution, then cooling to give a precipitate of chitosan salicylate, the temperature required for precipitation decreasing with decreasing molecular weight. Subsequent dissolution of the chitosan salicylate at room temperature or above, followed by adjusting the pH to ~9, yielded a precipitate of chitosan. The temperature dependence of the solubility of the salts formed by chitosan with 1-naphthylamine-4 sulphonic acid and 1-naphthol-4-sulphonic acid has been used for producing thermally reversible gels[65] (section 6.3.3).

Formation of polyelectrolyte complexes

Oppositely charged polyelectrolytes are known to interact rapidly in solution to form a polyelectrolyte complex which normally precipitates out from the solution.[70, 71] Katchalsky, in a study of the reaction between poly(methacrylic acid) and poly(lysine hydrochloride), found that precipitation occurred when the total concentration of positive groups equals the total concentration of negative groups,[72] and this stoichiometry is the basis of the colloid titration method previously developed by Terayama.[73] The resultant polyelectrolyte complexes may be neutral or have an excess of cationic or anionic groups, this excess being balanced by the presence of anionic or cationic counter ions respectively. Two possible general structures have been suggested for these complexes:[71] a ladder structure and a scrambled salt structure which are shown in Figure 6.2. Polyelectrolyte complexes are frequently soluble in multicomponent solutions consisting of water-miscible organic solvents together with an electrolyte such as NaBr.[71] Typical suitable organic solvents are acetone and dioxane.

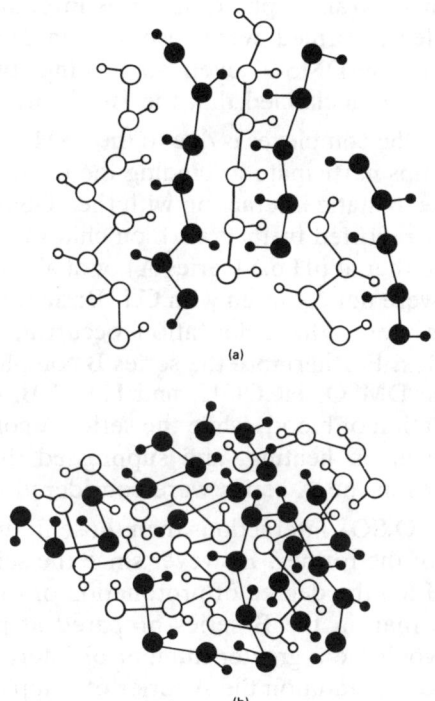

(a)

(b)

FIGURE 6.2   Proposed (a) ladder and (b) scrambled salt structures for polyelectrolyte complexes[71]

In many of the studies carried out on polyelectrolyte complexes of chitosan, or of hydroxyethyl chitosan which has been frequently used so that the studies can be extended to the alkaline pH region, the complexes are described in terms of the mixing ratio R which is defined as $R = P_{Cm}/(P_{Cm} + P_{Am})$, where P_{Cm} and P_{Am} are the molar concentrations of monosaccharide residues of the cationic polyelectrolyte and the anionic polyelectrolyte, respectively, in the mixture giving the maximum precipitate.

Thiele and Langmaach[74] appear to have been the first to attempt to prepare a chitosan-containing polyelectrolyte complex, using alginic acid as the polyanion and hydroxyethyl chitosan as the polycation. Kikuchi examined the formation of complexes by interaction of chitosan with heparin.[75] The complexes swelled extensively in water and were soluble on heating in H_2O–HCl–CH_3OH, H_2O–KBr–CH_3COCH_3 or HCOOH. They stained on treatment with C.I. Basic Blue 17 (Toluidine Blue), the staining occurring throughout the complex indicating an even distribution of heparin within the complex. The complex having a N:S mole ratio of 2.36:1, which had been formed from a mixture having a N:S mole ratio of 1.77:1, enhanced the coagulation of blood. Kikuchi and Fukuda[76] prepared complexes from chitosan–dextran sulphate mixtures in dilute aqueous acetic acid. All the complexes formed were insoluble in H_2O–HCl–CH_3OH, H_2O–KBr–CH_3COCH_3 or DMSO, even on heating, but were partially soluble in HCOOH. It was claimed that the IR absorption band at 1480 cm^{-1} in the spectra of the complexes is due to the $-\overset{+}{N}H_3$ group and that this shows that these groups participate in binding the dextran sulphate, most probably through electrostatic interaction with the $-OSO_3^-$ groups. These authors subsequently reported further work on chitosan–dextran sulphate complexes[77] formed either at pH 6.5 (series A) or at about pH 1 (series B). Series B complexes were not coloured with C.I. Basic Blue 17 whereas the series A complexes were, the colouration occurring fairly uniformly thoughout the complex. Furthermore the series B complexes were soluble or partially soluble in DMSO, HCOOH, and H_2O–KBr–CH_3COCH_3, and enhanced the coagulation of blood, while the series A complexes had very limited solubility, even on heating, and suppressed the coagulation of blood. The complexes in the A series were considered to have a greater number of $-\overset{+}{N}H_3$ . . . $\overset{-}{O}_3SO-$ interactions than those in the B series, hence the lower solubility of the former. However since the series A complexes were prepared at pH 6.5 the degree of protonation of the chitosan chains would be much less than in the B series prepared at pH 1, hence it is unlikely that there would be a greater number of interchain salt linkages formed in the A series. In addition the A series of complexes were stained with C.I. Basic Blue 17, indicating the presence of $-OSO_3^-$ ionic groups not involved in complex formation with $-\overset{+}{N}H_3$ groups, while the B series were not stained. This suggests that all the $-OSO_3^-$ groups are involved in salt links in the complex.

Shinoda and Nakajima[78-80] have made a detailed study of the complexes formed between hydroxyethyl chitosan[0.38] (HECh) and cellulose sulphate (CS), heparin (HEP), hyaluronic acid (HA), chondroitin sulphate A (CSA) and chondroitin sulphate C (CSC). Heparin and cellulose sulphate represent polyelectrolytes having a relatively high linear charge density while the other three (HA, CSA, CSC) represent relatively low linear charge density polyelectrolytes. These authors pointed out that the concentrations of both the $-\overset{+}{N}H_3$ and the $-COO^-$ groups in the polyelectrolytes will depend on the pH as well as on the polymer structure, and that for formation of a neutral complex

$$C_{(SO_3H)} + \alpha C_{(COOH)} = \beta C_{(NH_2)}$$

where C represents the molar concentrations of the groups, and α and β are the degrees of dissociation of the two pH-dependent groups. The values of α and β at any pH may be determined from the potentiometric titration curve. Thus R_s, the mixing ratio required to produce a neutral complex at any selected pH, may be calculated.

For the HECh–HEP complexes[78, 80] the plot of the mixing ratio R versus pH of mixing closely follows the theoretical curve for R_s versus pH of mixing, coinciding exactly at pH < 3.5 with $R = 0.65$. This led the authors to propose that the polyelectrolyte complex has the ladder structure with one HEP chain sandwiched between two HECh chains (Figure 6.3). Although R is constant at 0.65 over the pH range 3–5 for the HECh–CS complexes, again suggesting two HECh chains to each CS chain, it is very far below the theoretical R_s value of ~ 0.8, indicating that the complex would be anionic in character rather than neutral. Again an extended ladder-like structure was anticipated since the chains would be expected to be highly extended because of their high linear charge density.[78, 80]

In the complexes of HECh with the anionic polyelectrolytes of low linear charge density the curves for the experimentally determined R and the calculated R_s values, as functions of the pH, cross over at an R value of 0.5. This led the authors to propose that the complex contains an equal number of anhydro sugar residues from each polymer. However the discrepancy between the two curves for the HECh–HA system at pH 3.5, where the HECh is fully protonated while the HA is only partially dissociated, is large. Shinoda and Nakajima proposed[79] that the presence of protonated amine groups induces the dissociation of COOH groups that would otherwise be undissociated at the pH of complex formation

$$
\left|
\begin{array}{ll}
-\overset{+}{N}H_3 & HOOC- \\
& + \\
-\overset{+}{N}H_3 & HOOC-
\end{array}
\right|
\;\rightleftharpoons\;
\left|
\begin{array}{l}
-\overset{+}{N}H_3 \ldots \overset{-}{O}OC- \\
-\overset{+}{N}H_3 \ldots \overset{-}{O}OC-
\end{array}
\right|
\; + \; 2H^+
$$

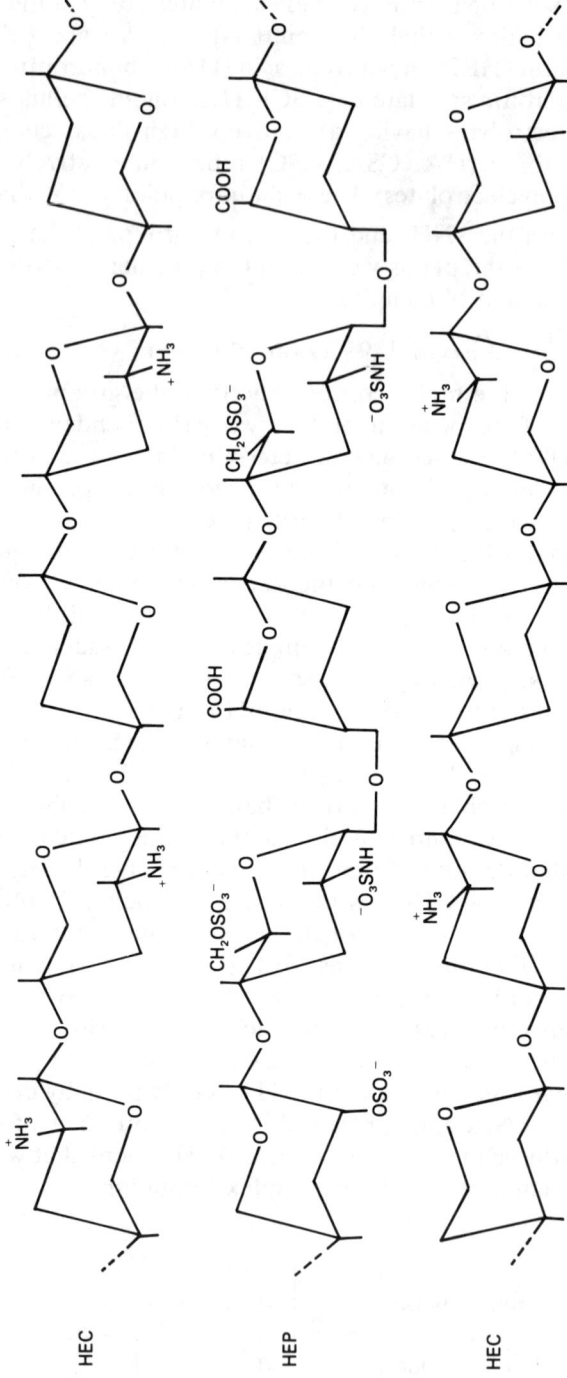

FIGURE 6.3 Schematic model for a HECh–HEP–HECh ladder complex as proposed by Shinoda and Nakajima[78]

Determination of the pH before and after mixing in the absence of buffer shows that at an initial pH of 3.53 there is a decrease in the final pH with increase in the HA:HECh ratio, and that about 70% of the $-NH_3^+$ groups in the HECh have been electrically neutralised, most through this mechanism of induced dissociation. The authors concluded that the two polyelectrolytes form a ladder-like complex enabling a close approach of the chains and development of the induced dissociation.

At alkaline pH values the HECh will only have a low degree of protonation whilst the HA will be almost fully dissociated. Induced dissociation was also proposed in this case

$$
\left|\begin{array}{cc} -NH_2 & \overset{-}{O}OC- \\ & + \\ -NH_2 & \overset{-}{O}OC- \end{array}\right| \quad \underset{}{\overset{2H_2O}{\rightleftharpoons}} \quad \left|\begin{array}{c} -\overset{+}{N}H_3\ldots.\overset{-}{O}OC- \\ \\ -\overset{+}{N}H_3\ldots.\overset{-}{O}OC- \end{array}\right| \quad + \ 2OH^-
$$

and an increase in pH on mixing at pH 7.5 was observed, in agreement with this suggestion, although the magnitude of the effect was much less than that of the decrease observed on mixing at pH 3.53. This was explained as due to the different conformations of the HECh prior to complex formation, the HA solution being added to the HECh solution in all the experiments. At pH 3.53 the HECh, being fully protonated, will have an extended conformation hence the added HA can form a ladder-like complex, so enabling induced dissociation to take place. At pH 7.5 the HECh will adopt a coiled conformation, leading to random interaction with the HA and a scrambled salt structure in which induced dissociation will be very limited because of the non-proximity of the groups. Scanning electron micrographs support this view, the complex formed at pH 3.53 being a fibrous network while that formed at pH 7.5 is a non-fibrous, amorphous material.[79]

Fukuda and Kikuchi[81] investigated complex formation between chitosan and carboxymethyl dextran. They concluded that the pH, mole ratio of N:COOH, and order of mixing are important in determining the composition of the complex, and that the number of $-\overset{+}{N}H_3 \ldots \overset{-}{O}OC-$ salt linkages in complexes prepared at pH 3 is smaller than in complexes prepared at pH 6.5; no complex formed on mixing at pH < 2. The greater salt linkage concentration in the pH 6.5 complexes is readily understood since at this pH the degree of dissociation of both groups will be appreciable while at pH 3, although the amine groups will be almost completely protonated, the extent of dissociation of the –COOH groups will be negligible with $\alpha < 0.05$. The complexes were soluble in H_2O–KBr–CH_3COCH_3 (60:20:20 wt-%) and H_2O–HCl–dioxane (5:45:50 wt-%).

Hirano et al.[82] used chitosan[0.12] and homogeneously N-acylated derivatives of it to form complexes at pH 2.8 and 4.5 with chondroitin sulphate A, chondroitin sulphate C and heparin. In all cases the mixing ratio R agreed with the composition ratio of the complex formed, which was determined from the N content. The R values increased with increase in the extent of N-acylation of the chitosan, because of the decrease in potential $-\overset{+}{N}H_3$ groups, but almost the same R value was obtained at the same extent of N-acylation regardless of the nature of the N-acyl group (N-acetyl-, N-propionyl-, N-hexanoyl-, and N-tetradecanoyl-) although the maximum level of N-acylation at which definite R values could be obtained decreased with increase in molecular weight of the N-acyl group. The distribution of amine groups in the partially N-acylated chitosans was shown to be random by periodate oxidation and Hirano et al.[82] concluded that the ladder-structure proposed by Shinoda and Nakajima[79] is incorrect, the random nature of the $-\overset{+}{N}H_3$ groups making matching up difficult. However Shinoda and Nakajima used heterogeneously deacetylated material and this has been shown[58, 59] to be a block copolymer, so that long segments of the chain would be fully deacetylated thereby allowing the formation of a ladder structure.

Srinivasan et al. have studied the dielectric properties and electrical conductivity of complexes formed between hydroxyethyl chitosan (HECh) and dextran sulphate (DS), carboxymethyl cellulose (CMC), poly(galacturonic acid) (GUA), and alginic acid (AA)[83, 84] and between HECh and chondroitin sulphate A (CSA), chondroitin sulphate C (CSC), and hyaluronic acid (HA).[85] Thin transparent films were formed from the complexes by dehydration under reduced pressure on mercury and these were used for measurement of the dielectric constant (ε') and loss factor (ε'') as functions of temperature and frequency.

The HECh–DS, HECh–CSA and HECh–CSC complexes showed little change in ε' and ε'' with change in temperature or frequency, while the other four complexes all showed qualitatively similar behaviour to each other, ε' and ε'' increasing with increase in temperature, the increase being greater the lower the frequency. Furthermore the rate of increase with rise in temperature showed an abrupt increase at about room temperature with all four complexes. However, quantitatively the increases are much more pronounced for the HECh–AA and HECh–HA complexes than for the HECh–CMC and HECh–GUA complexes.[83, 85] The behaviour of the latter two complexes could be explained on the basis of the Maxwell–Wagner interfacial polarisation mechanism[86] which postulates spheres of high conductivity occluded in a medium of low conductivity, the contribution from interfacial polarisation to the dielectric constant increasing substantially at temperatures above the glass transition temperature. The behaviour of the HECh–AA and HECh–HA complexes was explained by the Schwarz mechanism for dielectric behaviour of biocolloids,[87] previously invoked by Michaels et al.[88] to explain the effects of absorbed NaBr on the dielectric properties of a synthetic polyelectrolyte complex. The Schwarz mechanism postulates an occluded spherical phase containing fixed charges but with the surfaces surrounded by a very thin layer of mobile counterions. However the authors concluded[85] that a physically realistic picture can only be obtained if the microdomains are considered to be cylindrical rather than spherical. Such a microdomain structure would be in agreement with the ladder structure previously proposed[79] for the HEC–HA complex prepared at pH 3.45. Unfortunately Srinivasan et al. do not give details of the pH of mixing during formation of their complexes.

In recent years there have been a number of reports of encapsulation using polyelectrolyte complexes of chitosan with anionic polyelectrolytes to form the outer membrane or skin. Although complex coacervate capsules using other cationic polyelectrolytes had been reported previously, the first report of their preparation with chitosan is that by Rha et al.[68] who used chitosan and sodium alginate as the two components. Two routes for preparation of the capsules are described; in the first an alginate solution is added dropwise to a chitosan solution and the capsules collected and washed with water or a buffer solution. In the second the chitosan solution is added dropwise to an alginate solution and, after washing, the capsules are treated in a solution containing Ca^{2+} or similar divalent ions, so surrounding the interphasic membrane with an ionotropic gel layer. Comparisons were made of the growth rates of two test micro-organisms, Bacillus subtiles and Bacillus lichenformis, as free cells or as entrapped cells in either chitosan–polyphosphate globules or chitosan–alginate capsules. While the growth rates of the cells in the globules were only about 35% of the growth rate of the free cells, the growth rates for the cells in the capsules were 98% and 73%, respectively, of the rates for the two types of cells as free cells.

Rha's group has reported a number of studies on the permeability of the interphasic membranes in capsules formed by dropwise addition of a solution containing the anionic polyelectrolyte and a protein to a coagulating solution of chitosan.[89–91] In the first the diffusion of γ-globulin (molecular weight 1.56×10^5) and bovine serium albumin (BSA, molecular weight 6.9×10^4) through alginate–chitosan[0.2] interphasic membranes was examined.[89] The diffusion rates of the two proteins were found to decrease with decrease in pH of the coagulating chitosan bath over the range 5.9 to 3.5, with addition of $CaCl_2$ or $NaCl$ to the chitosan solution and, in the case of γ-globulin, with increase in the chitosan concentration from 1 g dm^{-3} to 5 g dm^{-3}. No diffusion of the larger protein fibrinogen (molecular weight 4.0×10^5) through the membrane was observed under any conditions. The decreases in the diffusion rates of γ-globulin and BSA on decreasing pH and increasing electrolyte concentration correlate with increases in mechanical strength of the capsules. The effects of variation of preparation conditions on the permeability of the capsules were attributed to changes in pore size and compactness of the membrane structure.

The permeability of carboxymethyl cellulose–chitosan[0.2] capsules has also been studied [90] using BSA as the diffusing protein, γ-globulin having too high a molecular weight to diffuse through the interphasic membrane. The permeability was found to decrease slightly with increase in reaction time for capsule formation but the change was not considered significant, while unlike the alginate–chitosan capsules the permeability increased with increase of electrolyte, either NaCl or phosphate buffer, in the chitosan solution. The permeability also increased with increase in the molecular weight of the chitosan sample used, most probably because of the ability of the lower-molecular-weight material to form a more compact, less porous structure.

A second study of protein diffusion from alginate–chitosan capsules using ovalbumin (molecular weight 4.6×10^4), BSA and γ-globulin, confirmed[91] the decrease in permeability with decreasing pH and increasing chitosan concentration of the coagulating bath found previously.[89] However, contrary to this earlier report the permeability was found to increase with increase in NaCl concentration in the coagulation bath, in agreement with the results for the carboxymethyl cellulose–chitosan capsules.[90]

Kim and Rha[92] have evaluated the effects of encapsulation on the growth of mammalian cells, the hybridoma cells ATCC CRL-1606 and ATCC HB-8852, and their production of monoclonal antibodies, using alginate–chitosan capsules. The maximum cell densities obtained in the encapsulated cultures were two orders of magnitude greater than those obtained in the free cell cultures, while the concentrations of the monoclonal antibodies produced by the encapsulated cells were some 20 times greater than the concentrations produced by the free cell cultures. These results demonstrate that encapsulation is an efficient method of increasing the cell density and concentration of final product in mammalian cell culture.

Knorr et al. have also reported on the production and properties of chitosan-based capsules using alginates or carrageenans as the anionic polyelectrolyte component.[93–96] Again capsules having either the anionic or the cationic polyelectrolyte in the liquid core were prepared and the effects of variation in a number of preparation parameters on the mechanical strength of the capsules examined.[93] Using a chitosan solution as the core liquid and sodium alginate as the counter ion, and hence the outer component of the interphasic membrane, it was found that the bursting strength of the capsules was increased by addition of $CaCl_2$ to the chitosan solution, and D-glucose to the alginate solution. Furthermore there was an optimum alginate concentration if maximum capsule strength was required (Figure 6.4).

Chitosan–heparin complexes have been studied to assess their potential as non-thrombogenic materials[75, 97, 98] and their compatability with blood

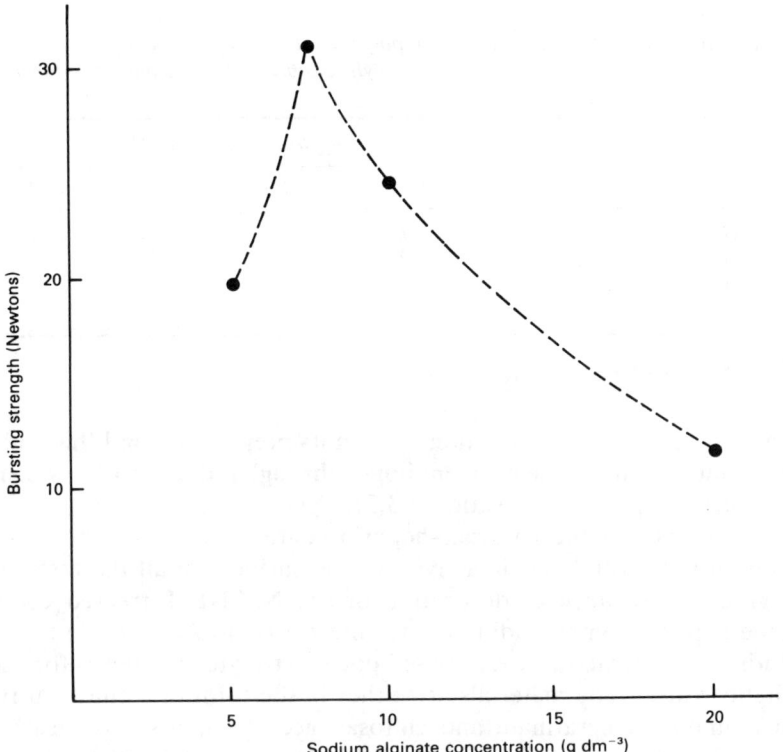

FIGURE 6.4   Effect of sodium alginate concentration (at a constant weight ratio of 3:1 for
D-glucose:sodium alginate) on the mechanical stability of chitosan–alginate
complex coacervate gel capsules[96]

appears to depend to some extent on their method of preparation.[75, 97] The
most detailed study to date is that of Hammer *et al.*[98] who prepared
supported chitosan–heparin complexes on the surfaces of polymeric materials such as poly(ethylene), poly(propylene), silicone rubber, poly(vinyl
chloride), and cellulose. The substrates were first primed either with
chromic acid solutions or by oxygen R_f plasma treatment, then immersed
in a chitosan solution, dried, neutralised with NH_4OH, then complexed by
steeping in a heparin solution at pH 7.0 with a phosphate buffer. Uptakes
on the substrates were approximately 4×10^{-5} g cm^{-2} for the chitosan and
2–3×10^{-5} g cm^{-2} for the heparin (calculated assuming that the heparin
used had an activity of 100 IU mg^{-1}. The extent of ionic binding between
the two electrolytes must be limited since complex formation was carried
out at pH 7.0, at which the degree of dissociation of the amine groups
would only be approximately 0.2 calculated from the data of Shinoda and
Nakajima.[78] Thus the heparin could be completely removed by soaking in
~4M NaCl solution. To improve the stability of the complex NaCNBH₃

TABLE 6.3   *Effect of a chitosan–heparin polyelectrolyte complex coating on the thrombogenic activity of polyethylene cathater implantations as a function of time*[98]

Material	Normalised total VTS[a]			
	0.5 h	2 h	24 h	96 h
Polyethylene	100	100	100	100
Polyethylene–chitosan–heparin	0	0	76	
Polyethylene–chitosan–heparin (treated with NaCNBH₃)	0	0	0	22

[a] VTS = visual thrombus score.

was added to the heparin solution used in its preparation and this had the effect of introducing some covalent bonds through reduction of any Schiff's base structures produced (section 4.3.3).

The presence of the chitosan–heparin coating led to a very large reduction in the thrombogenic activity of the surfaces in all the tests used. The greater resistance to desorption of the NaCNBH₃-treated coatings became apparent on extending the testing period to 24 h (Table 6.3).

Widra has patented chitosan-based polyelectrolyte complexes for use as synthetic skins.[99] The materials are either in the form of a bilayer formed from ammonium keratinate and chitosan acetate or a sandwich of ammonium keratinate between a layer of chitosan acetate and a layer of collagen acetate. These are not true polyelectrolyte complexes, complex formation only occurring at the interfaces between the layers, while the supported chitosan–heparin complexes of Hammer *et al.*[98] would be intermediate between these and true complexes formed in solution.

6.2 SOLUTION PROPERTIES

6.2.1 Chitin solutions

Vincendon[29] observed that in the ¹H NMR spectrum of methyl β-D-(di-N-acetyl)chitobioside in DMAc–LiCl the C(3′)OH group has a relatively low δ value and coupling constant ($\delta = 4.8$ ppm, $J = 0.9$ Hz) compared to that of the C(3)OH group ($\delta = 5.9$ ppm, $J = 2.75$ Hz) or of the C(4)OH group ($\delta = 5.96$ ppm, $J = 2.75$ Hz) (see Figure 1.9 for numbering scheme). The low δ value for the C(3′)OH group was attributed to its inability to interact with the LiCl molecules owing to the presence of an intramolecular C(3′)OH . . . O(5) hydrogen bond, such as exists in the crystalline state[100] and which has been shown to exist in cellulose oligomers in solution.[101] Further evidence supporting this was obtained from measurements of the temperature dependence of the proton chemical shift of the C(3′)OH signal, which was much less than that of the signals for the other OH groups or the amido NH groups. Similar behaviour was observed for the C(3′)OH groups of chitin in DMAc–LiCl solutions – low δ value, a low coupling constant value, and a low sensitivity to temperature change – indicating that the C(3′)OH . . . O(5) hydrogen bonds present in chitin in the solid state persist in the chains in solution. The presence of these hydrogen bonds would be expected to hinder rotation about the glycosidic bond and hence impart considerable rigidity to the chitin chain in this solvent system.

Evidence of such rigidity in chitin chains comes from a detailed study by Terbojevich et al.[37] using viscosity and light scattering measurements. They found high values for the second virial coefficient, although despite this the chitin chains were observed to associate in moderately concentrated solutions, and a linear expansion factor, α_s, of close to unity. These results imply that the chitin chain behaves as a rather stiff molecule in DMAc–LiCl, and this is further supported by values of 150–400 Å for the persistence length, the particular value obtained depending on the method of calculation.

However Saito et al. have found[102] using ^{13}C NMR spectroscopy that, in water, chitin oligomers up to a DP of 10 show behaviour similar to that of a random coil, with rapid conformational changes taking place.

6.2.2 Polyelectrolyte nature of chitosan in acid solution

In acid solution the amine groups of chitosan are protonated (section 5.1) and under these conditions chitosan would be expected to exhibit behaviour typical of a polyelectrolyte. Van Duin and Hermans[103] and subsequently Lee,[4] Kienzle-Sterzer et al.,[104] and Lyubina et al.,[105] showed that in dilute solutions in the absence of added electrolyte there is an abnormal increase in the viscosity number with decrease in chitosan concentration, but that normal straight line viscosity number versus concentration plots are obtained in the presence of added electrolyte. This is typical of polyelectrolyte solutions and is a result of the increase in coil dimensions that occurs on dilution owing to electrostatic repulsion between chain segments. This repulsion may be suppressed by addition of a low-molecular-weight electrolyte which functions by screening the electrical charges on the polymer chain. The viscosity number has also been found to decrease with increasing acetic acid concentration at a given chitosan concentration.[104]

If aqueous HCl is used as solvent, rather than aqueous acetic acid, linear viscosity number versus concentration plots are obtained without addition of added neutral electrolyte[105] provided the HCl concentration ⩾ 0.05M. In terms of ionic strength this solution is approximately equal to a dilute acetic acid solution containing 0.05M NaCl, which also gives a linear

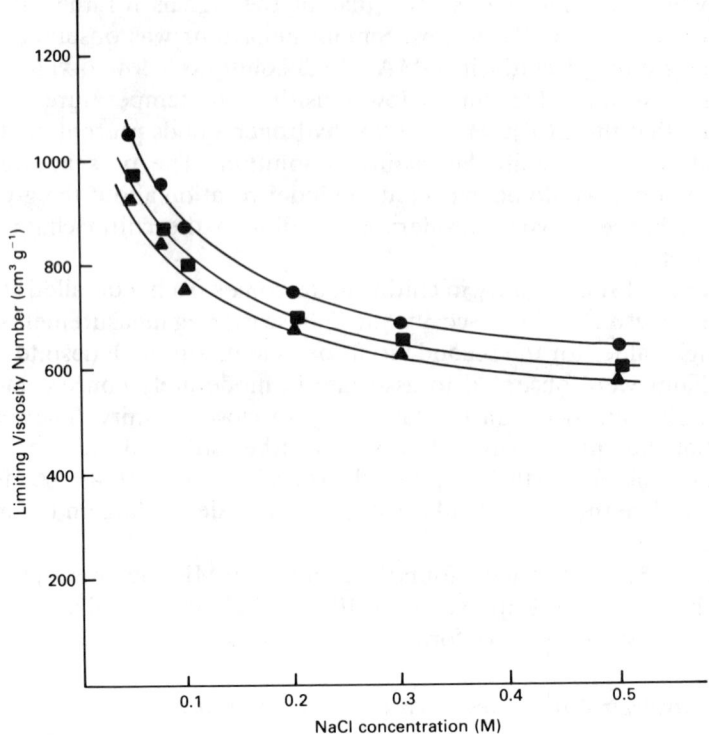

FIGURE 6.5   Limiting Viscosity Number values for chitosan as a function of concentration
of added NaCl at different degrees of ionisation (α)[107]: ●, $\alpha = 0.9$;
■, $\alpha = 0.7$; ▲, $\alpha = 0.6$

plot.[105] Linear plots may be obtained from viscosity data measured in
aqueous acetic acid without added neutral electrolyte by using[104] the
semi-empirical equation of Yuan *et al.*:[106] $\eta_{sp}/C = [\eta]_\infty(1 + Kc^{-0.5})$. The
intercept obtained, $[\eta]_\infty$, may be related to the LVN obtained in the
presence of a neutral electrolyte.

The LVN obtained for a chitosan sample depends on the concentration
of neutral electrolyte present, the value decreasing with increasing concentration of electrolyte. The effect of ionic strength and of the degree of
ionisation of the chitosan chain has been determined by Kienzle-Sterzer *et
al.*[107] and shows that the LVN decreases with both increasing electrolyte
concentration and decreasing degree of ionisation (Figure 6.5). These
results demonstrate the influence of electrostatic repulsion between
charged chain segments, and its suppression by the electrolyte, on the
hydrodynamic volume of chitosan molecules in solution.

Another characteristic of polyelectrolytes is that there is a linear relationship between the LVN values obtained at different electrolyte

concentration, and $I^{-0.5}$, where I is the ionic strength. This behaviour has been observed with chitosan solutions[105, 108, 109] although, while Lyubina *et al.*[105] found linearity down to an electrolyte concentration of 4×10^{-3}M, Rinaudo and Domard[109] found that linearity was only obtained down to a concentration of 8×10^{-3}M. Rodriguez-Sanchez *et al.*[108] found that the plots of [η] *versus* $I^{-0.5}$ for a number of 1:1 electrolytes fell on a single line. They concluded that the conformational changes of chitosan molecules are independent of the nature of the counter ion used, as is the LVN. This does not agree with the results of Domard and Rinaudo[110] who obtained an LVN of 374 cm³ g⁻¹ for a chitosan[0.0] in 0.2M NaCl and of 454 cm³ g⁻¹ in 0.2M NaBr. In this case the nature of the counter ion clearly has a considerable influence on the LVN. However there is a difference in the experimental technique used in the two studies. Domard and Rinaudo dissolved the chitosan by addition of the stoichiometric quantity of the appropriate acid, followed by addition of the necessary electrolyte, while Rodriguez-Sanchez *et al.*[108] used 0.7M aqueous acetic acid as solvent, the required amounts of electrolyte being added to these solutions. Thus in the former technique the specified counter ion is the only anion present while in the latter technique there will be acetate ions present in addition to the specified counter ion. This difference in procedure may account for the conflicting results.

6.2.3 Conformation and molecular dimensions of chitosan chains in acid solution

As chitosan is a β-(1→4)-linked glucan it would be expected to exist in solution as a random coil having a relatively extended conformation because of the low flexibility of the chain. In the analysis of the solution behaviour of stiff polymers use may be made of the equivalent freely jointed chain proposed by Kuhn[111] in which the real chain consisting of n links of length l is replaced by an equivalent chain having a smaller number of larger, freely jointed links or 'statistical chain elements' (Figure 6.6), both chains having the same contour length and chain end separation. The length of the statistical chain element, A_m, increases with increasing chain stiffness but the approach remains valid provided A_m does not exceed 1/10 of the chain length.[112] Thus it is applicable to many stiff polymers of high molecular weight. In the case of extremely stiff chains, particularly those of relatively low molecular weight, the value of A_m may approach that of the contour length, L, itself. In these circumstances a more appropriate model is the 'wormlike chain' model of Kratky and Porod.[113] In this model the angle between two successive rigid links is close to 180°, but there are no barriers to rotation. Stiffness is characterised by the persistence length P_L and for sufficiently long chains, where $L \gg P_L$, the model becomes similar to the randomly coiled equivalent chain of Kuhn.

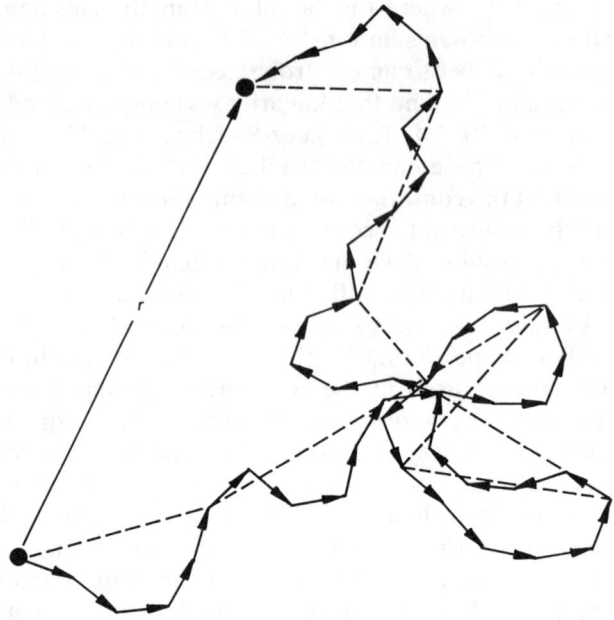

FIGURE 6.6   Two-dimensional representation of how a real polymer chain may be
approximated by an equivalent freely jointed chain. ($\rightarrow$, bond of real chain;
– – – –, statistical element of equivalent freely jointed chain)

Another measure of chain stiffness is the characteristic ratio, C_∞, which is the ratio of the unperturbed mean square end-to-end distance to the calculated mean square end-to-end distance of the random flight chain. It is a measure of how the short range interactions affect the chain flexibility.

The Kuhn statistical segment length has been determined for chitosan[0.18] by Pogodina et al.,[114] a value of 230 ± 30 Å being obtained from measurements in 0.33M CH$_3$COOH–0.3M NaCl, while Kienzle-Sterzer et al.[115] have calculated values of 73 Å for chitosan in 0.167M CH$_3$COOH–0.46M NaCl and of 122 Å in 0.2M CH$_3$COOH–O.1M NaCl–4M urea. It is to be expected that the chain would become more flexible with increasing ionic strength of the solution since electrostatic repulsion between chain segments will be reduced, and this has been shown to be case with sodium carboxymethyl cellulose ($DS = 1.06$) where the value of A_m was found to decrease from 640 Å in 0.005M NaCl to 180 Å in 0.2M NaCl and to 95 Å at infinite ionic strength.[116]

The persistence length has also been found to depend on the ionic strength of the medium and also on the degree of ionisation.[107] The values obtained by Kienzle-Sterzer et al.[107] range from 80 Å ($I = 0.05$M, $\alpha = 0.90$) to 52 Å ($I = 0.5$M, $\alpha = 0.60$) while Rinaudo and Domard[109] report a value of 42 Å for chitosan[0.17] at infinite ionic strength. This compares very well with the value of 44 Å which may be calculated from the experimental data[116] for sodium carboxymethyl cellulose ($DS = 1.06$), also at infinite ionic strength, using the Yamakawa–Fujii equation [117] in the form given by Kienzle-Sterzer et al.[107] The value of the characteristic ratio for chitosan has been calculated to be 14.8 compared with 14.2 for sodium carboxymethyl cellulose.[107]

Another measure of chain stiffness is the empirical coefficient B proposed by Smidsrød and Haug.[118] This is obtained from the slope, S, of the plot of LVN versus $I^{-0.5}$, together with the value for the LVN at $I = 0.1$M. The value of $I = 0.1$M was selected as the reference ionic strength since it had been included in most of the relevant published work. The Smidsrød–Haug parameter is given by $B = S/[\eta]^{1.3}_{I = 0.1}$ and has been shown[118] to be inversely correlated with a number of fundamental parameters of chain stiffness.

Rha et al.[68] report a value of 0.08 for B, which is similar to that calculated[118] for sodium carboxymethyl cellulose from the data of Brown and Henley,[116] or to 0.043–0.091 obtained by Terbojevich et al.[119] for samples of chitosan ranging from chitosan[0.52] to chitosan[0.12]. These latter results indicate that there is a tendency towards increasing flexibility with increase in the extent of deacetylation, and Terbojevich et al. have suggested that this arises from intramolecular hydrogen bonding between $C(6')OH$ and $C(2)NHCOCH_3$ groups, which stabilises the more extended chain conformations, decreasing as deacetylation increases.

Rinaudo and Domard[109] have calculated a value of 5.7×10^{-3} for B, which is similar to that quoted for DNA[118] and apparently an order of magnitude lower than the other values reported for chitosan. However this discrepancy is not as large as would at first appear since Smidsrød and Haug, in deriving their empirical equation, expressed the LVN values in dl g^{-1} whereas Rinaudo and Domard expressed them in cm^3 g^{-1}. Thus although the latter's value for S will be increased by a factor of 10^2 relative to the S values of Smidsrød and Haug and other workers, their value for the $[\eta]^{1.3}_{I = 0.1}$ term will be greater by a factor of 3.98×10^2. Hence the correct value of B from their data is 0.023 which is of the same order as the other values in the literature, although indicating a somewhat stiffer molecular chain.

The above results show that in solution chitosan behaves similarly to other β-(1→4)-glucans, particularly sodium carboxymethyl cellulose. In media of low ionic strength it adopts an extended conformation owing to the stiffening of the chain because of electrostatic repulsion between chain segments. In media of high ionic strength the chain becomes more flexible and adopts a random coil conformation. The chain flexibility is also temperature dependent, increasing with increasing temperature as is shown by the steady decrease in $[\eta]$ as the solution temperature is raised.[4, 114] The values obtained for $d(\ln[\eta])/dT$ were similar despite the differences in solvent systems, being 4×10^{-3} in 0.2M CH_3COOH–0.1M CH_3COONa[4], 3.2×10^{-3} in 0.2M CH_3COOH–0.1M NaCl–4M urea[4], and 5.3×10^{-3} in 0.33M CH_3COOH–0.3M NaCl.[114]

Another model of the chitosan molecule in solution has been proposed by Berkovich *et al.*[120] to account for the anomalously low values of between 0.147 and 0.296 obtained for the constant *a* in the Mark–Houwink equation. They argue that the chitosan chains adopt a dense quasiglobular conformation stabilised by extensive intramolecular hydrogen bonding between chain segments. However globular proteins, to which this chitosan structure was likened, exhibit low LVN values that are usually less than 5 cm^3 g^{-1} and are more or less independent of the molecular weight. Neither of these two characteristics is typical of chitosan solutions and, to explain the high viscosity values obtained, Berkovich *et al.* suggest that there is a high degree of intermolecular hydrogen bonding between the quasiglobular molecules. This concept has been supported by Kienzle-Sterzer and coworkers[68, 115, 121] who have concluded[115] that the chain conformation changes from a compact spherical structure in acetic acid-–NaCl to a random coil conformation in acetic acid–NaCl–urea, because of the capacity of urea to break hydrogen bonds. It has also been supported by Lyubina *et al.*[105] but has been criticised by Maghami and Roberts[122].

The conclusions of Lyubina *et al.* are somewhat contradictory as they state both that the chitosan molecule attains a swollen coil conformation (Gaussian, for $I \geqslant 0.1M$) and that its hydrodynamic properties comply with the general rules characteristic for polymers, while concluding that the proposal that chitosan exists in solution in the form of compressed, internally hydrogen bonded coils is verified by their results. Furthermore there are a number of contradictions in the paper of Berkovich *et al.*[120] who were unable to obtain a physically realistic value for the Kuhn statistical chain element using either a non-permeable or semipermeable coil model, although a value of $A_m = 80$ Å was obtained using a Gaussian coil model. In addition the decrease in LVN with increase in LiCl concentration was taken as evidence of disruption of the interglobular hydrogen bonds although they must also have assumed either an increase, or at the least no change, in the extent of intraglobular hydrogen bonding with increase in LiCl concentration. Kienzle-Sterzer *et al.* refer to the 'collapse' of the chitosan molecule with increasing electrolyte concentration[108, 115] but plotting their reported viscosity results[108] in the form of LVN *versus* log[NaCl] gives a linear relationship (Figure 6.7) which indicates[104] that no conformational transition of the chitosan molecules occurs, although one would be expected if the chains were to 'collapse'. Furthermore if the chain conformation were to change from a compact quasiglobular one to a random coil conformation on addition of urea, as has been suggested,[115] the presence of urea would be expected to cause a dramatic increase in

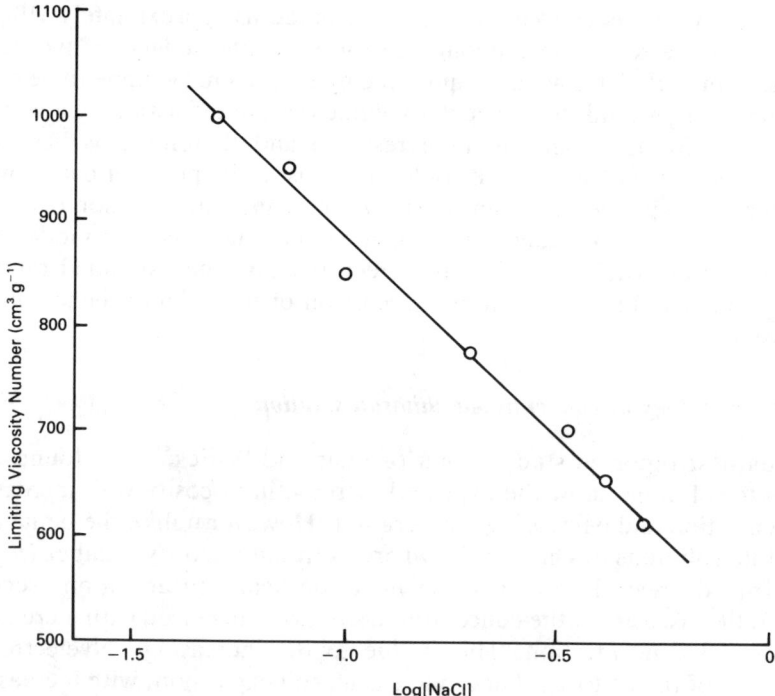

FIGURE 6.7   Limiting Viscosity Number *versus* log(NaCl concentration) plotted from the data of Rodriguez-Sanchez *et al.*[108]

solution viscosity. That this is not the case is shown by the results of Domard and Rinaudo[110] who report a LVN of 374 cm³ g⁻¹ in 0.2M NaCl and of 390 cm³ g⁻¹ in 0.1M NaCl–4M urea, the small increase being readily accounted for by the reduction in NaCl concentration, and those of Lee[4] and Lyubina *et al.*[105] which actually indicate that the addition of urea causes a decrease in the LVN.

Although Berkovich *et al.* attribute the high viscosity of chitosan solutions to interglobular hydrogen bonding, the lack of any concentration dependence in the molecular weight values obtained is cited by them as evidence of the absence of any macromolecular association, the chitosan molecules behaving 'like self-contained kinetically independent particles'.[120] Finally, at the concentrations used in viscosity measurements for determining LVN values the chitosan chains would have to have hydrodynamic volumes considerably in excess of their actual molecular volumes if they are to approach each other closely enough to form a hydrogen-bonded network as proposed.[120] For a sample having $\bar{M}_v = 2.62 \times 10^5$ the maximum concentration used[123] was 1.8 g dm⁻³ and at this concentration each chain would need to occupy a sphere of radius approximately 310 Å if the chains are to be close enough to form an intermolecular hydrogen-bonded network. This would require the hydrodynamic volume to be some 600 times larger than the molecular volume (calculated assuming a volume of 150 Å^3 for the D-glucosamine residue) and therefore, as has been pointed out,[122] such a network could only exist if the polymer chains were to adopt an expanded random coil network. Thus consideration of all the currently available evidence leads to the conclusion that the model proposed by Berkovich et al.[120] is incorrect and that the expanded random coil model should be used in any discussion of the solution behaviour of chitosan.

6.2.4 Rheology of concentrated chitosan solutions

The earliest reported study is that of Filar and Wirick[51] who found that chitosan solutions show the expected increase in viscosity with increasing concentration and decreasing temperature. However unlike the behaviour of dilute solutions of chitosan,[104] where both the viscosity number (n_{sp}/C) and $[\eta]_\infty$ decreased with increase in acetic acid concentration (section 6.2.2), the viscosity of the concentrated solutions increased with increase in acetic acid concentration. This is due to the increasing polyelectrolyte character of the chitosan, hence increasing coil expansion, with increase in acetic acid concentration since not all the amine groups will be protonated at the higher pH values because of their much higher solution concentration. A more detailed study by Rinaudo and Domard, referred to in a review paper,[109] has shown that for a series of chitosans the points for log η_{sp} versus log $\{ C[\eta] + k'(C[\eta])^2 \}$ fall on a single straight line, up to a value for log η_{sp} of approximately 0.77, regardless of the molecular weight, degree of deacetylation or concentration (Figure 6.8). Above this point the slope increases from its original value of 1 because of the molecular overlap that occurs between the coiled polymer chains as their concentration in the solution increases. The concentration at which molecular overlap begins is referred to as the critical concentration, C^*, and may be determined from the value of $C[\eta]$ at which divergence from rectilinearity in Figure 6.8 occurs. This has been found[109] to be 2.7, with the value of k' being 0.43, so that $C^* = 2.7/[\eta]$ for any chitosan sample. The samples examined, which had $\overline{DP}_v$ values between 1.39×10^3 and 11.35×10^3, had C^* values between 1.57×10^{-2} g cm^{-3} and 2.23×10^{-3} g cm^{-3} when calculated using the factor 2.7, which is considerably larger than the value of 1 which is customarily used[37] for determining C^*.

Above C^* the slope of the plot in Figure 6.8 increases and at the highest concentrations examined ($C[\eta] = 11.5$, $C = 6.69 \times 10^{-2}$ g cm^{-3} and $9.5 \times$ g cm^{-3} for the lowest and highest molecular weight materials respectively) the relationship between viscosity and molecular weight was found

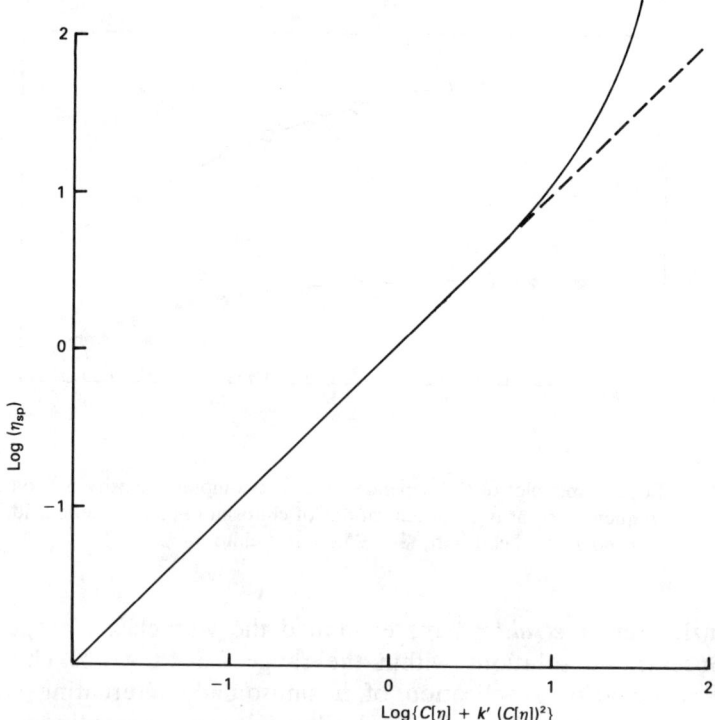

FIGURE 6.8   Master plot of $\log(\eta_{sp})$ as a function of $\log\{C[\eta] + k'\ (C[\eta])^2\}$ for chitosans having different N-acetylation values and different $\overline{DP}_v$ values.[109] Departure from linearity occurs at a value of approximately 0.77 on the abscissa

to be $\eta_{sp} \propto \overline{M}_v.^3$ This is in reasonable agreement with the value of 3.4 for the exponent that has been found for a large number of polymers.[124] Although shear effects normally become apparent in the region none were observed by Rinaudo and Domard, although Kienzle-Sterzer et al.[125] had previously reported them for chitosan solutions in the same concentration range. Rinaudo and Domard suggested that the conflicting observations could be due to the higher pH value used by Kienzle-Sterzer et al. in their studies. A higher pH would reduce the linear charge density of the chitosan chains and promote chain entanglement, thereby increasing the likelihood of pronounced shear effects.

Rinaudo and Domard also examined the temperature dependence of the viscosity for an 8 g dm^{-3} solution of chitosan. Conversion of their results to approximately the same form as that used by Lee[4] and by Pogodina et al.[114] gives $d\eta_{rel}/dT = 23 \times 10^{-3}$, which is 4–5 times larger than the values obtained in the earlier studies. Rinaudo and Domard attributed this difference to the higher chitosan concentration used in their work.

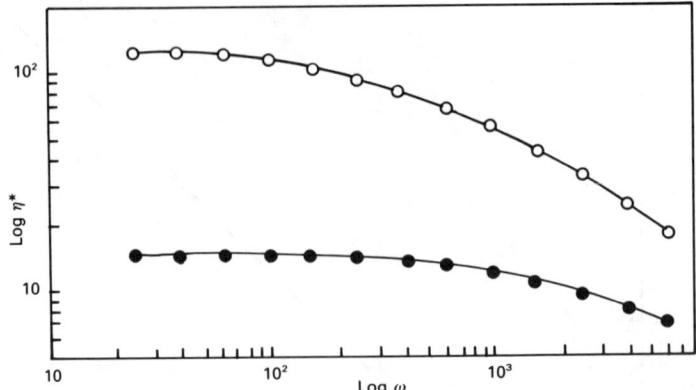

FIGURE 6.9   Logarithmic plot of the dependence of the complex viscosity, η^*, on the frequency, ω, at two concentrations of chitosan in 0.66M acetic acid: $\bigcirc = 40$ g dm^{-3} chitosan; $\bullet = 25$ g dm^{-3} chitosan[125]

Kienzle-Sterzer et al.[125] have examined the viscoelastic properties of semiconcentrated solutions within the range 2.5–4.0 wt-% chitosan in 0.66M acetic acid. Application of a sinusoidally alternating stress of frequency ω to a viscoelastic material will result in an alternating strain that will be out of phase with the stress, the phase angle between stress and strain being $\delta(\omega)$. It may be shown[126] that

$$\text{Tan } \delta = G'' / G'$$

where $G'(\omega)$ is the shear storage modulus and $G''(\omega)$ is the shear loss modulus. The phase relationships can be expressed either as a complex modulus

$$G^* = G' + i\,G''$$

or as a complex viscosity

$$\eta^* = \eta' - i\,\eta''$$

with $\eta' = G''/\omega$ and $\eta'' = G'/\omega$. As the frequency approaches zero the real component η' tends towards the steady flow viscosity η_0.

Kienzle-Sterzer et al.[125] found that η^* increases with increase in chitosan concentration but is frequency-dependent above a frequency value which decreases with increasing chitosan concentration (Figure 6.9). This was considered due to the increase in importance of segmental motions relative to the importance of translational motions with increasing chitosan concentration, leading to an increase in the energy stored by the chains and hence increasing η'' and/or decreasing η'.

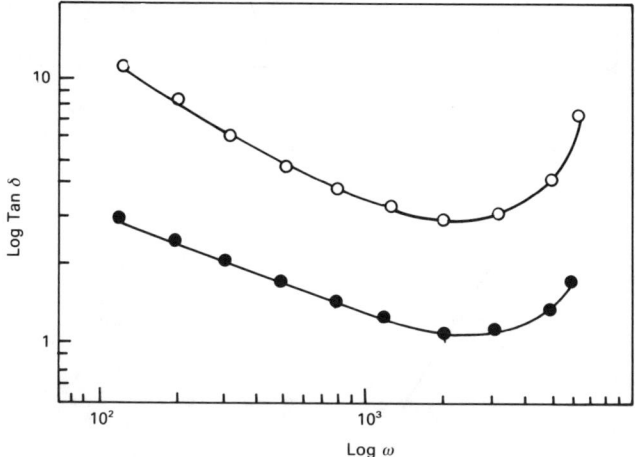

FIGURE 6.10   Logarithmic plot of the dependence of Tan δ on the frequency, ω, at two concentrations of chitosan in 0.66M acetic acid: $\bullet$ = 40 g dm^{-3} chitosan; $\bigcirc$ = 25 g dm^{-3} chitosan[125]

The frequency dependence of the storage modulus and the loss modulus increases with increase in the frequency, crossing over at high frequency in the case of the more concentrated solutions owing to entanglement coupling which prolongs molecular rearrangement times. A plot of log G'_R versus log ω_R where G'_R is the reduced storage modulus, G'/C, and ω_R is the reduced frequency, gives a single master curve for all the samples, indicating that the relaxation mechanisms are the same for all concentrations. Figure 6.10, which is the plot of log Tan δ versus log ω, indicates that the interactions giving rise to entanglements are the same at all concentrations, although the modulus of the entanglement network increases with increase in chitosan concentration.

6.2.5   Chitosan-induced metachromasy

Although polyelectrolyte-induced metachromasy in anionic polyelectrolyte–cationic dye systems had been well known for over 50 years, the first report[127] of analogous behaviour in chitosan–anionic dye systems was not recognised as such, the spectral changes shown by Eosin (C.I. Acid Red 87), Erythrosin (C.I. Acid Red 51) and Uranine (C.I. Acid Yellow 73) in the presence of chitosan hydrochloride being claimed to be non-metachromatic .[127] More recently the interaction between chitosan and anionic dyes in solution has been the subject of a series of papers by Gummow and Roberts[66, 128–130] who showed that the spectral changes

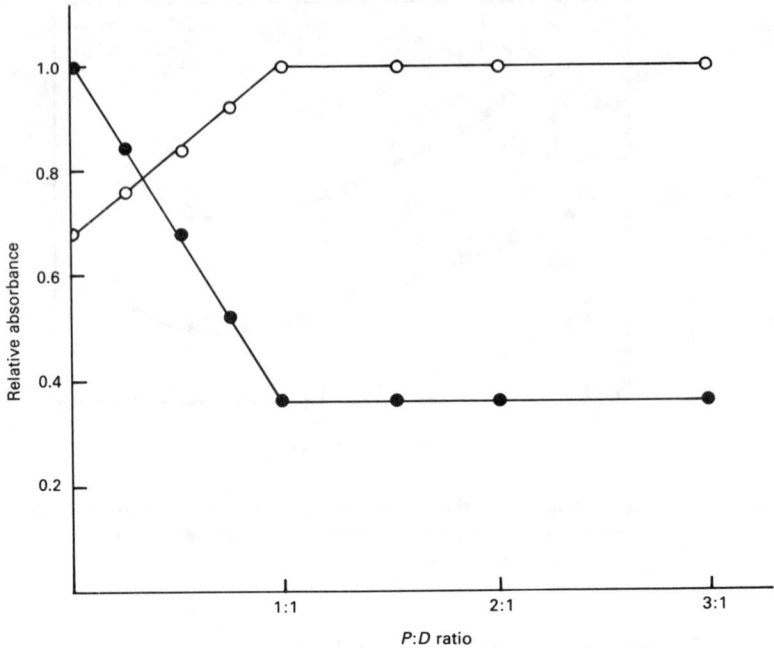

FIGURE 6.11　Change of absorbance of C.I. Acid Red 88 as a function of the $P{:}D$ ratio – the ratio of $\overset{+}{N}H_3$ groups (on chitosan):SO_3^- groups (on dye). Absorbance is expressed relative to the maximum value at the two wavelengths: ●, relative absorbance at 505 nm (λ_{max} for monomeric dye); –O–, relative absorbance at 450 nm (λ_{max} for aggregated dye)[128]

induced by chitosan could be reversed by the addition of neutral salts, alcohols, urea or excess chitosan, and by raising the temperature.[66] This confirmed that these spectral changes are indeed metachromatic and similar to those observed in anionic polyelectrolyte–cationic dye systems. The limiting value for the decrease in absorbance at λ_{max} was found to occur at a chitosan amine group:dye ion ratio of 1:1 (Figure 6.11), and this stoichiometry enables metachromatic titrations with chitosan and Orange II (C.I. Acid Orange 7), or any other suitable dye, to be used to determine either the degree of acetylation, using a chitosan solution of known concentration, or the concentration in solution of a chitosan having a known degree of N-acetylation.[128] This latter technique has been utilised in studies of the adsorption of chitosan on cellulose.[131, 132]

There is a much greater diversity of structure among anionic dyes compared with cationic dyes, both in the chemical types and in the number of charged groups in the dye ion, all cationic dyes carrying a single charged group. Gummow and Roberts examined the metachromatic behaviour of some 40 dyes and observed that both hypsochromic and bathochromic shifts could be induced by chitosan, and that the direction of shift could be correlated with the general structure of the dye.[130] These authors also proposed a new mechanism for polyelectrolyte-induced metachromasy based on their studies of the chitosan–anionic dye system. Instead of the mechanism of Michaelis[133] who suggested that it is due to two or more dye ions binding to adjacent sites on the polyelectrolyte chain, thereby forming aggregates which give rise to the characteristic spectral changes, Gummow and Roberts proposed[129] that metachromasy arises when dye ions are exchanged for the counter ions held within the electrostatic domain of the randomly coiled polyelectrolyte. This exchange results in a highly concentrated dye solution within the electrostatic domain of the chitosan chain, leading to aggregation of the dye ions in this solution and to the observed spectral changes. The reduction in the total energy of the system and the increase in entropy resulting from hydrophobic bonding, that occur on aggregation of the dye ions, constitute the driving forces behind the exchange of dye ions for counter ions.

6.3   GELATION

The formation of gels from chitosan has been referred to previously in discussing N-acylation of chitosan (section 4.2.1), Schiff's base formation (section 4.4), adsorption of proteins (section 5.3.2), and chromatographic applications of chitin and chitosan (section 5.3.5). However it is convenient to discuss gel formation as a coherent topic rather than treating it merely as a means of preparing chitosan derivatives.

Chitosan-based gels may be broadly divided into thermally non-reversible gels and the very much smaller group of thermally reversible gels. Within the first group a further subdivision into those formed by N-acylation and those produced by Schiff's base (aldimine) formation is useful, and this division of the topic is used here.

6.3.1   N-Acylchitosan gels

These were first described by Hirano et al. in 1975,[134] the gels being prepared by treating a 20 g dm^{-3} solution of chitosan[0.0] in 10 vol.-% aqueous acetic acid with acetic anhydride, the volume ratio of chitosan solution:acetic anhydride being 2:1. A rigid gel formed within 30 minutes at room temperature and after dialysis against distilled water the gel was colourless, transparent and rigid; soluble in formic acid but insoluble in water, 50 vol.-% HCOOH, 10 vol.-% CH$_3$COOH, alcohols and acetone. It was also infusible. Analysis of the dehydrated gel showed that there was an average of 2.36 acetyl groups per anhydro-D-glucosamine residue, the presence of both N- and O-acetyl groups being confirmed by IR spectroscopy and the substitution level by NMR spectroscopy in deuterated formic acid. This was expanded in a second paper[135] in which N-acetyl-, N-propionyl- and N-butyrylchitosan gels were prepared using 10 vol.-% aqueous acetic, propionic and butyric acid as solvents for treatment with the appropriate acyl anhydride. In all cases O-acylation accompanied N-acylation but the extent decreased from an O-acetyl DS of 1.4, to 0.3 for O-propionyl and 0.1 for O-butyryl. De-O-acylation was carried out by treatment overnight at room temperature with 0.5M KOH in 95 vol.-% ethanol. The gels were all only soluble in formic acid.

Selective N-acylation during gelation was first demonstrated by Hirano *et al.*[136] using a 1:5 mixture of 10 vol.-% acetic acid:methanol as solvent and a range of acyl anhydrides up to dodecanoic anhydride and also benzoic anhydride. Although selective N-acylation was also achieved with tetradecanoic, hexadecanoic and octadecanoic anhydrides, gelation did not occur with these and instead precipitates were obtained. Similar behaviour was observed with homogeneously prepared mixed N-acylchitosans[137] where all those derivatives containing a major proportion of N-tetradecanoyl, N-hexadecanoyl or N-octadecanoyl groups, formed precipitates rather than gels. In the absence of methanol both N- and O-acylation was again observed with acetic, propionic and butyric anhydrides (Table 4.6).[136] The use of cosolvents other than methanol has been examined and with formamide, ethylene glycol, diethylene glycol (Digol), and triethylene glycol (Trigol), rapid gelation occurs. However the use of ethanol and 1-butanol leads to formation of gelatinous lumps while acetone and 2-propanol cause precipitation of the polymer.[138]

The effects of variation in the acetic anhydride:–NH$_2$ group mole ratio has been investigated by Hirano and Yamaguchi[139] using both 10 vol.-% aqueous acetic acid and a 1:5 mixture of 10 vol.-% acetic acid: methanol as solvents. In the former solvent, gel formation requires a mole ratio greater than 13:1, rigid gels being produced at a mole ratio of 20:1. In the aqueous acetic acid–methanol system, a rigid gel is obtained with a mole ratio of only 1.7:1. Syneresis occurs with the gels formed in the methanol-containing system using mole ratios of 13:1 or greater, but did not occur with gels formed at mole ratios of up to 50:1 in the absence of methanol, although it did occur with the gel formed at a mole ratio of 170:1. This requirement of a high anhydride concentration if gels are to be produced in the absence of methanol was also found by Roberts and Taylor in their study of the preparation of uniform chitin beads by gel formation in the discontinuous phase of an emulsion.[140] Beads which neither coagulated nor deformed under low shear stress were only obtained at a mole ratio of 30:1 or higher.

The importance of other variables has been studied by determining their effects on the rate of gelation, as measured by the time to the onset of

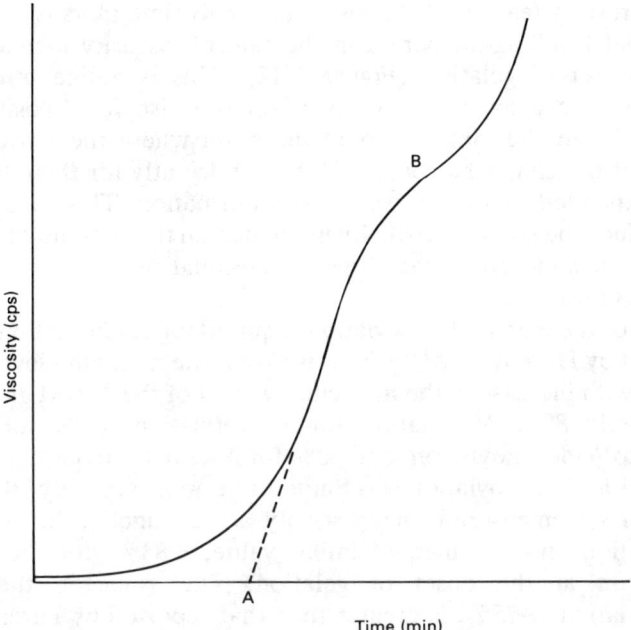

FIGURE 6.12   Typical plot of viscosity *versus* time during formation of *N*-acylchitosan gels,
showing the time to onset of gelation (A) and the step in the curve
(B)[138, 141]

gelation.[138, 141] The rate of gelation was found to increase with increase in
chitosan concentration, acyl anhydride concentration and temperature,
and with decrease in the molecular weight of the acyl anhydride. The time
to onset of gelation, measured as indicated in Figure 6.12, increased from
about 7 minutes with acetic anhydride to about 260 minutes with decanoic
anhydride for the same molar concentration of anhydride and amine
group. This increase in gelation time was attributed to increasing steric
hindrance at the reaction site leading to a reduction in the fraction of
effective intermolecular collisions and hence a reduction in the rate of
N-acylation.[138] The energies of activation for *N*-acetylation and *N*-hexa-
noylation were found to be 68 ± 7 kJ mol^{-1} and 57 ± 6 kJ mol^{-1} respectively,[141] and this dependence on molecular weight supports the suggestion
of increasing steric hindrance at the reaction centre with increase in the
molecular size of the anhydride. The other observed effects are in agreement with the conclusion[139] that gelation occurs through aggregation of the
polysaccharide chains, presumably[141] through hydrophobic bonding rather
than helix formation of the type which occurs with carrageenans, and that
this is aided by the reduction in electrostatic repulsion as *N*-acylation
proceeds. Thus the rate of gelation depends upon the rate of *N*-acylation.

An interesting feature of the viscosity *versus* time plots is that they all show a brief levelling-off period in the rate of viscosity increase shortly after the onset of gelation (Figure 6.12). This is unlike other gelling systems such as agar, which show a smooth rise in viscosity. It was suggested[141] that the kink occurs at the point where the polyelectrolyte character of the chains has been reduced sufficiently for them to collapse from an expanded to a more compact conformation. This collapse would tend to reduce the solution viscosity in opposition to the trend of increasing viscosity with build-up of the three-dimensional network through chain segment aggregation.

Values for the extent of *N*-acylation required for gel formation were first determined by Hirano *et al.*[137] who found that the minimum level required decreased with increase in the molecular weight of the *N*-acyl group, being approximately 80% *N*-acylation for *N*-acetylation, 60% for *N*-propionylation to *N*-decanoylation, and 40% for *N*-dodecanoylation. The value required with *N*-benzoylation was found to be approximately 70%. Moore and Roberts[141] measured the viscosity as a function of the extent of *N*-acylation and obtained a similar value, ~83%, for the extent of *N*-acetylation at the onset of gelation. The value in the case of *N*-hexanoylation, ~75%, is greater than that reported by Hirano *et al.*[137] but their work was carried out using chitosan[0.0] whereas Moore and Roberts used chitosan[0.24]. The figure of 75% obtained by the latter workers is the value of the total *N*-acylation and hence contains 24% of *N*-acetyl groups which are less effective than *N*-hexanoyl groups in promoting gelation. This factor would not apply in the case of *N*-acetyl-chitosan gels and hence there is good agreement between the two sets of figures. It was also found that the level of *N*-acylation required for gelation decreases with increase in the chitosan concentration (Table 6.4) but is independent of the reaction temperature.

The gel produced in the aqueous acetic acid–methanol system using a mole ratio for acetic anhydride:amine group of 2:1 gave, after lyophilisation, an IR spectrum similar to that of chitin, and the ^{1}H NMR spectrum confirmed that it was chitin[1.0].[139] Furthermore the CP/MAS ^{13}C NMR spectra of native α-chitin and lyophilised *N*-acetylchitosan show identical ^{13}C shifts, thus the chain conformation in the gel must be similar to that in native chitin.[142] However the ^{13}C signals are considerably sharper in the spectrum of native chitin, indicating that it has a higher degree of crystallinity.[142] This is in agreement with the previous observation[139] that the lyophilised *N*-acetylchitosan is soluble in formic acid while native chitin is not, although this was originally attributed[139] to differences in molecular conformation and molecular weight rather than to differences in crystallinity. The ^{13}C chemical shifts of both *N*-decanoylchitosan and *N*-octadecanoylchitosan (the latter being presumably obtained by lyophilisation of the precipitate obtained on reaction with the anhydride since *N*-octa-

TABLE 6.4   *Effects of chitosan concentration and temperature on the N-acylation level required for gelation[141]*

Chitosan[a] conc. (g dm^{-3})	Temperature (°C)	Total % N-acylation at onset of gelation on	
		N-acetylation	N-hexanoylation
30	25		60
20	25		62
10	25	82.3	73.5
10	35	85.0	79.0
10	45	82.5	74.5

[a] Starting material was chitosan[0.24].

decanoylchitosan does not gel[137]) are similar to those of *N*-acetylchitosan, although the signals are considerably broader, which indicates a similarity in chain conformation but a considerably reduced level of crystallinity.[142] High resolution ^{13}C NMR spectroscopy of the gels themselves, together with ^{13}C CP/MAS NMR spectroscopy of the lyophilised material, indicates that the crosslinks or junction points giving rise to the gel structure are formed by association of the polymer chains in an extended conformation similar to that occurring in native chitin. However the extent of this association or crystallisation is less than in native chitin since only a limited amount of interchain association is permissible if a gel, rather than a precipitate, is to be formed.

The ultrastructure of the lyophilised *N*-acetylchitosan gels has been examined by SEM and found to be microporous, with non-uniform pores in the range 30–50 μm diameter and a pore wall thickness of <1.5 μm. Increase in the chitosan concentration gives a decrease in pore diameter. The structure has a honeycomb-like appearance and a characteristic feature is that the junctions in the pore assemblies always contain three pore wall membranes.[143]

An account of a number of applications for *N*-acylchitosan gels has been given by Hirano *et al.*[144]

6.3.2   N-Alkylidene- and N-arylidenechitosan gels

An *N*-alkylidene gel, *N*-methylidenechitosan, the first reported chitosan-based gel, was originally prepared by Broussignac[145] using dilute aqueous acetic acid as the solvent system. Gelation occurred in approximately 5 minutes with a 30 g dm^{-3} solution of chitosan and 40 minutes with a 20 g dm^{-3} solution, with no gelation occurring within 24 h using a 10 g dm^{-3} solution. In all three cases the formaldehyde concentration used was approximately 14 g dm^{-3}.

Hirano *et al.* subsequently reported[146] the formation of *N*-alkylidene-

and N-arylidenechitosan gels using aqueous acetic acid–methanol as the reaction medium. The aldehydes used, all of which produced rigid gels, were formaldehyde, acetaldehyde, propionaldehyde, acrolein, salicylaldehyde, p-tolualdehyde, cinnamaldehyde and glutaraldehyde. No mention was made of higher aliphatic aldehydes than propionaldehyde but Moore has reported[147] that straight chain C_4–C_{10} aliphatic aldehydes do not form gels under the conditions used by Hirano et al.[146]

The effect of varying the formaldehyde:amine group mole ratio was examined using a final chitosan[0.0] concentration of 16.7 g dm^{-3} and it was found that the minimum mole ratio required for gel formation was between 4:1 and 8:1, the product obtained using the former value being a highly viscous solution, but not a gel, and that obtained with the latter value being a rigid gel. No intermediate values were examined. Elemental analysis of the dried products produced using mole ratios of 8:1–64:1 indicated that the extent of reaction of the amine groups was relatively constant at about 80% over this wide range of formaldehyde concentrations.

The minimum amount of aldehyde required for gel formation has been found to increase with increase in the aldehyde molecular weight in the series HCHO < CH_3CHO < CH_3CH_2CHO.[148] The gels are colourless, rigid, and infusible up to 200°C. They are insoluble in, and stable towards, cold and boiling water, 2M NaOH, DMF, DMSO, and common organic solvents, but are degraded by 0.5M HCl and by 10 vol.-% CH_3COOH. The structure of N-methylidenechitosan gels was found to be that of a polyphasic xerogel,[148] similar to that of the N-acetylchitosan gels[143] but with a range of pore size dimensions smaller than those found in the latter gels. A second difference is that there is no fixed number of pore wall membranes at the junction zones of pore assemblies, whereas N-acetylchitosan gels always show three such membranes per junction zone.[143]

The most extensively studied gel system involving Schiff's base formation is that using glutaraldehyde,[146, 149, 150] which is unlike the N-alkylidene- and N-arylidenechitosan gels discussed above in involving covalent crosslink formation between the chains. Roberts and Taylor, using viscosity measurements to determine the rate of gelation, have found it to be proportional to both the chitosan and the glutaraldehyde concentrations and to the temperature, while it is inversely proportional to the acetic acid concentration.[150] The activation energy of gelation was found to be approximately 63 kJ mol^{-1}, which is close to the activation energies for formation of N-acylchitosan gels.[141] The similarity in activation energies was considered to be reasonable since both gelation mechanisms involve nucleophilic attack by an $-NH_2$ group on a $>C=O$ group. However, despite this similarity, plots of viscosity versus time do not show the inflection point or kink previously observed[141] during the formation of N-acylchitosan gels. Instead the plot shows an initial period where there is very little change in viscosity, followed by an extremely rapid rise, up to the limit of the viscometer, in which no evidence of any kink could be observed.

The addition of neutral electrolytes was also found to increase the rate of reaction between amine and aldehyde groups and hence the rate of gelation. The effectiveness was found to increase in the order $Cs^+ < Rb^+ < K^+ < Na^+$ for monovalent cations, $Ba^{2+} < Sr^{2+} < Ca^{2+} < Mg^+ <$ for divalent cations, and $NO_3^- < F^- < Cl^- < Br^- < SCN^-$ for anions.[150] Two possible explanations for the observed variation in effectiveness of the electrolytes were suggested:

(a) Gel formation requires chain segments from different chains to approach each other sufficiently closely for crosslinking to occur and this would be facilitated by screening the positively charged chain segments by specific ion adsorption, so that the effects of specific ions should parallel their effects in the flocculation of positively charged sols, where Na^+ is more effective than K^+ and Cl^- than NO_3^-.

(b) Increasing the ionic strength of the reaction medium should decrease the activity coefficient of the transition-state-activated complex and

$$\ldots -NH_2 + O=CH- \ldots \rightleftharpoons \ldots \left[-\overset{+}{N}H_2 - \overset{\overset{\displaystyle O^-}{|}}{C}H- \right] \ldots \rightarrow$$

$$\ldots -NH-CH(OH)- \ldots \rightarrow \ldots -N=CH- \ldots + H_2O$$

increase the rate of reaction,[151] and in this case the effectiveness of a specific electrolyte will depend on its activity coefficient.

A correlation between the activity coefficient of the added electrolyte and the rate of gelation was in fact found, but this does not of itself rule out the first explanation.

The actual structure of the crosslinks responsible for gelation in the chitosan–glutaraldehyde system is the subject of some controversy. Muzzarelli *et al.* have claimed[149] that the reaction is not simply one of Schiff's base formation (**6.1** in Figure 6.13) and Muzzarelli has suggested[152] that the mechanism of Richards and Knowles for crosslinking proteins with glutaraldehyde[153] also applies to the chitosan–glutaraldehyde system. Richards and Knowles concluded, on the basis of ^1H NMR spectroscopy, that in aqueous solution glutaraldehyde is largely polymeric and contains significant quantities of α,β-unsaturated aldehyde groups, such as **6.2a,b**, formed by the aldol condensation reaction. They suggested[153] that these unsaturated aldehyde groups form Michael-type adducts with amine groups, rather than simple Schiff's base derivatives, and Muzzarelli *et al.*[149] have extended this concept to the chitosan–glutaraldehyde system and proposed that gelation is due to the formation of crosslinks such as **6.3** (Figure 6.13) rather than **6.1**. However Roberts and Taylor[150] have shown

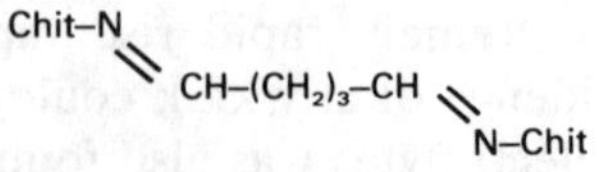

Chit–N
        CH–(CH₂)₃–CH
                    N–Chit

6.1

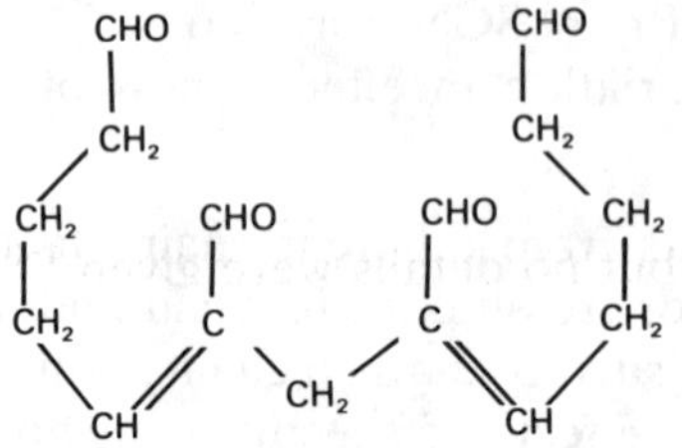

CHO   CHO   CHO   CHO
| | | |
CH₂   CH₂   C   CH₂
CH₂   CHO   CHO   CH₂   CH₂   CH   CHO   CH₂
CH₂   C   C   CH₂   CH₂   C   C   CH₂
CH   CH₂   CH   CH   CH₂   CH

6.2a 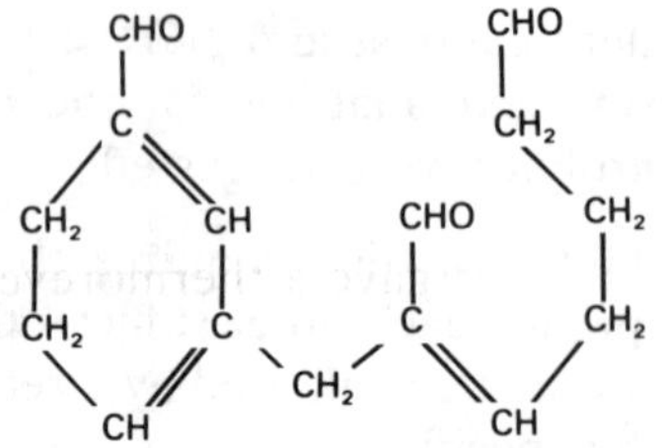   **6.2b**

CHO   CHO
| |
CH₂   CH₂
CH₂   CHO   CHO   CH₂
CH₂   CH   CH   CH₂
CH   CH₂   CH
Chit–NH   NH–Chit

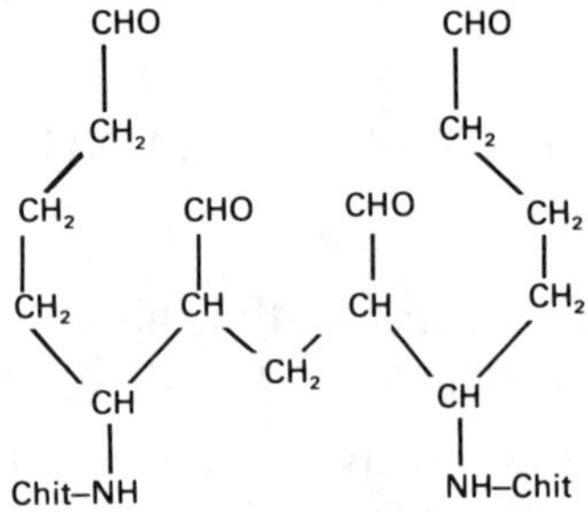

6.3

FIGURE 6.13   Proposed structures for glutaraldehyde–crosslinked chitosan[150, 152]

by UV, ¹H NMR, and ¹³C NMR spectroscopy of glutaraldehyde, purified[154] glutaraldehyde, and the 2,4-dinitrophenylhydrazine derivatives of glutaraldehyde, octanal and crotonaldehyde, that the concentration of α,β-unsaturated aldehyde groups in aqueous glutaraldehyde is very low, less than 0.15% of the total aldehyde group concentration. Furthermore glutaraldehyde purified by the method of Gillett and Gull[154] induces gelation in chitosan solutions at the same rate as does unpurified glutaraldehyde having the same aldehyde group concentration. These results have been claimed as support for a simple Schiff's base structure (**6.1**) for the crosslinks.[150]

A major argument against a simple Schiff's base structure for such crosslinks is their stability towards acid treatment,[153] whereas Schiff's bases are known to be susceptible to acid hydrolysis. However Blanco and

Guisán have shown[155] that while attachment of an enzyme to an insoluble support through a single Schiff's base structure is unstable and readily reversible, attachment involving two or more Schiff's base structures is apparently irreversible. In the case of the chitosan–glutaraldehyde product, hydrolysis of a –CH=N– group would give an amine and an aldehyde group that would be held, by adjacent crosslinks, in a very favourable position to interact and reform the Schiff's base structure. This is similar to the proposal of cooperative crosslink formation in which it was suggested[150] that formation of the initial network aids formation of further crosslinks since the chain segments adjacent to an existing crosslink will be held at approximately the correct separation for the formation of additional crosslinks.

Gels related to the N-alkylidenechitosan gels have been produced by reductive N-alkylation (section 4.3.3) using D-galactose, D-glucose, lactose, and melibiose as the aldehydes.[156] The ultra-structures of the xerogels formed varied with variation in the carbohydrate used; that formed using D-galactose had a smooth, non-porous surface with a highly ordered microfibrillar substructure, while that formed using D-glucose had a polyphasic microporous structure with non-uniform pore dimensions and very thin membrane walls. The non-porous character of the D-galactose product was in agreement with the previous observation[157] that it did not act as a ligand for binding metal ions. The gel produced using lactose was similar to that formed from D-glucose, with a polyphasic microporous structure provided it was prepared by extensive dialysis, while that from melibiose had a highly ordered microfibrillar ultrastructure. ESR spectroscopy of nitroxide spin-labelled products shows[157] that the latter has a more extended, ordered, and less mobile arrangement of side chains than the lactose-based derivative. No simple relationship between chemical structure and the observed ultrastructure was found.

6.3.3   Thermally reversible gels

The first report of a chitosan-based thermally reversible gel is that of Hayes and Davies[158] who found that solutions of chitosan in 1.1M oxalic acid gradually gelled on standing at room temperature. The length of time required for gelation was considerable and depended on the chitosan concentration, being approximately 21 days for a 30 g dm^{-3} solution and 24 h for a 70 g dm^{-3} one. The gels melted on heating and gelled on cooling, the time required to gel after melting being considerably less than that required in the initial formation. The melting points were reasonably high and increased with increase in chitosan concentration, being 88°C for a 50 g dm^{-3} solution and 92°C for a 100 g dm^{-3} one. The authors found that addition of iodine to the solution produced a blue colouration in the gel, the rate of colour formation being comparable to the rate of gel formation, and suggested that in the presence of oxalic acid the polymer chains form double helices which act as the junction points in the gel network. Solutions of chitosan in 2M dichloroacetic acid were reported to behave similarly,[56] although no details were given beyond the statement that the solutions gel on standing.

Hirano and co-workers[159] subsequently reported on this type of gel and proposed that the mechanism of formation of the junction points is one of salt formation similar to that for formation of alginate gels in the presence of Ca^{2+} ions. In a study of partial N-succinylation of chitosan[0.0] the same authors noted that the product having 35% of the amine groups acylated gave a thermoreversible gel, but no details were given.[160]

A method of preparing thermoreversible chitosan gels through the use of large organic counter ions has been described.[65] The process involves mixing heated solutions of chitosan acetate and of the sodium salt of either 1-naphthol-4-sulphonic acid (NSA) or 1-naphthylamine-4-sulphonic acid (NASA), the mixture gelling on cooling. The gels have reasonably sharp melting points (T_m) that remain constant over a number of melting/cooling cycles and are independent of the mixing temperature, provided it is above T_m, the time of heating the solution after mixing, or the length of time the gels are held at ambient prior to determining T_m. However T_m is strongly dependent on the concentration of NSA or NASA, increasing from about 40°C to 85°C as the concentration of NSA is increased from 0.04M to 0.18M, or from 30°C to 85°C with increase in the NASA concentration from 0.03M to 0.12M.

Unlike the chitosan–oxalic acid gels the chitosan concentration required for gel formation is low, 2–5 g dm^{-3} being typical, and similar to the concentrations used for gel formation with other polysaccharides such as the carrageenans. Surprisingly the melting temperatures in the chitosan–NSA system are independent of the chitosan concentration within the above limits of 2–5 g dm^{-3}, provided that the chitosan:acetic acid ratio remains constant. Even if the acetic acid concentration is kept constant so that this ratio decreases with increasing chitosan concentration, T_m is independent of the chitosan concentrations except at the lowest concentration (Table 6.5). However decreasing the chitosan concentration causes a decrease in the maximum NSA concentration that may be used without forming an insoluble non-melting precipitate rather than a thermally reversible gel. Therefore, since the main factor controlling T_m is the NSA concentration, there is a decrease in the highest T_m value obtainable with decreasing chitosan concentration. This does not appear to be the case when NASA is used, although the values of T_m show a slightly greater dependence on chitosan concentration (Table 6.5).

Addition of neutral electrolytes causes a reduction in T_m in these systems although the order of effectiveness is reversed, being NaCl < NaBr < NaI for chitosan–NSA gels and NaI < NaBr < NaCl for chitosan–NASA

TABLE 6.5   *Effect of chitosan concentration on the melting temperatures of NSA- and NASA-based thermally reversible gels*[65]

Chitosan concentration (g dm^{-3})	5.0	3.5	2.5	2.0
T_m (°C) – NSA-based gels[a]	57.5	58.5	57.0	50.5
T_m (°C) – NASA-based gels[b]	57.0	57.0	53.0	53.5

[a] Chitosan dissolved in 0.1M acetic acid–0.12M NSA.
[b] Chitosan dissolved in 0.07M acetic acid–0.12M NASA.

gels. No mechanism has been proposed for these thermoreversible gels but empirical tests show that the nature of the counter ion is critical, no gels being formed by addition of either naphthalene-1-sulphonic acid (Na$^+$) or naphthalene-2-sulphonic acid (Na$^+$).

6.4 ADSORPTION OF CHITOSAN FROM SOLUTION

The adsorption of chitosan may be considered under two main sub-headings based on the substrate dimensions. The first deals with adsorption onto macroscopic surfaces and in this flocculation is not normally an outcome of the adsorption process. The second deals with adsorption onto suspended particles of colloidal dimensions and here flocculation or stabilisation will normally occur. Recovery of soluble protein is included under this second heading since this depends on coagulation or flocculation of the insoluble polyelectrolyte complex (section 6.1.4) initially formed between the chitosan and the protein.

6.4.1 Adsorption of chitosan on cellulose

The adsorption of chitosan and water-soluble chitosan derivatives on cellulose has been studied by several workers,[131, 132, 161–164] primarily because of the possible utilisation of chitosan in paper making. Numerous other studies have dealt with the effectiveness of chitosan in this role but have not examined the actual adsorption process.

Cellulose acquires a negative surface charge on immersion in water[165] and hence electrostatic interaction between chitosan and the cellulose substrate would be expected to play an important role in the adsorption mechanism. This was first suggested by Allan *et al.*[161] who attributed the low equilibrium adsorption of chitosan on a high α-cellulose pulp to the high linear charge density and extended conformation of the chitosan chains, which makes them very efficient at neutralising anionic sites on the fibre surface. Allan *et al.* argued that adsorption will be complete once the anionic sites are all occupied or screened and hence the α-cellulose, having a low concentration of anionic (–COOH) groups, will have a low adsorption value. In contrast an unbleached sulphite pulp having approximately

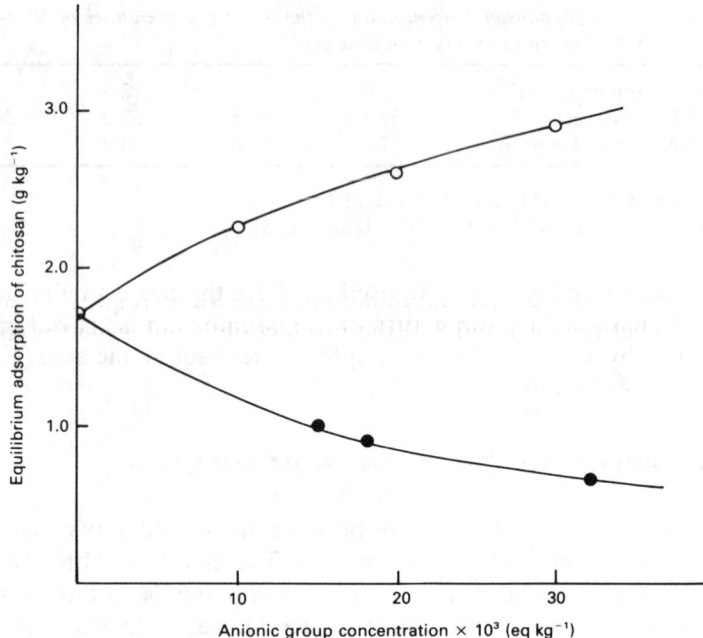

FIGURE 6.14   Equilibrium adsorption of chitosan as a function of anionic group concentration: $\bigcirc$, carboxylic acid groups on oxidised cellulose; $\bullet$, sulphonic acid groups on reactive dyed cellulose[132]

three times the concentration of anionic groups showed increased adsorption of chitosan. Domszy et al.[131] found an increase in the amount of chitosan[0.16] adsorbed with increase in the $-COOH$ group content of cellulose substrates modified by oxidation by $NaIO_4$ followed by acidified $NaClO_2$. They attributed this to increasing electrostatic interactions but subsequently Domszy and Roberts[132] suggested that the increase might arise in part from higher surface area values of the oxidised cellulose substrates due to their greater hydrophilic character, increasing adsorption of chitosan with increasing surface area due to mechanical treatments having previously been reported.[131, 163]

That a higher anionic group content will not in itself necessarily give rise to higher adsorption values was shown by the decrease in the adsorption values obtained with increase in the dye content of reactive-dyed cotton[132] (Figure 6.14). This was attributed to steric effects causing a reduction in the surface areas of chitosan and cellulose in contact, thereby reducing the Van der Waals' forces and nullifying the increase in electrostatic interaction. The fact that direct-dyed cotton shows an increase in chitosan adsorption[132] was considered to support this explanation since direct dyes, unlike reactive dyes, are known to be adsorbed onto cellulose chains in a planar conformation[165] and hence would not disrupt the close approach of the two polysaccharide chains.

If the level of chitosan adsorption is primarily governed by electrostatic interaction then the amount of chitosan adsorbed should increase with decrease in the linear charge density of the chitosan, since each chain will neutralise fewer anionic sites the lower their linear charge density. Decreasing this may also affect uptake by reducing both the coil dimensions and the electrostatic repulsion between chains, and both these effects would lead to less effective screening of the surface by chitosan chains already *in situ*. This reduction in the charge density may be achieved either by raising the pH or increasing the degree of *N*-acetylation and both approaches have been shown to raise the adsorption level of chitosan on cellulose. Tanaka *et al.*[162] found that the adsorption of *O*-hydroxyethyl chitosan[0.22] on a bleached kraft pulp was at a maximum at approximately pH 8.2, the amount adsorbed decreasing sharply with either increase or decrease of pH, while Domszy *et al.*[131] found a linear relationship between the degree of *N*-acetylation and the amount of chitosan adsorbed. Tanaka *et al.* calculated that maximum adsorption was obtained when the *O*-hydroxyethyl chitosan had approximately 1 protonated amine group for every 100 anhydro sugar residues.[162]

The conformation of the adsorbed chitosan chains has not yet been established but an attempt has been made to determine the frequency of ionic interaction by measuring the amount of the cationic dye Methylene Blue (C.I. Basic Blue 9) displaced from the anionic sites of a carboxyl-containing cellulose pretreated with Methylene Blue, on adsorption of chitosan.[132] The results indicate that only 1 amine group in every 800, approximately, is directly involved in ionic interactions with the carboxylic acid groups on cellulose. This corresponds to an average of 2–3 ionic bonds for each adsorbed chitosan chain and suggests that loop adsorption[166] is occurring (Figure 6.15).

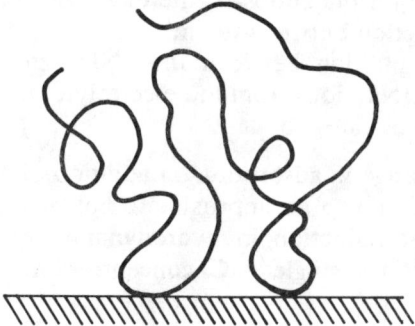

FIGURE 6.15   Loop adsorption model for the conformation of an adsorbed polymer chain[166]

Other factors have been found to influence the extent to which chitosan may be adsorbed on cellulose. Since water-swollen cellulose will act as a porous substrate the molecular weight of the chitosan would be expected to have a considerable effect on the amount adsorbed, since high-molecular-weight chains would be confined to the macroscopic surface whereas low-molecular-weight chains would be able to penetrate the substrate and adsorb onto the internal surfaces. An inverse relationship between the amount of chitosan adsorbed and its molecular weight has been demonstrated,[131] while the effect of increasing temperature has been found to increase the adsorption.[131, 164] The effect of temperature has been explained[131] as arising from both a proposed decrease in the solubility of chitosan with increasing temperature, by analogy with the known inverse relationship between solubility and temperature for water-soluble cellulose derivatives, and from a reduction in the hydrodynamic volume of the chains with increasing temperature which has the effect of increasing the surface area available for adsorption.

The effect of added electrolyte has also been examined.[132, 164] Rippon[164] found no change in the amount of chitosan adsorbed on addition of 0.34 mol dm^{-3} NaCl but Domszy and Roberts found an initial increase in adsorption, followed by a decrease, with increasing NaCl concentration.[132] This latter behaviour is similar to that reported by Tanaka et al.[162] for the adsorption of a synthetic cationic polyelectrolyte, poly(dimethyldiallylammonium chloride), on cellulose. The initial rise in the adsorption is due to the reduction in the hydrodynamic volume of the chitosan chains brought about by the addition of neutral electrolyte (section 6.2.3). This decrease in coil size has a similar effect to that of a reduction in molecular weight, increasing the surface area accessible to the chitosan chains.[131] Opposing this are two effects that cause a decrease in the amount of chitosan adsorbed. These are:

(i) suppression of the negative surface charge on the cellulose and of the positive charge on the chitosan, thereby reducing the extent of electrostatic interaction between them;

(ii) increasing competition between the $-\overset{+}{N}H_3$ groups on the chitosan chains, and the Na$^+$ ions from the electrolyte, for the anionic sites on the cellulose substrate.

The lack of any change in adsorption on addition of NaCl, as reported by Rippon,[164] may be due to an approximate balancing between these two effects and that of reduction in hydrodynamic volume, since Rippon examined the effect at a single NaCl concentration.

6.4.2 *Adsorption on inorganic dispersions*

Although kaolin is a widely used substrate for studies of the adsorption and flocculating properties of polymers, very little use has been made of it in studying these aspects of the behaviour of chitosan. The major study is that of ·Domard et al.[167] who do refer to a previous study by Jiang et al., unfortunately published in a relatively inaccessible journal.[168] Domard et al. studied the adsorption behaviour of four chitosan samples, together with a quaternised derivative N,N,N-trimethylchitosan[0.05] chloride.

The equilibrium adsorption isotherm for the quaternised derivative shows an initial increase in the amount adsorbed, which then levels off, with the height of this plateau increasing slightly with increase in pH from 4 to 6. That for the unmodified chitosan is very similar at pH 4 but differs significantly at pH 6, showing no evidence of any levelling off at even the highest chitosan concentrations used. This difference in behaviour was attributed to the large decrease in the linear charge density of chitosan on going from pH 4 to pH 6, whereas this remains relatively constant for the quaternised derivative. Furthermore the equilibrium adsorption curves for all four chitosan samples failed to reach this adsorption plateau, regardless of molecular weight or degree of N-acetylation, but showed that the equilibrium adsorption at any solution concentration increases with increase in molecular weight and increase in the amine group concentration. Although this is the reverse of that observed with the chitosan on cellulose,[131] it agrees with the increase in adsorption with increase in molecular weight usually observed for the adsorption of polymers on non-porous adsorbents, but the increase in adsorption with decreasing N-acetyl content is less ameanable to explanation. However the saturation values for chitosan[0.05] and chitosan[0.15], calculated using the Langmuir equation, are similar, being 25.0 and 24.3 g kg^{-1} respectively. Furthermore the value for the chitosan[0.05] sample drops to 17.0 g kg^{-1} at pH 4, where its linear charge density will be much higher, in agreement with the results obtained for adsorption on cellulose. However the adsorption intensity is much higher at the lower pH, showing the importance of electrostatic interactions between the cationic polyelectrolyte and the negatively charged kaolin surface.

The flocculating capability of chitosan was also found to be much more pH dependent than that of the quaternised derivative. Maximum floccula-tion, as determined by measuring the sedimentation volume and light transmission, occurs with a lower concentration of chitosan at pH 4 than at pH 6, while the sediment volume at any chitosan concentration is much greater at pH 6 than pH 4 (Figure 6.16). In the case of the quaternised chitosan derivative maximum flocculation occurs with a lower polymer concentration at pH 6 than at pH 4, owing presumably to there being only a relatively small change in linear charge density over this pH range while there is a considerable increase in the negative surface charge on the kaolin particles, the zeta potential changing from −18 mV at pH 4 to −35 mV at pH 6. Furthermore there is a very much smaller difference between the sediment volumes obtained at pH 4 and pH 6 for any concentration of the quaternised chitosan. With both polymers and at both pH values

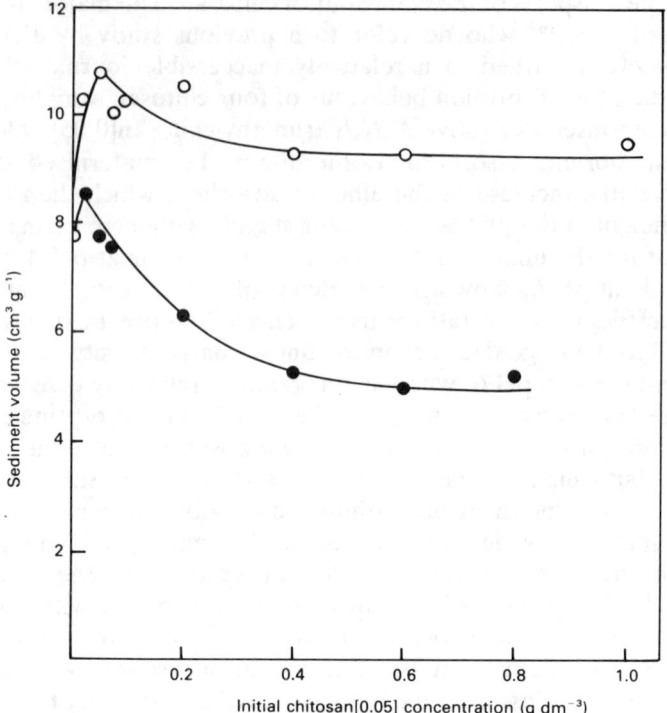

FIGURE 6.16   Variation in sediment volume as a function of chitosan[0.05] concentration:[167] ○, pH 6; ●, pH 4

the adsorption process leads to a reversal of the net charge of the solid and this stabilises the dispersion.

6.4.3   Adsorption on organic dispersions

There have been numerous reports on the use of chitosan as a flocculant for organic solids, particularly for the recovery of suspended organic matter from waste liquors in the food processing industry. The work, much of which has been carried out by Bough and co-workers, includes studies of the treatment of waste liquors from egg,[169] poultry,[170] vegetable,[171] cheese,[172] shellfish,[173–175] and meat[173, 176] processing plants. In general reductions of 80–98% in the suspended solids may be achieved using coagulation and settling to recover the solids, with lower reductions being achieved using dissolved air flotation.[177]

Most of the studies have been directed towards optimising the particular flocculation process rather than obtaining an understanding of the underlying principles, but Wu and Bough[178] have attempted to correlate characteristics of chitosan, as controlled by the deacetylation process, with its effectiveness as a flocculant in wastewater systems. They concluded that the extent of deacetylation and the molecular weight distribution, as measured by HPLC, are the best parameters for characterising the effectiveness of a sample. Viscosity, measured using a rotating spindle viscometer, was not a reliable indicator of performance. However the optimum values of these two parameters were found to vary with the particular wastewater system used and Wu and Bough concluded that a case-by-case survey is required for each type of application.

More recently an evaluation of the coagulating abilities of 24 chitosan samples, produced from different crustacean and fungal sources, has been reported.[176] Their performances were compared with those for a synthetic cationic polyelectrolyte, for $FeCl_3$, and for $Al_2(SO_4)_3$, using wastewater from a meat processing plant. The general order of effectiveness for reduction of turbidity and of suspended solids level was found to be crustacean-derived chitosan > synthetic cationic polyelectrolyte > fungi-derived chitosan > inorganic coagulants. However the chitosans were not characterised and those derived from fungal sources were used as the chitosan–glucan complex (section 2.5.5), so that it is difficult to form definite conclusions from this work. Bough[172, 173] had previously reported chitosan to be more effective than a range of synthetic polymer flocculants for reducing the turbidity of shrimp processing wastes and cheese whey, but evaluation of all the polymers was carried out under the optimum conditions established for chitosan and it is possible that some of the synthetic polymers would have been more effective than chitosan if their application conditions were also optimised.

The use of inorganic salts as co-flocculants has given conflicting results. Moorjani et al.[179] found that the use of 1.5×10^{-4} g dm^{-3} chitosan plus 1×10^{-2} g dm^{-3} $Al_2(SO_4)_3$ gave a slightly greater reduction in turbidity than did the use of 6×10^{-4} g dm^{-3} chitosan alone, and a considerably greater reduction than using 30×10^{-2} g dm^{-3} $Al_2(SO_4)_3$ alone. A similar small improvement in the recovery of suspended organic solids in meat packing wastewater was obtained by Bough using 4×10^{-2} g dm^{-3} $FeCl_3$ in conjunction with 1×10^{-2} g dm^{-3} chitosan, compared with using chitosan alone.[173] In contrast to this Bough et al.[170, 171] found that the addition of $FeCl_3$ gave no improvement with poultry chiller and vegetable processing wastes, and a similar lack of improvement in efficiency on addition of $FeCl_3$ has been reported by No and Meyers[175] for reduction of turbidity in crawfish waste using chitosan.

Chitosan has been found to be a good conditioning agent for dewatering activated sludges from brewing and vegetable processing wastes.[177] A study of its effectiveness with municipal wastes – anaerobically digested sludge, primary sludge plus excess activated sludge, and excess activated sludge – has been reported.[180] Chitosan was found to be an excellent dewatering aid for raw sludges, whose pH values are relatively low, but to be relatively ineffective with anaerobically digested sludges which have alkaline pH values. In this it is at a disadvantage compared with synthetic cationic polyelectrolytes which are effective over a much wider pH range. However the biodegradation rate, in soil environments, of the recovered solids is considerably higher when chitosan is used as the dewatering aid rather than either using a synthetic polymer or no aid at all.

Chitosan has been used as a flocculant for the recovery of microalgae cultures grown in both sea and fresh water,[181-183] its effectiveness being less in the former. This has been attributed to the ionic strength of sea water, and confirmation of this comes from the observation that reduction of the ionic strength below 0.1M increases the effectiveness of chitosan considerably. However chitosan was found to be effective at an ionic strength of 0.7M when used in conjunction with $FeCl_3$, provided that the chitosan was added prior to the $FeCl_3$. Chitosan has also been used as a flocculant to aid the collection of cells from fermentation broths.[184] A study using *Candida utilis* as a typical yeast culture, and the dissolved air flotation process for recovery of the floc, found that maximum recovery was obtained at pH 6.5. A dosage level of 8×10^{-2} g dm^{-3} chitosan gave 98% recovery at pH 6.5 compared with less than 20% recovery in the absence of chitosan. The optimum recovery ratio, which is the ratio of the volume of air-saturated water introduced to the volume of broth sample, was found to be 0.5. Increasing the dosage level caused a decrease in the recovery level down to approximately 50% at a dosage level of 15×10^{-2} g dm^{-3} chitosan, owing to stabilisation caused by charge reversal. Further increase in the dosage level to 17×10^{-2} g dm^{-3} caused the recovery level to rise to approximately 90%, but no explanation was offered for this effect.[184]

Agerkvist *et al.*[185] have reported a detailed study of the use of chitosan as a selective flocculant for cell wall fragments in *E. coli* homogenates, a heterogeneous system whose solution phase contains the required product β-galactosidase, other proteins, and nucleic acids. Chitosan-induced flocculation removes up to 98% of the cell debris by sedimentation under gravity, compared with only 70% by centrifugation at 15,000 g without addition of chitosan, but may also cause a considerable reduction in the other components as well. Both the amount of chitosan required for the optimal flocculation dosage (OFD), and the selectivity of flocculation, increase with increase in the pH (Table 6.6). Although flocculation of all the components takes place to some extent, the correct choice of pH enables up to 50% of the proteins to remain in solution even in the presence of excess chitosan. Since chitosan appeared to interact preferentially with high-molecular-weight proteins the authors suggested that the selectivity could be raised for proteins having a lower molecular weight than β-galactosidase.

TABLE 6.6 *Effects of pH on the selectivity of chitosan as a flocculant with E.* coli *cell homogenates, determined at the optimum flocculation dosage (OFD)[185]*

pH	OFD (g kg^{-1})	Material remaining in the clarified solution (%)			
		Cell debris	β-Galactosidase	Proteins	Nucleic acids
4.0	12	10	0	5	10
5.0	22	2	64	45	12
6.0	38	3	41	63	14
6.9	43	3	59	56	15

The nucleic acids present in the solution phase interact with the chitosan thereby preventing it interacting with cell debris and leading to higher OFD values. Pretreatment of the homogenate with a nuclease breaks down any DNA or RNA present to oligonucleotides which do not interact in this way. Thus pretreatment with a DNase reduced the OFD by approximately 25%, while pretreatment with both a DNase and an RNase reduced it by 55%. The authors concluded that flocculation-enhanced separation is suitable as an upstream purification process in the production of intracellular products from bacteria.

In some of the studies discussed above it was noted that although the suspended solids content could be reduced by up to 98%, much lower reductions were generally achieved in the Chemical Oxygen Demand[175, 177] which may be assumed to arise mainly from proteins present in solution in the waste waters. In some wastewaters the dissolved protein may represent over 80% of the total solids in the system[175] so that its recovery is worthwhile both from the point of view of its value as a by-product and the reduction in effluent discharge costs.

Senstad and Almas investigated the recovery of protein from shrimp processing wastewater, determining the effect of pH on the extent of protein removal at different ratios of chitosan to protein.[186] Most of the proteins present had isoelectric points ≤ pH 5 so that at pH values below this the addition of chitosan resulted in the removal of less protein than could be removed in the absence of chitosan by adjusting the solution pH to the same value. At pH 5 itself very little difference was found in the amount of protein removed, either in the absence of chitosan or in its presence with chitosan:protein weight ratios of 0.005:1 to 0.25:1. Above pH 5 both the amount of protein removed, and the pH giving maximum removal, were found to increase with increase in the chitosan:protein ratio, except for the highest ratio of 0.25:1 where restabilisation may be occurring. No apparent effect of electrolyte on the amount of protein removed was observed.

Holland *et al.*[187–189] have examined the recovery of protein from liquid mussel protein concentrate, obtaining higher yields of recovered protein using chitosan as the flocculant than could be obtained using two anionic flocculants, lignosulphonic acid or sodium hexametaphosphate (Calgon). Contrary to the results of Senstad and Almas,[186] Holland and Shahbaz[187] found that even at pH 4.5 the addition of chitosan gave an increase in the amount of protein recovered compared with that obtained by acidification to pH 4.5 alone. Some 40–65% of the soluble protein was recovered and these values are similar to the values reported for the reduction in COD achieved by treating a number of wastewaters with chitosan.[175, 177] All the suspended solids were removed in these treatments.[187–189]

Baxter[190] has studied the recovery of soluble protein by flocculation with chitosan using model systems containing α-lactalbumin, β-lactoglobulin, and bovine serum albumin, the proteins having isoelectric points of 4.8, 5.1–5.4 and 5.1 respectively. The proteins were precipitated by chitosan at pH values above their isoelectric points, but not below them. The amount of protein precipitated above the isoelectric point increased with increase in the amount of chitosan added up to a maximum, the OFD, then decreased with further increase owing to restabilisation. Both the OFD and the total amount precipitated at the OFD increased with increase in pH, the maximum amounts precipitated at pH 6.4 being 80% for α-lactalbumin, 83% for β-lactoglobulin, and 72% for bovine serum albumin.

Addition of NaCl at pH 6.4 caused a decrease in the amount precipitated, that for bovine serum albumin decreasing from 72% in the absence of NaCl, to approximately 10% in 0.04M NaCl. The OFD was not noticeably affected by the increase in ionic strength and Baxter concluded that the reduction in the precipitate is due to a reduction in the size of the coiled chitosan chains, leading to a reduction in the extent of bridging between protein molecules and a resultant increase in the fraction of smaller, more soluble, chitosan–protein complexes.

The effects of structural changes in the chitosan were also studied. Increasing the extent of deacetylation gave an increase in the amount of protein precipitated, with no noticeable change in the OFD, and Baxter suggested that the persistence length of the chitosan chain increases with increase in the deacetylation level, making it a more effective bridge between protein molecules. At the same level of N-acetylation a chitosan prepared directly from chitin by heterogeneous deacetylation was found to precipitate more protein at the OFD than did one prepared by homogeneous N-acetylation of a highly deacetylated precursor, although the OFD values were similar. Baxter proposed that the block copolymer pattern of the former (section 2.5.8) gave rise to either more effective ionic interaction with proteins or a more extended chitosan chain. However the possibility of molecular weight effects cannot be ruled out at this stage as Poole[191] has shown that low-molecular-weight chitosan – having a viscosity < 25 cps for a 10 g dm^{-3} solution in 1 vol.-% acetic acid – interacts with proteins at acid pH values that are above the isoelectric point of the protein, without causing precipitation. That interaction has taken place is demonstrated by the enhanced foaming properties of the solution, even in the presence of lipids, leading Poole to suggest the use of such systems in the preparation of aerated food products. Presumably, although charge neutralisation has taken place, the chitosan chains are not long enough to bridge the electrically 'neutralised' globular proteins.

REFERENCES

1. P.P. von Weimarn, *Ind. Eng. Chem.*, **19** (1927) 109.
2. G.L. Clark and A.F. Smith, *J. Phys. Chem.*, **40** (1936) 863.
3. R.H. Hackman and M. Goldberg, *Carbohydr. Res.*, **38** (1974) 35.
4. V.F. Lee, *University Microfilms (Ann Arbor)*, *74/29446*, 1974.
5. R.C. Capozza, *German Patent 2,505,305* (1975).
6. K. Nagasawa, Y. Tohira, Y. Inoue and N. Tanoura, *Carbohydr. Res.*, **18** (1971) 95.
7. K. Nagasawa and Y. Inoue, *Chem. Pharm. Bull.*, **19** (1971) 2617.
8. K. Nagasawa and N. Tanoura, *Chem. Pharm. Bull.*, **20** (1972) 157.
9. P.R. Austin, in *Chitin, Chitosan, and Related Enzymes*, J.P. Zikakis (ed.), Academic Press, New York, 1984, p. 227.
10. D. Gagnaire, J. Saint-Germain and M. Vincendon, *Makromol. Chem.*, **183** (1982) 593.
11. P.R. Austin, J.E. Castle and C.J. Albisetti, in *Chitin and Chitosan*, G. Skjåk-Braek, T. Anthonsen and P. Sandford (eds), Elsevier, London, 1989, p. 749.
12. J. Noguchi, S. Tokura and N. Nishi, in *Proceedings 1st International Conference on Chitin/Chitosan (1977)*, R.A.A. Muzzarelli and E.R. Pariser (eds), MIT Sea Grant Program Report MITSG 78–7, 1978, p. 315.
13. P.R. Austin, *US Patent 3,892,731* (1975).
14. T. Ando and S. Kataoko, *Kobunshi Ronbunshu*, **37** (1980) 1.
15. N. Nishi, J. Noguchi, S. Tokura and H. Shiota, *Polym. J.*, **11** (1979) 27.
16. O. Somorin, N. Nishi, S. Tokura and J. Noguchi, *Polym. J.*, **11** (1979) 391.
17. K. Kaifu, N. Nishi, T. Komai, S. Tokura and O. Somorin, *Polym. J.*, **13** (1981) 241.
18. K. Kaifu, N. Nishi and T. Komai, *J. Pol. Sci., Pol. Chem. Ed.*, **19** (1981) 2361.
19. N. Nishi, H. Ohnuma, S. Nishimura, O. Somorin and S. Tokura, *Polym. J.*, **14** (1982) 919.
20. G.G. Allan, P.G. Johnson, Y-z. Lai and K.V. Sarkanen, *Chem. Ind.*, (1971) 127.
21. L.A. Buffington and E.S. Stevens, *J. Amer. Chem. Soc.*, **101** (1979) 5159.
22. P.R. Austin, in *Methods in Enzymology*, W.A. Wood and S.T. Kellogg (eds), Academic Press, New York, 1988, vol. 161, p. 403.
23. P.R. Austin, *German Patent 2,707,164* (1977).
24. F.A. Rutherford and P.R. Austin, in ref. 12, p. 182.
25. D. Balasubramanian and R. Shaikh, *Biopolymers*, **12** (1973) 1639.
26. A. Ciferri and S. Russo, *Amer. Chem. Soc., Pol. Preprints*, **18(2)** (1977) 87.
27. M. Paner and L.F. Beste, *Amer. Chem. Soc., Pol. Preprints*, **17(1)** (1976) 65.
28. S.M. Aharoni and E. Wasserman, *Macromolecules*, **14** (1981) 454.
29. M. Vincendon, in *Chitin in Nature and Technology*, R.A.A. Muzzarelli,

C. Jeuniaux and G.W. Gooday (eds), Plenum Press, New York, 1986, p. 343.
30. C.J. Brine and P.R. Austin, *Amer. Chem. Soc., Sympos. Ser.*, **18** (1975) 505.
31. P.R. Austin, *US Patent 4,029, 727* (1977).
32. K. Kifune, K. Inoue and S. Mori, *US Patent 4,431,601* (1984).
33. M. Nakajima, K. Atsumi and K. Kifune, in ref. 9, p. 407.
34. T. Uragami, Y. Ohsumi and M. Sugihara, *Polymer*, **22** (1981) 1155.
35. F.A. Rutherford and W.A. Dunson, in ref. 9, p. 135.
36. S. Aiba, M. Izume, N. Minoura and Y. Fujiwara, *Carbohydr. Polym.*, **5** (1985) 285.
37. M. Terbojevich, C. Carraro, A. Cosani and E. Marsano, *Carbohydr. Res.*, **180** (1988) 73.
38. M. Terbojevich, A. Cosani, C. Carraro and G. Torri, in ref. 11, p. 407.
39. C.L. McCormick and D.K. Lichatowich, *J. Pol. Sci., Pol. Lett. Ed.*, **17** (1979) 479.
40. C.L. McCormick, D.K. Lichatowich, J.A. Pelezo and K.W. Anderson, *Amer. Chem. Soc., Pol. Preprints*, **21** (1980) 109.
41. C.L. McCormick and K.W. Anderson, in ref. 9, p. 41.
42. P.A. Small, *J. Appl. Chem.*, **3** (1953) 71.
43. K.L. Hoy, *J. Paint Technol.*, **42** (1970) 76.
44. J.L. Gardon, in *Encyclopedia of Polymer Science and Technology*, H.F. Mark, N.G. Gaylord and N.M. Bikales (eds), Interscience, New York, 1965, p. 834.
45. H. Burrell, in *Polymer Handbook*, J. Brandrup and E.H. Immergut (eds), Wiley–Interscience, 2nd edn, 1975, p. 337.
46. P.R. Austin and S. Sennet, in ref. 26, p. 279.
47. S.T. Horowitz, S. Roseman and H.J. Blumenthal, *J. Amer. Chem. Soc.*, **79** (1957) 5046.
48. E.R. Hayes and D.H. Davies, in ref. 12, p. 406.
49. J.G. Domszy and G.A.F. Roberts, *Makromol. Chem.*, **186** (1985) 1671.
50. P. Gross, E. Konrad and H. Mager, in *Chitin and Chitosan*, S. Hirano and S. Tokura (eds), The Japanese Society of Chitin and Chitosan, Tottori, 1982, p. 205.
51. L.F. Filar and M.G. Wirick, in ref. 12, p. 169.
52. R. Yamaguchi, S. Hirano, Y. Arai and T. Ito, *Agric. Biol. Chem.*, **42** (1978) 1981.
53. G.W. Rigby, *US Patent 2,040,879* (1936).
54. G.W. Rigby, *US Patent 2.040,880* (1936).
55. R.A.A. Muzzarelli, *Chitin*, Pergamon, Oxford, 1977, p. 104.
56. E.R. Hayes, D.H. Davies and V.G. Munroe, in ref. 12, p. 103.
57. M.L. Wolfrom, G.G. Maher and A. Chaney, *J. Org. Chem.*, **23** (1957) 1990.
58. T. Sannan, K. Kurita and Y. Iwakura, *Makromol. Chem.*, **176** (1975) 1191.
59. T. Sannan, K. Kurita and Y. Iwakura, *Makromol. Chem.*, **177** (1976) 3589.
60. G.G. Maghami and G.A.F. Roberts, unpublished work.
61. Lion Corporation, *Japan Patent 142,710* (1985).
62. G.K. Moore and G.A.F. Roberts, *Int. J. Biol. Macromol.*, **2** (1980) 73.
63. H. Brunswick, *Biochem. Zeit.*, **113** (1921) 11.
64. J. Doczi, *US Patent 2,795,579* (1957).
65. G.A.F. Roberts, in ref. 11, p. 479.
66. B.D. Gummow and G.A.F. Roberts, *Makromol. Chem.*, **186** (1985) 1245.
67. K.D. Vorlop and J. Klein, *Biotech, Lett.*, **3** (1981) 9.
68. C. Rha, D. Rodriguez-Sanchez and C.A. Kienzle-Sterzer, in *Biotechnology of Marine Polysaccharides*, R.R. Colwell, E.R. Pariser and A.J. Sinskey (eds), Hemisphere, New York, 1985, p. 283.

69. D. Knorr, in ref. 29, p. 428.
70. H. Bungenberg de Jong, in *Colloid Science*, H. Krurzt (ed.), Elsevier, Amsterdam, 1949, vol. 2, chapters 8, 9 and 11.
71. A.S. Michaels, *Ind. Eng. Chem.*, **57** (1965) 32.
72. A. Katchalsky, *Pure Appl. Chem.*, **26** (1971) 327.
73. H. Terayama, *J. Pol. Sci.*, **8** (1952) 243.
74. H. Thiele and L. Langmaach, *Zeit. Naturforsch.*, **12** (1957) 14.
75. Y. Kikuchi, *Makromol. Chem.*, **175** (1974) 2209.
76. Y. Kikuchi and H. Fukuda, *Makromol. Chem.*, **175** (1974) 3593.
77. H. Fukuda and Y. Kikuchi, *Makromol. Chem.*, **178** (1977) 2895.
78. K. Shinoda and A. Nakajima, *Bull. Inst. Chem. Res., Kyoto Univ.*, **53** (1975) 392.
79. K. Shinoda and A. Nakajima, *Bull. Inst. Chem. Res. Kyoto Univ.*, **53** (1975) 400.
80. A. Nakajima and K. Shinoda, *J. Colloid Interfac. Sci.*, **55** (1976) 126.
81. H. Fukuda and Y. Kikuchi, *Bull. Chem. Soc. Japan*, **51** (1978) 1142.
82. S. Hirano, C. Mizutani, R. Yamaguchi and O. Miura, *Biopolymers*, **17** (1978) 805.
83. R. Srinivasan and R. Kamalam, *Biopolymers*, **21** (1982) 251.
84. R. Srinivasan and R. Kamalam, *Biopolymers*, **21** (1982) 265.
85. S. Pushpa and R. Srinivasan, *Biopolymers*, **23** (1984) 59.
86. S.B. Dev, A.M. North and J.C. Reid, in *Dielectric Properties of Polymers*, F.E. Kraasz (ed.), Plenum Press, New York, 1972, p. 217.
87. G. Schwarz, *J. Phys. Chem.*, **66** (1962) 2636.
88. A.S. Michaels, G.L. Falkenstein and N.S. Schneider, *J. Phys. Chem.*, **69** (1965) 1456.
89. C. Hwang, C. Rha and A.J. Sinskey, in ref. 29, p. 389.
90. T. Shioya and C. Rha, in ref. 11, p. 627.
91. S-K. Kim and C. Rha, in ref. 11, p. 635.
92. S-K. Kim and C. Rha, in ref. 11, p. 617.
93. M.M. Daly and D. Knorr, *Biotechnol. Progress*, **4** (1988) 76.
94. D. Knorr and M.M. Daly, *Process Biochem.*, **48** (1988) 48.
95. M.D. Beaumont, Y. Pandya and D. Knorr, *Food Biotechnol.*, **2** (1988) 137.
96. D. Knorr, in ref. 11, p. 101.
97. Y. Kikuchi and A. Noda, *J. Appl. Pol. Sci.*, **20** (1976) 2561.
98. W.J. Hammer, H.V. Mendenhall, R.L. Vigdahl, R.H. Ferber and L.C. Haddad, in ref. 50, p. 213.
99. A. Widra, *US Patent 4,570,629* (1986).
100. R. Minke and J. Blackwell, *J. Mol. Biol.*, **120** (1978) 167.
101. D. Gagnaire, J. Saint-Germain and M. Vincendon, *J. Appl. Pol. Sci., Polymer Symposia*, **37** (1983) 261.
102. H. Saito, R. Tabeta and S. Hirano, in ref. 50, p. 71.
103. P.J. Van Duin and J.J. Hermans, *J. Pol. Sci.*, **36** (1959) 295.
104. C.A. Kienzle-Sterzer, D. Rodriguez-Sanchez and C. Rha, *J. Appl. Pol. Sci.*, **27** (1982) 4467.
105. S. Ya. Lyubina, I.A. Strelina, L.A. Nud'ga, E.A. Plisko and I.N. Bogatova, *Vysokomol. Soyed.*, **A25** (1983) 1467.
106. L. Yuan, T.J. Dougherty and S.S. Stivala, *J. Pol. Sci., Pt A2*, **10** (1972) 171.
107. C.A. Kienzle-Sterzer, D. Rodriguez-Sanchez and C. Rha, in ref. 9, p. 383.
108. D. Rodriguez-Sanchez, C.A. Kienzle-Sterzer and C. Rha, in ref. 50, p. 30.
109. M. Rinaudo and A. Domard, in ref. 11, p. 71.
110. A. Domard and M. Rinaudo, *Int. J. Biol. Macromol.*, **5** (1983) 49.
111. W. Kuhn, *Kolloid-Zeit.*, **68** (1934) 2.

112. P.J. Flory, *Principles of Polymer Chemistry*, Cornell University Press, Ithaca, 1953, p. 412.
113. O. Kratky and G. Porod, *Recueil Trav. Chim. Pays-Bas*, **68** (1949) 1106.
114. N.V. Pogodina, G.M. Pavlov, S.V. Bushin, A.B. Mel'nikov, Ye. B. Lysenko, L.A. Nud'ga, V.N. Marsheva, G.N. Marchenko and V.N. Tsvetkov, *Vysokomol Soyed.*, **A28** (1986) 232.
115. C.A. Kienzle-Sterzer, D. Rodriguez-Sanchez and C. Rha, in ref. 29, p. 338.
116. W. Brown and D. Henley, *Makromol. Chem.*, **79** (1964) 68.
117. H. Yamakawa and M. Fujii, *Macromolecules*, **7** (1974) 128.
118. O. Smidsrød and A. Haug, *Biopolymers*, **10** (1971) 1213.
119. M. Terbojevich, A. Cosani, M. Scandola and A. Fornasa, in ref. 26, p. 349.
120. L.A. Berkovich, G.I. Timofeyeva, M.G. Tsyurupa and V.A. Davankov, *Vysokomol. Soedin.*, **A22** (1980) 1834.
121. E.R. Lang, C.A. Kienzle-Sterzer, D. Rodriguez-Sanchez and C. Rha, in ref. 50, p. 34.
122. G.G. Maghami and G.A.F. Roberts, *Makromol. Chem.*, **189** (1988) 195.
123. G.A.F. Roberts and J.G. Domszy, *Int. J. Biol. Macromol.*, **4** (1982) 374.
124. T.G. Fox, S. Gratch and S. Loshack, in *Rheology*, F.E. Eirich (ed.), Academic Press, New York, 1956, vol. 1.
125. C.A. Kienzle-Sterzer, D. Rodriguez-Sanchez and C. Rha, in ref. 50, p. 26.
126. J.D. Ferry, *Viscoelastic Properties of Polymers*, Wiley, New York, 3rd edn, 1980.
127. N. Mataga and M. Koizumi, *Bull. Chem. Soc. Japan*, **28** (1955) 51.
128. B.D. Gummow and G.A.F. Roberts, *Makromol. Chem.*, **186** (1985) 1239.
129. B.D. Gummow and G.A.F. Roberts, *Makromol. Chem., Rapid Commun.*, **6** (1985) 381.
130. B.D. Gummow and G.A.F. Roberts, *Makromol. Chem.*, **187** (1986) 995.
131. J.G. Domszy, G.K. Moore and G.A.F. Roberts, in *Cellulose and its Derivatives*, J.F. Kennedy, G.O. Phillips, D.J. Wedlock and P.A. Williams (eds), Ellis Horwood, Chichester, 1985, p. 463.
132. J.G. Domszy and G.A.F. Roberts, in ref. 29, p. 331.
133. L. Michaelis, *Cold Spring Harbor Symp. Quant. Biol.*, **12** (1947) 131.
134. S. Hirano, S. Kondo and Y. Ohe, *Polymer*, **16** (1975) 622.
135. S. Hirano and Y. Ohe, *Agric. Biol. Chem.*, **39** (1975) 1337.
136. S. Hirano, Y. Ohe and H. Ono, *Carbohydr. Res.*, **47** (1976) 315.
137. S. Hirano, O. Miura and R. Yamaguchi, *Agric. Biol. Chem.*, **41** (1977) 1755.
138. G.K. Moore and G.A.F. Roberts, *Int. J. Biol. Macromol.*, **2** (1980) 73.
139. S. Hirano and R. Yamaguchi, *Biopolymers*, **15** (1976) 1685.
140. G.A.F. Roberts and K.E. Taylor, in ref. 11, p. 577.
141. G.K. Moore and G.A.F. Roberts, *Int. J. Biol. Macromol.*, **2** (1980) 78.
142. H. Saito, R. Tabeta and S. Hirano, in ref. 50, p. 71.
143. S. Hirano, R. Yamaguchi and N. Matsuda, *Biopolymers*, **16** (1977) 1987.
144. S. Hirano, H. Senda, Y. Yamamoto and A. Watanabe, in ref. 9, p. 77.
145. P. Broussignac, *Chim. Ind. Genie Chim.*, **99** (1968) 1241.
146. S. Hirano, R. Yamaguchi, N. Matsuda, O. Miura and Y. Kondo, *Agric. Biol. Chem.*, **41** (1977) 1547.
147. G.K. Moore, Ph.D. Thesis (CNAA), Trent Polytechnic, UK, 1978.
148. S. Hirano, N. Matsuda, O. Miura and T. Tanaka, *Carbohydr. Res.*, **71** (1979) 344.
149. R.A.A. Muzzarelli, G. Barontini and R. Rocchetti, *Biotechnol. Bioeng.*, **18** (1976) 1445.
150. G.A.F. Roberts and K.E. Taylor, *Makromol. Chem.*, **190** (1989) 951.
151. E.S. Gould, *Mechanism and Structure in Organic Chemistry*, Holt, Rinehart

and Winston, New York, 1959, p. 186.
152. R.A.A. Muzzarelli, in ref. 55, p. 136.
153. F.M. Richards and J.R. Knowles, *J. Mol. Biol.*, **37** (1968) 231.
154. R. Gillett and K. Gull, *Histochemie*, **30** (1972) 162.
155. R.M. Blanco and J.M. Guisán, *Enzyme Microb. Technol.*, **11** (1989) 360.
156. L.D. Hall, M. Yalpani and N. Yalpani, *Biopolymers*, **20** (1981) 1413.
157. L.D. Hall and M. Yalpani, *J. Chem. Soc., Chem. Commun.*, (1980) 1153.
158. E.R. Hayes and D.H. Davies, in ref. 12, p. 193.
159. R. Yamaguchi, S. Hirano, Y. Arai and T. Ito, *Agric. Biol. Chem.*, **42** (1978) 1981.
160. R. Yamaguchi, Y. Arai, T. Ito and S. Hirano, *Carbohydr. Res.*, **88** (1981) 172.
161. G.G. Allan, G.D. Crosby and K.V. Sarkanen, *Int. Paper Phys. Confer.*, (1975) 109.
162. H. Tanaka, K. Tachiki and M. Sumimoto, *Tappi*, **62** (1979) 41.
163. F. Onabe, I. Osawa, M. Usuda and T. Kadoya, *Mokuzai Gakkaishi*, **30** (1984) 839.
164. J.A. Rippon, *J. Soc. Dyers Colourists*, **100** (1984) 298.
165. I.D. Rattee and M.M. Breuer, *The Physical Chemistry of Dye Adsorption*, Academic Press, London, 1974.
166. E. Jenkel and B. Rumbach, *Zeit. Elektrochem.*, **55** (1951) 612.
167. A. Domard, M. Rinaudo and C. Terrassin, *J. Appl. Pol. Sci.*, **38** (1989) 1799.
168. T. Jiang, E. Luan, E. Wu, M. Suping and L.U. Depei, *Huanjing Huaxue*, **3** (1984) 31.
169. W.A. Bough, *Poultry Sci.*, **54** (1975) 1904.
170. W.A. Bough, A.L. Shewfelt and W.L. Salter, *Poultry Sci.*, **54** (1975) 992.
171. W.A. Bough, *J. Food Sci.*, **40** (1975) 297.
172. W.A. Bough and D.R. Landes, *J. Diary Sci.*, **59** (1976) 1874.
173. W.A. Bough, *Process Biochem.*, **11** (1976) 13.
174. R.A. Johnson and S.M. Gallanger, *J. Water Pollut. Control Fed.*, **56** (1984) 970.
175. H.K. No and S.P. Meyers, *J. Agric. Food Chem.*, **37** (1989) 580.
176. N. Castellanos-Perez, M. Maldonado-Vega, G. Fernandez-Villagomez and S. Caffarel-Mendez, in ref. 11, p. 567.
177. W.A. Bough and D.R. Landes, in ref. 12, p. 218.
178. A.C.M. Wu and W.A. Bough, in ref. 12, p. 88.
179. M.N. Moorjani, D.I. Khasim, S. Rajalakshmi, P. Puttarajappa and B.L. Amla, in ref. 12, p. 210.
180. T. Ansano, in ref. 12, p. 231.
181. M.V. Rao and K.P. Krishnamoorthi, *Indian J. Environ. Health*, **29** (1979) 183.
182. D. Bilanovic, G. Shelef and A. Sukenik, *Biomass*, **17** (1988) 65.
183. P. Gualtieri, L. Bersanti and V. Passarelli, *Ann. Inst. Pasteur/Microbiol.*, **139** (1988) 717.
184. C.R. Holland, in ref. 11, p. 559.
185. I. Agerkvist, L.B. Eriksson and S-O. Enfors, in ref. 11, p. 543.
186. C. Senstad and K.A. Almas, in ref. 26, p. 568.
187. C.R. Holland and M. Shahbaz, *Irish J. Food Sci. Technol.*, **9** (1985) 107.
188. C.R. Holland and M. Shahbaz, *J. Food Eng.*, **5** (1986) 135.
189. C.R. Holland and P. McComiskey, *J. Food Technol.*, **21** (1986) 763.
190. A.D. Baxter, Ph.D. Thesis (CNAA), Humberside Polytechnic, UK, 1990.
191. S. Poole, in ref. 11, p. 523.

Subject Index

Author Index